Y0-AAJ-266

COSMIC PLASMA PHYSICS

COSMIC PLASMA PHYSICS

Proceedings of the Conference on Cosmic Plasma Physics
Held at the European Space Research Institute (ESRIN),
Frascati, Italy, September 20-24, 1971

Edited by Karl Schindler
European Space Research Institute
Frascati, Italy

PLENUM PRESS • NEW YORK—LONDON • 1972

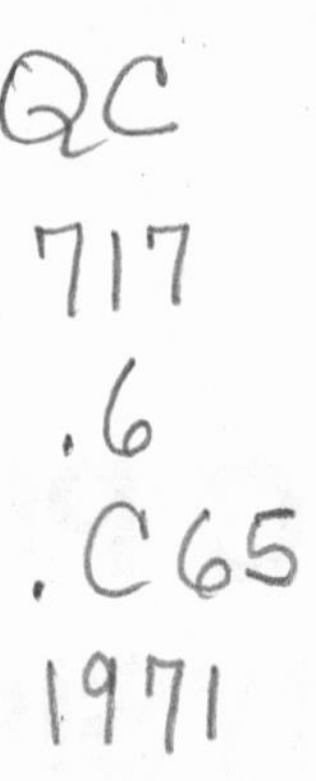

Library of Congress Catalog Card Number 78-188924
ISBN 0-306-30582-8

A Division of Plenum Publishing Corporation
227 West 17th Street, New York, N. Y. 10011

United Kingdom edition, published by Plenum Press, London
A Division of Plenum Publishing Company, Ltd.
Davis House (4th Floor), 8 Scrubs Lane, NW10 6SE, London, England

Printed in the United States of America

6/21/73

FOREWORD

The plan to hold a conference on cosmic plasma physics originated in the Plasma Physics Division of the European Physical Society, whose chairman, B. Lehnert, took the first steps towards its realization.

ESRIN readily adopted this idea, and preliminary contacts with a number of other groups showed that there was a good deal of interest in bringing together people working in different areas of the field of cosmic plasma physics. It was clearly felt that an exchange of views and experience, and an attempt to define problem areas, would be profitable. In this spirit a programme was devised which covered a large variety of topics, ranging from ionospheric to galactic structures.

A diversified programme of this kind runs the risk that the communication between the various fields of specialization remains insufficient. It was gratifying to find that within the wide field of cosmic plasma physics a lively dialogue was in fact possible.

The Conference was sponsored by the European Physical Society. Financial support was provided by ESRO.

It is a pleasure to acknowledge the excellent suggestions of the programme committee members L. Biermann, N. D'Angelo, R. Gendrin, and B. Lehnert. I should like to thank my colleagues B. Bertotti, K. Lackner, and J.F. McKenzie, and numerous other ESRIN staff members, for their valuable help. I feel particularly indebted to the conference secretary, Miss Sachs, who did the real work while I just signed the letters.

ESRIN, Frascati
November 1971

K. Schindler

CONTENTS

INTRODUCTION

PLANETARY ENVIRONMENTS

SOLAR WIND

SOLAR WIND INTERACTION WITH PLANETS AND COMETS

SOLAR PHYSICS

STELLAR AND INTERSTELLAR PLASMA

PULSARS

PULSARS (cont'd)

GENERAL THEORY

SHOCK WAVES, TURBULENCE

COSMIC RAYS

† *IS* Invited survey lecture
IT Invited topical lecture
C Contributed paper

RELATIONS BETWEEN COSMIC AND LABORATORY PLASMA PHYSICS

Hannes Alfvén

Division of Plasma Physics

Royal Institute of Technology, Stockholm

ABSTRACT

Like all other fields of physics, plasma physics cannot be developed without an intimate contact between theory and experiments. Theories which have not been checked by experiments often describe a hypothetical medium which has little similarity with a real plasma.

Some general remarks are made about the present state of astrophysics.

EXPERIMENTAL AND THEORETICAL APPROACH TO PLASMA PHYSICS

Plasma physics has started along two parallel lines. One is the hundred years old investigations in what was called "electrical discharges in gases". To a high degree this approach was experimental and phenomenological, and only very slowly it reached some degree of theoretical sophistication. Most theoretical physicists looked down on this field, which was complicated and awkward. The plasma exhibited striations and double-layers, the electron distribution was non-Maxwellian, there were all sorts of oscillations and instabilities. In short, it was a field which was not at all suited for mathematically elegant theories.

The other approach came from the highly developed kinetic theory of ordinary gases. It was thought that with a limited amount of work this field could be extended to include also ionized gases. The theories were mathematically elegant and claimed to derive all the properties of a plasma from first principles. The theories had very little contact with experimental plasma physics, and all the awkward and complicated phenomena which had been observed in the study of discharges in gases were simply neglected.

Fig.1 Terrella experiment (Block, 1955). When a magnetized sphere is immersed in a plasma, luminous rings, corresponding to the auroral zones, are often produced.

In cosmical plasma physics the experimental approach was initiated by Birkeland, who was the first one to try to connect laboratory plasma physics and cosmic plasma physics. (Neither of the terms was used at that time!) (Fig.1.) Birkeland observed aurorae and magnetic storms in nature, and tried to understand them through his famous terrella experiment. He found that when his terrella was immersed in a plasma, luminous rings around the poles were produced (under certain conditions). Birkeland identified these rings with the auroral zones. As we know today this was essentially correct. Further he constructed a model of the polar magnetic storms, supposing that the auroral electrojet was closed through vertical currents (along the magnetic field lines). Also this idea is essentially correct. Hence although Birkeland could not know very much about the complicated structure of the magnetosphere, research today follows essentially Birkeland's lines, of course supplemented by space measurements, see Dessler (1968) and Boström (1968).

Unfortunately, the progress along these lines was disrupted. Theories about plasmas - at that time called ionized gases - were developed without any contact with the laboratory plasma work. In spite of this - or perhaps because of this - the belief in them was so strong that they were applied directly to space. The result was the Chapman-Ferraro theory, which soon got generally accepted to such an extent that Birkeland's approach was almost completely forgotten. For thirty or forty years it was often not even mentioned in text books and surveys, and all attempts to revive it and develop it were neglected. Similarly, the Chapman-Vestine current system, according to which magnetic storms were produced by currents exclusively flowing in the ionosphere, took the place of Birkeland's three-dimensional system.

CONFRONTATION BETWEEN THEORY AND EXPERIMENTS

The smashing victory of the theoretical approach over the experimental approach lasted as long as a confrontation with reality could be avoided. However, from the theoretical approach, it was concluded that plasmas could easily be confined in magnetic fields and heated to such temperatures as to make thermonuclear release of energy possible. When attempts were made to construct thermonuclear reactors, a confrontation between the theories and reality was unavoidable. The result was catastrophic. Although the theories were "generally accepted" the plasma itself refused to believe in them. Instead the plasma showed a large number of important effects, which were not included in the theory. It was slowly realized that one had to build up new theories but this time in close contact with experiments.

The thermonuclear crisis did not affect cosmical plasma physics very much. The development of the theories went on because they largely dealt with phenomena in regions of space where no real check was possible. The fact that the basis of several of the theories had been proved to be false in the laboratory had very little effect: One said that this did not necessarily prove that they must be false also in cosmos!

The second confrontation, however, came when space missions made the magnetosphere and interplanetary space accessible to physical instruments. The first results were interpreted in terms of the generally accepted theories, or new theories were built up on the same basis. However, when the observational technique became more advanced, it became obvious that these theories were not applicable. The plasma in space was just as complicated as laboratory plasmas. Today very little is left of the Chapman-Ferraro theory and nothing of the Chapman-Vestine current system. Many theories which have been built on a similar basis may have to share their fate.

THE FIRST AND SECOND APPROACH TO COSMIC PLASMA PHYSICS

The result is that the "first approach" has been proved to be leading into a dead end street and we have to make a "second approach" (see Alfvén 1968). The characteristics of these approaches are shown in Table 1.

Table 1

Cosmical Electrodynamics

First Approach	Second Approach
Homogeneous models	Space plasmas have often a complicated inhomogeneous structure
Conductivity $\sigma = \infty$	σ depends on current and often suddenly becomes 0.
Electric field $E_{\parallel} = 0$	$E_{\parallel}$ often $\neq 0$
Magnetic field lines are "frozen-in" and "move" with the plasma	Frozen-in picture often completely misleading
Electrostatic double layers neglected	Electrostatic double layers are of decisive importance in low-density plasmas
Instabilities neglected	Many plasma configurations unrealistic because they are unstable
Electromagnetic conditions illustrated by magnetic field line picture	It is equally important to draw the current lines and discuss the electric circuit
Filamentary structures and current sheets neglected or treated inadequately	Currents produce filaments or flow in thin sheets
Maxwellian velocity distribution	Non-Maxwellian effects often decisive
Theories mathematically elegant and very well developed	Theories still not very well developed and partly phenomenological

SOME RESULTS OF LABORATORY PLASMA PHYSICS

The first laboratory experiment with reference to cosmical physics had the character of "scale model experiments", see Block (1955, 1956, 1967), Danielsson and Lindberg (1964, 1965), Schindler (1969), Podgorny and Sagdeev (1970). It was soon realized, however, that no real scaling of cosmical phenomena down to laboratory size is possible, among other things because of the large number of involved parameters, which obey different scaling laws. Hence laboratory experiments should rather aim at clarifying a number of basic phenomena of importance in cosmical physics.

More specifically, laboratory experiments have demonstrated that a plasma often exhibits the following properties which earlier had been neglected:

1. Quite generally magnetized plasma exhibits a large number of instabilities. Lehnert (1967) lists 32 different types, but there may be still a few more.

2. A plasma has a tendency to produce electrostatic double layers, in which there is a strong electric field over a small distance. Such layers may be stable, but very often they produce oscillations. The phenomenon is basically independent of magnetic fields. If a magnetic field is present, the double layer "cuts" the "frozen-in" field lines. A survey of the laboratory results and their application to cosmic phenomena (especially in the ionosphere) has been given by Block (1971).

3. If a current flows through an electrostatic double layer (which often is produced by the current itself) the layer may cut off the current. This means that the

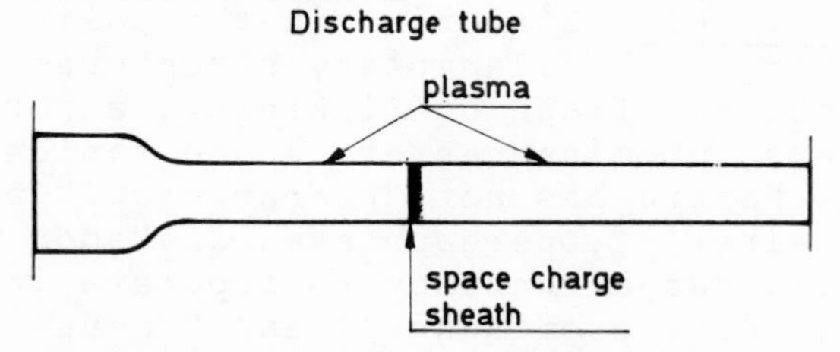

Fig. 2 Electrostatic double sheaths are often produced in a plasma. The figure shows a discontinuity produced spontaneously by the plasma in a thin tube. Over the double sheath a voltage drop is produced which sometimes becomes very large (10^5 volts) and may disrupt the discharge.

voltage over the double layer may reach any value necessary to break the circuit (in the laboratory say 10^5 or 10^6 volt - in solar flares even 10^{10} volt). The plasma "explodes", and a high vacuum region is produced, see Carlqvist (1969), Babic, Sandahl, Torvén (1971); Fig.2.

4. Currents parallel to a magnetic field (or in absence of magnetic fields) have a tendency to "pinch", i.e. to concentrate to filaments and not flow homogeneously (H.Alfvén and C.-G.Fälthammar 1963); Fig.3. This is one of the reasons why cosmic plasmas so often exhibit filamentary structures. The beautiful space experiments by Lüst and his group are important in this connection (although not completely understood; see Völk and Haerendel 1971).

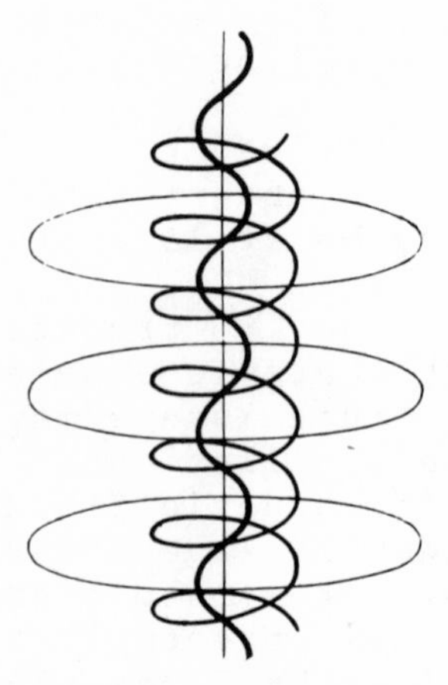

Fig. 3 Simple model of a filamentary structure. The lines depict magnetic field lines. The current flows parallel to these lines.

5. The result of 1-4 is that homogeneous models are usually not applicable. Striation in the positive column of a glow discharge and filamentary structures (arc at atmospheric pressure, flash of lightning, auroral rays, coronal streamers, prominences etc.) are typical non-homogeneities. Nature has not "horror vacui" but a "horror homogeneitatis", perhaps even an "amor vacui". In fact, a plasma has a tendency to separate into high density regions (e.g. prominences) and "vacuum" regions or low density regions (the surrounding corona).

6. If the relative velocity between a magnetized plasma and a non-ionized gas surpasses a certain "critical

velocity"

$$v_{crit} = (2eV_{ion}/m_a)^{1/2}$$

the interaction becomes very strong and leads to a quick ionization of the gas. The phenomenon is of importance to the problem of the origin of the solar system and may also be decisive to the theory of cometary tails. L. Danielsson (1970, 1971) will discuss this in a lecture later at this symposium.

7. The transition between a fully ionized plasma and a partially ionized plasma, and vice versa, is often discontinuous (Lehnert 1970). When the fed-in energy changes gradually, the degree of ionization jumps suddenly from 0.1% to 100%. The border between a completely ionized and a partially ionized plasma is normally very sharp.

8. Borders of this kind act as semipermeable membranes, enriching elements with low ionization voltage in the hot region and elements with high ionization voltage in the cool regions. This may be an efficient chemical differentiation process in cosmos (Lehnert 1968, 1969, 1970). Similar differentiation can also be produced in rotating plasmas (Bonnevier 1966).

9. Flux amplification. If in an experimentally produced plasma ring the toroidal magnetization exceeds the poloidal magnetization, an instability is produced by which the poloidal magnetization increases (Lindberg, Witalis, and Jacobsen 1960, Lindberg and Jacobsen 1964). This phenomenon may be of basic importance to the understanding of how cosmic magnetic fields are produced (H. Alfvén 1961); Figs. 4 and 5.

10. When a plasma moving parallel to a magnetic field reaches a point where the field lines bend, a laboratory plasma may bend in the opposite direction to the bend of the field lines, contrary to what would be natural to assume in most astrophysical theories (Lindberg and Kristoferson 1971); Fig.6.

11. Furthermore, shock and turbulence phenomena in low-pressure plasmas have to be studied in the laboratory before it is possible to clarify the cosmic phenomena (see Podgorny and Sagdeev 1970).

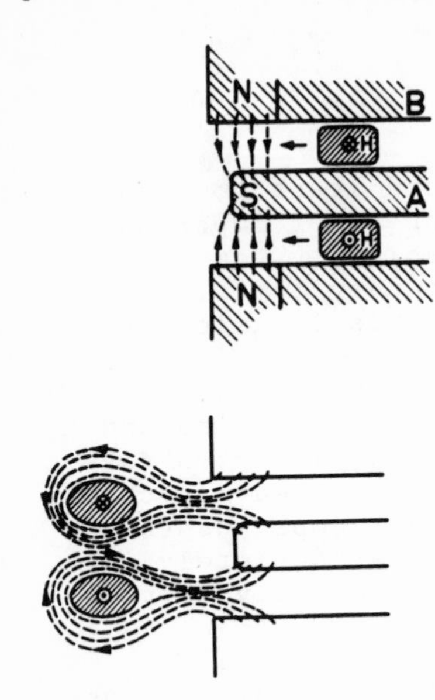

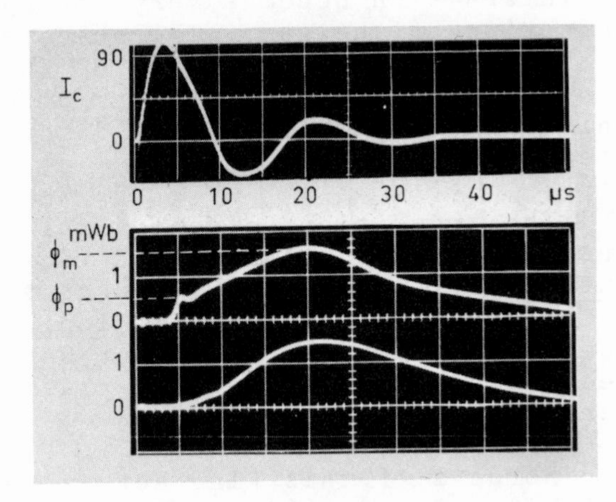

Fig. 5

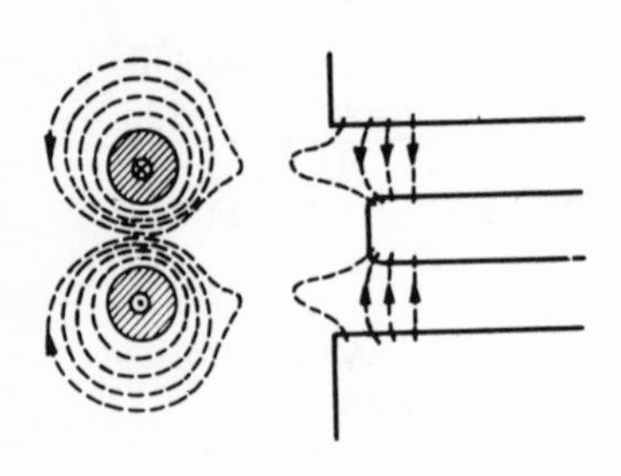

Fig. 4

Fig. 4 By shooting a plasma ring with toroidal magnetization through a magnetic field, a free ring with both toroidal and poloidal magnetization is produced. If the toroidal magnetic energy is too large, a part of it is transferred to poloidal magnetic energy (through kink instability of the current).

Fig. 5 Discharge current I_c in the gun, and poloidal magnetic field. The middle curve shows how the ring, when shot out from the gun, first gets a poloidal flux ϕ_p. An instability of the ring later transforms toroidal energy into poloidal energy, thus increasing the flux from ϕ_p to ϕ_m.

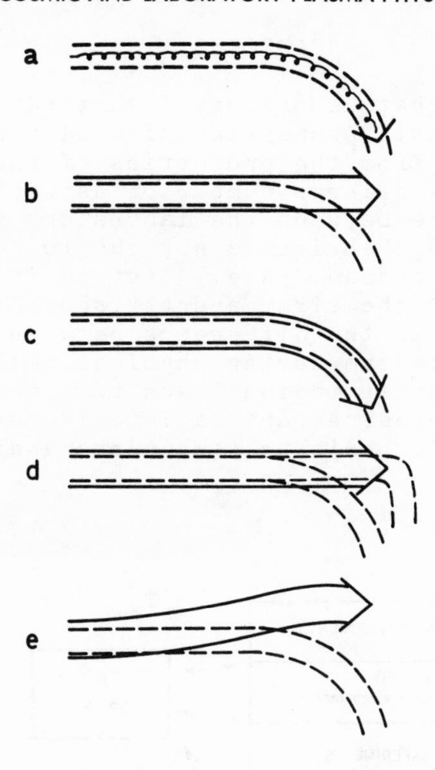

Fig. 6 a) In a magnetic field which has a downward bend, charged particles shot parallel to the field will follow the bend.

b) If instead a plasma beam is shot one should expect that it either

c) follows the bend (like a), or

d) continues to move straight forward bringing the "frozen-in" field lines with it, or gets electrocally polarized and moves straight forward without bringing the field lines with it.

e) In reality it does neither. In the experiment by Lindberg and Kristoferson it bends instead upwards!

12. Further experiments of interest include studies of magnetic conditions at neutral points (Bratenahl and Yeates 1970).

CONCLUSIONS

Thus we find that laboratory investigations begin to elucidate the basic properties of a plasma. These differ drastically from the properties of the medium which is treated in a large number of astrophysical theories. The difference between the laboratory plasma and the plasma of these theories is not due to the difference between laboratory and space. Instead it reflects the controversy between the first and the second approach. In other words, it is the difference between a hypothetical medium and a medium having physical reality. The treatment of the former medium leads to speculative theories of little interest except as intellectual excercise. The latter medium is basic to the understanding of the world we live in.

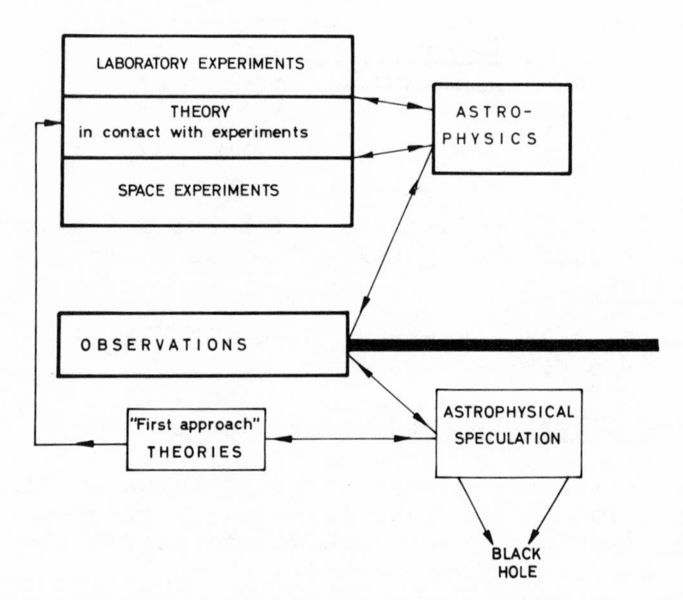

Fig. 7 A combination of "first approach" theories with observations leads to astrophysical speculation, which often has little contact with reality. A check of postulated plasma processes by laboratory (and space) experiments is necessary before these are applied to astrophysics. A healthy development of astrophysics requires a clear distinction between speculation and scientific theories.

Fig. 7 is meant to illustrate the present strategic position in astrophysics. Before we are allowed to combine them with observations, the "first approach" theories must be processed through the laboratory where many of their ingredients no doubt will be filtered away. This is the only way of building up astrophysics with a minimum of speculation. Again we have to learn that science without contact with experiments is a very dangerous enterprise, which runs a large risk of going astray.

THE SPECULATIVE CHARACTER OF PRESENT ASTROPHYSICS

When we thus have found that what has been sacrosanct astrophysical theory for 30 or 40 years has now turned out to be nothing but unfounded speculation, we cannot avoid asking ourselves if the situation is similar in other fields of astrophysics. I believe the answer is yes and would like to submit the following list of fields which sooner or later may be ripe for drastic revisions. For each field the present "sacrosanct" theory is listed together with a "heretic" theory which often is so heretic that it is not even mentioned in textbooks or review articles.

Galactic Intensity of C.R.

Sacrosanct: The intensity of the common C.R. (energy 10^{10}-10^{11}ev) is essentially the same in the whole galaxy as measured near the earth. The particles move freely in the whole galaxy, filling it to the same intensities.

Heretic: There may very well be a magnetic field, e.g. in the heliosphere, which forms a screen between the solar system and the galaxy. In the same way as the van Allen radiation in the earth's magnetic field differs by orders of magnitude from the radiation outside, the interplanetary intensity may be orders of magnitude larger than the interstellar intensity. Recent observations of large C.R. gradients in interplanetary space (Krimigis and Venkatesan 1969, O'Gallagher 1967) support this view.

Possible test: Both views depend on speculations about the structure of magnetic fields in the surrounding of the solar system. Nothing is known about this. A decision has probably to wait until spacecraft pass the outer shock-front of the solar wind.

Solar Magnetic Fields

Sacrosanct: The sun's general magnetic field varies in strength and sign from day to day. Sunspots are produced by mechanisms close to the solar surface.

Heretic: It is known that the readings of "solar magnetographs" do not give the magnetic field but a complicated function of magnetic field, turbulence, temperature etc. The rapid changes in the "magnetograph" readings may as well be due to other factors than changes in the magnetism. Hence there is no evidence that the general field varies. It may be a dipole field. Sunspots may be caused by disturbances in the core, travelling up as hydromagnetic waves.

Test: New methods of measurements and theoretical analysis of magnetic fields in the solar atmosphere are necessary.

Type of Matter in Space

Sacrosanct: All the celestial objects we observe consist of ordinary matter. This view was generally accepted before the antiproton was discovered and it was realized that we at present have no way of distinguishing a koinostar from an antistar. When this is understood, however, the claim that all celestial objects consist of koinomatter is nothing but a hypothesis, which may be true or false.

Heretic: Some of the celestial objects we observe may consist of antimatter. If this is true drastic revisions of theories of the structure, dynamics, and evolution of galaxies will be necessary.

Test: The investigations necessary for a general decision about the existence of antimatter have been listed in a recent paper (Alfvén 1971). A decision whether a certain star consists of koinomatter or antimatter can probably not be made unless a spacecraft is sent there.

The quoted examples, which easily could be increased in number, demonstrate how speculative much of modern astrophysics is, and that many of the most popular theories run the risk of having no permanent value or - to put it more bluntly - being nothing but a sink of theoretical work and machine time.

REFERENCES

Alfvén, H., 1961, On the Origin of Cosmic Magnetic Fields, Ap. J. 133, 1049.

Alfvén, H., and Fälthammar, C.-G., 1963, Cosmical Electrodynamics, Fundamental Principles, Clarendon Press, Oxford, p.192.

Alfvén, H., 1968, Second Approach to Cosmical Electrodynamics, Ann. d. Geophysique 24, 1.

Alfvén, H., 1971, Plasma Physics Applied to Cosmology, Physics Today 24, 28.

Babic, M., Sandahl, S., and Torvén, S., 1971, The Stability of a Strongly Ionized Positive Column in a Low Pressure Mercury Arc, Xth Internat. Conf. on Phenomena in Ionized Gases, Oxford, Sept. 13-17, 1971, p. 120.

Block, L.P., 1955, Model Experiments on Aurorae and Magnetic Storms, Tellus 7, 65.

Block, L.P., 1956, On the Scale of Auroral Model Experiments, Tellus 8, 234.

Block, L.P., 1967, Scaling Considerations for Magnetospheric Model Experiments, Planet. Space Sci. 15, 1479.

Block, L.P., 1971, Potential Double Layers in the Ionosphere, TRITA-EPP-71-14; Cosmic Electrodynamics, in press.

Bonnevier, B., 1966, Diffusion Due to Ion-Ion Collisions in a Multicomponent Plasma, Arkiv f. Fysik 33, 255.

Boström, R., 1968, Currents in the Ionosphere and Magnetosphere, Ann. Géophysique 24, 681.

Bratenahl, A., and Yeates, G.M., 1970, Experimental Study of Magnetic Flux Transfer at the Hyperbolic Neutral Point, Physics of Fluids 13, 2696.

Carlqvist, P., 1969, Current Limitations and Solar Flares, Solar Physics 7, 377.

Danielsson, L., 1970, Review over the Critical Velocity, Part II Experimental Observations, to be published in Cosmic Electrodynamics.

Danielsson, L., 1971, this volume.

Danielsson, L., and Lindberg, L., 1964, Plasma Flow through a Magnetic Dipole Field, Physics of Fluids 7, 1878.

Danielsson, L., and Lindberg, L., 1965, Experimental Study of the Flow of a Magnetized Plasma through a Magnetic Dipole Field, Arkiv f. Fysik 28, 1.

Dessler, A.J., 1968, Solar Wind Interactions, Ann. Géophysique 24, 333.

Krimigis, S.M., and Venkatesan, D., 1969, The Radial Gradient of Interplanetary Radiation Measured by Mariners 4 and 5, J. Geophys. Res. 74, 4129.

Lehnert, B., 1967, Experimental Evidence of Plasma Instabilities, Plasma Physics 9, 301.

Lehnert, B., 1968, Screening of a High-Density Plasma from Neutral Gas Penetration, Nuclear Fusion 8, 173.

Lehnert, B., 1969, On Separation Effects in the Boundary Region of a Plasma Surrounded by Neutral Gas, TRITA-EPP-69-24, Roy. Inst. Technol., Stockholm.

Lehnert, B., 1970, Rotating Plasmas, TRITA-EPP-70-37; Nuclear Fusion, in press; see also Cosmic Electrodynamics 1, 1970, 397.

Lindberg, L., Witalis, E., and Jacobsen, C.T., 1960. Experiments with Plasma Rings, Nature 185, 452.

Lindberg, L., and Jacobsen, C.T., 1964, Studies of Plasma Expelled from a Coaxial Plasma Gun, Physics of Fluids Supplement S44, 844.

Lindberg, L., and Kristoferson, L., 1971, Reverse Deflection and Contraction of a Plasma Beam Moving Along Curved Magnetic Field Lines, TRITA-EPP-71-13; Cosmic Electrodynamics, in press.

O'Gallagher, J.J., 1967, Cosmic-Radial Density Gradient and Its Rigidity Dependence Observed at Solar Minimum on Mariner IV, Ap. J. 150, 675.

Podgorny, I.M., and Sagdeev, R.Z., 1970, Physics of Interplanetary Plasma and Laboratory Experiments, Soviet Physics Uspekhi 98, 445.

Schindler, K., 1969, Laboratory Experiments Related to the Solar Wind and the Magnetosphere, Rev. Geophys. 7, 51.

MAGNETOSPHERIC SUBSTORMS

F.V. Coroniti and C.F. Kennel

Plasma Physics Group, Department of Physics

University of California, Los Angeles, California

1. Introduction

During geomagnetically disturbed times magnetospheric substorms periodically occur every one to three hours. Energetically, substorms represent the principal dissipation process for the energy coupled into the magnetosphere by the solar wind, and provide the energy source for geomagnetic storms. Many theories for the rapid dissipation phase, known as substorm breakup, have been advanced[1-7], but there remains the question of how the magnetosphere temporally evolves into a breakup configuration. Recent experimental work has proven that a coherent substorm growth phase[8], lasting about one hour, is initiated by a southward shift in the solar wind magnetic field[9-11]. A complex pattern of events then follows all of which are consistent with the gradual development of internal magnetospheric convection driven by enhanced field-line reconnection[12-14] at the front-side magnetopause.

The magnetosphere is a bounded hydromagnetic system with both external, the magnetopause, and internal, the ionosphere, plasma-neutral sheet, plasmapause, and particle precipitation boundaries. These boundaries interact via electric fields and plasma currents to control the internal convection rate self-consistently. At the sharp boundary gradients the hydromagnetic description often breaks down, and dissipation rates must then be determined by microscopic plasma turbulence theory. For time dependent convection, the boundary structure and the intensity of turbulent dissipation rates temporally evolve.

In this paper we develop a model of substorm growth phase which logically proceeds from a time dependent convective flow. We draw heavily on previous theoretical work[12, 2,3,15,5,16,17,18]

2. Dayside Magnetopause-Convection Onset

Let's assume that the magnetosphere is in a very quiet, slowly convecting state. At time zero, the solar wind magnetic field shifts abruptly southward and holds steady, and field-cutting at the dayside magnetopause begins at a steady rate. As the reconnection electric field begins to penetrate the magnetopause, a convective flux flow toward the boundary is established. The magnetopause, however, connects to the highly conducting dayside auroral oval ionosphere[19] where large dissipative Pederson currents flow in response to convection electric fields. Since the Pederson current is not divergence-free in the ionosphere, field-aligned currents, inward (outward) on the morning (evening) side, flow into the auroral oval ionosphere from the magnetopause (Figure 1).

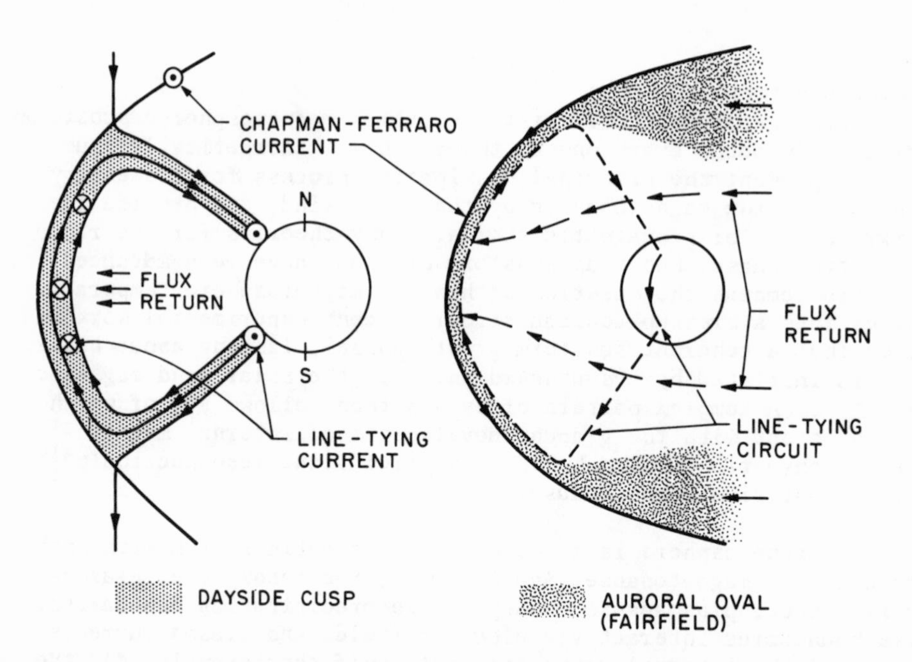

Figure 1. The line-tied magnetopause can be thought of as the superposition of two equivalent current circuits; a Chapman-Ferraro current, and a line-tying current which flows westward through the ionosphere as a Pederson current and closes as an equivalent eastward current on the magnetopause. The stress from the equivalent eastward line-tying current inhibits convection toward the boundary, and, in steady state, just balances the external acceleration stresses arising from magnetopause distortion.

Hence the magnetopause not only must generate sufficient current to stand-off the solar wind dynamic pressure, but must also supply the ionospheric Pederson current required by the internal convective flow. The magnetopause can easily generate more current by distorting its shape so as to increase its angle of attack to the solar wind.

Initially, the boundary distortion is small, and only a small convective flux return to the magnetopause is permitted. Since the initial flux return rate is less than the field-cutting rate, the magnetopause further distorts and shrinks in size until a new equilibrium, with balanced field-cutting and convective flux rates, is established. The rate of approach to equilibrium is determined by the ionospheric electrical conductivity. A rough theoretical analysis[20] which treats the magnetopause as a line-tied equivalent Chapman-Ferraro boundary predicts a 20 minute e-folding time for internal convection to build-up, during which the boundary shrinks by one or more earth radii (R_E) as a consequence. Since there is a time lag in convective flux return, flux is added to the geomagnetic tail which, if not immediately field-cut in the tail, increases the tail flux by about 10%. The convection build-up time constant [9], the inward boundary displacement[21], and the development of a westward electric field[22] are consistent with recent ground and satellite observations during substorm growth phase.

3. Configurational Changes in the Geomagnetic Tail

The decrease in size of the front-side geomagnetic cavity and the increase in tail flux during convection onset forces the tail magnetopause boundary to increase its angle of attack to the solar wind in order to balance the increase of tail normal stress; i.e., the tail intercepts more solar wind dynamic pressure. A rough model for a flaring tail[16,17], which includes only a thin plasma sheet current, was used[23] to estimate the change in tail magnetic field strength expected from the front-side magnetopause shrinkage and increased tail flux. Tail field increases of 40-50%, resulting primarily from magnetopause shrinkage, were calculated for the near tail region, approximately $30R_E$ behind the earth, consistent with satellite measurements in this region during substorm growth phase.[24,25]

The increased tail flaring implies that the solar wind tangential drag on the tail is enhanced. For the tail to remain in quasi-static force balance with no large scale hydromagnetic accelerations, the tangential stress must be balanced by the force of attraction between the tail current system and the earth's dipole[5,15]. The force balance relation can be used to estimate the downstream distance of the tail currents from the earth[5]. The increase of tangential flaring stress consistent with the front-side magnetopause shrinkage produces an earthward motion of the tail currents of about 2 R_E[23]. The combination of tail current inward motion and enhanced current strength produces an additional 30-40 gamma magnetic depression at the synchronous satellite orbit, in agreement with growth phase observations at ATS-1 [26,27].

Plasma sheet electrons of energies .2-5 KeV exhibit a sharp spatial decrease or inner edge located at a radius of about 10 R_E behind the earth[28,29]. This sharp decrease has been interpreted as a flow-precipitation boundary [30-32] If electrons are maintained isotropic by wave turbulent pitch angle scattering, the precipitation lifetime depends only on the electron energy and the magnetic mirror ratio. When this lifetime becomes comparable to the flow time across a typical gradient of magnetic mirror ratio, electrons are rapidly lost to the atmosphere thus creating a spatial temperature boundary. During convection build-up, the enhanced depression of the nightside magnetic field arising from tail current inward motion permits electrons to flow deeper into the magnetosphere before being lost. Adiabatic flow compression also heats the plasma. Again employing estimates consistent with frontside magnetopause shrinkage, we find that the plasma sheet inner edge moves 1-2 R_E earthward during growth phase.[23] Inward displacements of this magnitude have been observed by satellites[28], and are consistent with growth phase equatorward drift of auroral arcs[33] into which the inner edge maps[34].

From the above discussion we draw several conclusions. The observed increased tail magnetic field, inward motion of the tail current system, and earthward displacement of the plasma sheet inner edge are qualitatively and quantitatively consistent with the line-tied inward motion of the dayside magnetopause and the slow establishment of magnetospheric convection. During growth phase, the tail evolves into a configuration of enhanced stress which must be relaxed during the substorm.

4. Nightside Auroral Oval Ionosphere

As a first approximation, let's assume that a dense, highly conducting nightside auroral oval ionosphere exists only where maintained by electron precipitation from the magnetosphere. Hence the polar oval boundary is the last closed tail field line and the equatorward boundary maps into the plasma sheet inner edge. The ionospheric electron density maximizes at the equatorward boundary since plasma sheet electrons, having undergone flow compression, are denser and hotter at the inner edge (Figure 2). In addition, the ratio of the height-integrated Hall to Pederson conductivities Σ_H/Σ_P maximizes at the equatorward boundary since the hotter electron precipitation fluxes more deeply penetrate to the Hall conducting

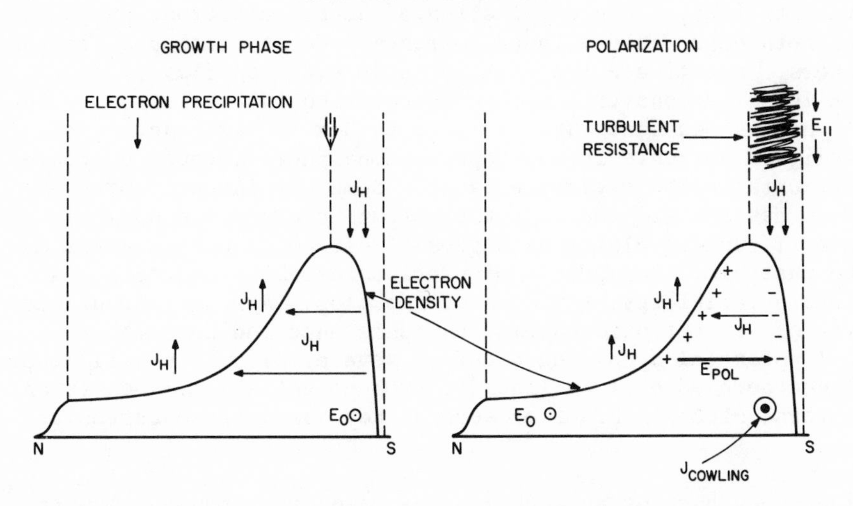

Figure 2. The westward convection electric field E_0 drives a northward Hall current J_H. The gentle northward and sharp southward decrease in electron density forces J_H to close in space. At break-up, the field-aligned closure currents at the southern edge of the oval become unstable, and a parallel anomalous resistance and electric field $E_{\parallel}$ develop. J_H then polarizes, and the polarization electric field E_{POL} drives an intense westward cowling electrojet.

ionospheric E region. Hence both the magnitudes of Σ_P and Σ_H and the ratio Σ_H/Σ_P maximize at the southern edge of the auroral oval.

The increasing westward convection electric field drives (in the northern hemisphere) a westward Pederson current, which must close in the magnetosphere, and a northward Hall current. Since Σ_H decreases northward, the Hall current either generates a southward polarization electric field or also closes via field-aligned currents in the magnetosphere (current out of the ionosphere at the northern edge of the oval, in at the southern edge). As long as the ionosphere and magnetosphere maintain near perfect electrical communication, the highly mobile plasma sheet electrons will short circuit any polarization electric fields, and permit the Hall current to close freely in the magnetosphere. The mapping of the electron plasma sheet inner edge into the oval produces a sharp conductivity gradient at the southern edge. Hence inward field-aligned currents are largest at the southern edge while the outward currents to the north are more distributed. Note that the southern field-aligned current is carried into space by cold ionospheric electrons.

During growth phase the convection electric field build-up coupled with increased precipitation of hotter electrons enhances the southern edge field-aligned currents. For typical pre-breakup parameters, we estimate the parallel cold electron flux to be $J \approx 5 \times 10^9$ electrons cm^{-2} sec^{-1} [35]. Detailed instability calculations[18] have shown that when J exceeds 3×10^9 cm^{-2} sec^{-1}, the field-aligned currents in the topside ionosphere become unstable to electrostatic ion cyclotron waves. Somewhat larger fluxes are needed to destabilize ion acoustic waves; however, the possible presence of runaway electrons suggests that ion acoustic waves may also be unstable. Regardless of precise instability details, an anomalous parallel resistance should develop. A rough turbulence theory[36] of ion cyclotron anomalous resistance indicates that parallel potential drops the order of several kilovolts will occur on the southern edge field lines[35]; such potentials are consistent with auroral electron beams of several KeV which are observed at breakup[37].

The appearance of an anomalous parallel resistance disrupts the perfect electrical communication between the ionosphere and magnetosphere thus preventing the free closure of the Hall current system. When the anomalous resistance becomes comparable to the total north-south Pederson resistance, a southward polarization electric field develops[35]. The polarization electric field drives a westward Cowling current which we identify as the breakup auroral electrojet. Initially the electrojet is localized to local midnight which is the maximum precipitation and tail stress region. Such a rapid development of a southward electric field at breakup has been observed in 19 substorms by Mozer[22].

Thus following the logic of convection build-up determined by the line-tied dayside magnetopause, the resulting changes in the tail configuration, and the tail-imposed structure of the ionosphere and its convection driven current systems, we have arrived at the formation of the auroral electrojet which magnetically is the ground signature of substorm breakup.

5. Substorm Breakup

From observation we know that substorm breakup detected on the ground is accompanied by a rapid flux flow toward the earth[38]. After a 10-20 minute delay the plasma sheet, which has continued thinning after breakup, rapidly expands[39], and the tail magnetic field simultaneously rotates to a less stressed, more dipole-like configuration[24,25]. Here our theoretical understanding is less certain, but we offer some speculation on the tail's response to breakup without claiming that our arguments are self-consistent.

The change of ionospheric electrical boundary conditions at breakup must affect the convective flow and tail current system.

The southward polarization electric field, initially localized to the southern edge, undoubtedly maps, at least in part, into the magnetosphere. The sense of this electric field is to drive the flow toward dawn, removing flux from and reducing plasma pressure gradients in the midnight breakup region. The spatially localized (both east-west and north-south) Cowling current electrojet is probably closed by field-aligned currents from the magnetosphere, although some ionospheric closure undoubtedly also occurs. This large additional demand for current by the ionosphere is satisfied by diverting a portion of the tail current into the electrojet region.

The flow acceleration toward dawn and the reduction of tail current resembles a hydromagnetic piston, which launches a rarefaction wave into the tail. The boundary conditions for the rarefaction wave are that the flow behind the wave must be accelerated toward the earth, the current density behind must be reduced, and the magnetic tension of the stretched tail field lines must be relaxed so that the field behind has a larger component normal to the neutral sheet. These boundary conditions can be satisfied by the slow hydromagnetic rarefaction wave[40]. Thus ionospheric breakup launches a slow rarefaction wave which propagates into the tail accelerating plasma and flux toward the nightside, relaxing the tail magnetic configuration, and producing an expansion of the plasma sheet. The rarefaction wave spreads in east-west extent until the tail boundary is reached. The northward motion of the auroral arcs follow breakup, which is not a consequence of an electric field[41], probably is the ionospheric image of the tail-ward propagating rarefaction wave.

At some point in the substorm, enhanced tail reconnection must field-cut the additional flux transported into the lobes of the tail during growth phase. Since wave propagation times in the plasma sheet or solar wind flow times to the distant tail neutral line exceed typical growth phase durations, enhanced reconnection is unlikely until after breakup. Furthermore, the enhanced adverse plasma sheet pressure gradient required by increased tail flaring tends to choke the reconnection driven tail flow. Two possibilities for triggering reconnection exist: the slow rarefaction wave, after having relaxed the adverse tail pressure gradient, permits enhanced reconnection upon arrival at the distant neutral line; or a new neutral line, perhaps stimulated by the rarefaction wave, forms much closer to the earth and then propagates down the tail.

ACKNOWLEDGEMENTS

It is a pleasure to acknowledge many informative discussions with W. I. Axford, M. P. Aubry, R. L. McPherron, G. L. Siscoe, and R. M. Thorne. C. F. Kennel acknowledges the partial support of the

Alfred P. Sloan Foundation. This research was supported under National Aeronautics and Space Administration, Grant NGR 05-007-190.

REFERENCES

1. Alfven, H., 1955, Tellus, 7, 50.
2. Axford, W.I., 1969, Rev. of Geophys., 7, 421
3. Atkinson, G, 1966, J. Geophys. Res., 71, 5157; 1967, ibid, 72, 1491; 5373.
4. Swift, D.W., 1967, Planet. Space Sci., 15, 1225.
5. Siscoe, G. L., and W.D. Cummings, 1969, Planet Space Sci., 17, 1795.
6. Oguti, T., 1971, Cosmic Electrodynamics.
7. Liu, C.S., 1970, J. Geophys. Res., 75, 3789.
8. McPherron, R.L., 1970, J. Geophys. Res., 75, 5592.
9. Nishida, A., 1968, J. Geophys. Res., 73, 1795; 5549.
10. Fairfield, D.H., and L.J. Cahill, 1966, J. Geophys. Res., 71, 155.
11. Arnoldy, R.L., 1971, J. Geophys. Res., 76, 5189.
12. Dungey, J.W., 1961, Phys. Rev. Letters, 6, 47.
13. Levy, R.H., H.E. Petschek, and G.L. Siscoe, 1964, AIAA J., 2, 2065.
14. Axford, W.I., H.E. Petschek, and G.L. Siscoe, 1965; J. Geophys. Res., 70, 1231.
15. Siscoe, G.L., 1966, Planet. Space Sci., 14, 947.
16. Tverskoy, B.A., 1968, Dynamics of the Earth's Radiation Belts, Publishing House, Physical-Mathematical Literature, Moscow.
17. Spreiter, J.R., and A.Y. Alksne, 1969, Rev. of Geophysics, 7, 11.
18. Kindel, J.M., and C.F. Kennel, 1971, J. Geophys. Res., 76, 3055.
19. Fairfield, D.H., 1968, J. Geophys. Res., 73, 7329.
20. Coroniti, F.V., and C.F. Kennel, 1971, submitted to J. Geophys. Res.
21. Aubry, M.P., C.T. Russell, and M.G. Kivelson, 1970, J. Geophys. Res., 75, 7018.
22. Mozer, F.S., 1971, J. Geophys. Res., 76.
23. Coroniti, F.V., and C.F. Kennel, 1971, submitted to J. Geophys. Res.
24. Fairfield, D.H., and N.F. Ness, 1970, J. Geophys. Res., 75, 7032.
25. Russell, C.T., R.L. McPherron, and P.J. Coleman, 1971, J. Geophys Res., 76, 1823.
26. Cummings, W.D., J.N. Barfield, and P.J. Coleman, 1968, J. Geophys Res., 73, 6687.
27. Coleman, P.J., and R.L. McPherron, 1970, in Particles and Field in the Magnetosphere, ed. B.M. McCormac, D. Reidel Publishing Co., Dordrecht, Holland.
28. Vasyliunas, V.M., 1968, J. Geophys. Res., 73, 2839; 7519.
29. Schield, M.A., and L.A. Frank, 1970, J. Geophys. Res., 75, 137.
30. Petscheck, H.E., and C.F. Kennel, 1966, Trans. Amer. Geophys. Union, 47, 137.

31. Kennel, C.F., 1969, Rev. of Geophys., 7, 379.
32. Vasyliunas, V.M., private communication.
33. Akasofu, S.I., 1969, Polar and Magnetospheric Substorms, Springer, New York.
34. Feldstein, Y.I., 1970, International Symposium on Solar-Terrestrial Physics, program abstract MY-2, Leningrad.
35. Coroniti, F.V., and C.F. Kennel, invited paper IAGA Symposium on Magnetospheric Substorms, Moscow, 1971.
36. Kindel, J.M., 1970, thesis, UCLA.
37. Evans, D.S., 1968, J. Geophys. Res., 73, 2315.
38. McPherron, R.L., and P.J. Coleman, 1970, EOS, Trans. Am. Geophys. Union, 51, 403.
39. Hones, E.W., J.R. Asbridge, and S.J. Bame, 1971, J. Geophys. Res., 76, 4402.
40. Kantrowitz, A.R., and H.E. Petschek, 1966, Plasma Physics in Theory and Application, ed. W.B. Kunkel, 148, McGraw-Hill, New York.
41. Kelly, M.C., J. A. Star, and F. S. Mozer, 1971, J. Geophys. Res., 76, 5269.

PARAMETRIC INSTABILITIES GENERATED IN THE IONOSPHERE BY INTENSE RADIO WAVES

C. Oberman, F. Perkins, and E. Valeo

Plasma Physics Laboratory, Princeton University

Princeton, N. J.

The earth's ionosphere is, from energetic grounds, one of the few regions where man can apply sufficient energy to modify the environment significantly. The energy content of 100 cubic kilometers of ionospheric plasma is of the order of 10 megawatt-secs. This estimate, coupled with an electron-ion temperature relaxation time $\sim$ 10 secs, suggests the H.F. frequency transmitters in the megawatt power range can significantly alter the ionosphere provided the electromagnetic energy can be absorbed. Normally, radio waves incident on the ionosphere are reflected. But, as Perkins and Kaw [1] have shown, the power densities needed to produce heating of the ionosphere are also sufficient to trigger parametric instabilities [2] which cause an anomalous ac resistivity.

The parametric instability is essentially a decay type instability — a transverse electromagnetic wave splits into an electron plasma (Langmuir) wave and an acoustic wave. The consequences of parametric instabilities in the ionosphere are:

1. Creation of short wavelength fluctuations: $10^4 > \lambda > 10^1$ cm. The properties of such fluctuations can be ascertained by Thomson scatter radars.
2. A two-component structure in frequency. The ratio between the electron density

fluctuations in the Langmuir wave $\tilde{n}_L$ to the acoustic wave $\tilde{n}_A$ is

$$\frac{|\tilde{n}_A|^2}{|\tilde{n}_L|^2} = \frac{1}{k^2 D^2} \frac{\nu_e}{\omega_o} \frac{E^2}{E_T^2} ,$$

where ν_e is the electron collision frequency and E_T the threshold field.

3. Heating of energetic electrons.

There is rapidly growing experimental evidence[3] that all these effects occur.

The anomalous ac resistivity depends on the saturation level of the instability. Our nonlinear calculations indicate that this level is determined by nonlinear Landau damping.

ACKNOWLEDGMENT

This work was supported by the Air Force Office of Scientific Research and the Advanced Research Projects Agency.

REFERENCES

1. F. W. Perkins and P. K. Kaw
J. Geophys. Res. 76, 282 (1971)
2. D. F. DuBois and M. V. Goldman
Phys. Rev. Letters 14, 544 (1965);
P. K. Kaw and J. M. Dawson
Phys. Fluids 12, 2586 (1969)
3. Radio-Wave Modification of the Ionosphere
J. Geophys. Res. 75, 6402-6452 (1970)
W. Utlant and R. Cohen
Science (in press)
A. Y. Wong and R. J. Taylor
Phys. Rev. Letters 27, 644 (1971).

STUDY OF A JOVIAN PLASMASPHERE AND THE OCCURRENCE OF JUPITER RADIOBURSTS

L. Conseil[†], Y. Leblanc[†], G. Antonini[††], and D. Quemada[††]

[†] Observatoire de Paris, Meudon

[††] Groupe de Plasmas Stellaires et Planetaires, Université de Paris

INTRODUCTION

The modulation by Io of the rate of occurrence of radiobursts from Jupiter is well known (Bigg 1964), and many authors have tried to explain this influence of Io on the emission of the radiobursts. Duncan (1970) proposed that Io stimulates the emission when it crosses the "magnetopause" of Jupiter. Melrose (1967) suggested that the motion of Io through a "corotating region" inside the magnetosphere of Jupiter may be the cause.

The suggestion of Melrose has led us to identify this "corotating region" with a plasmasphere similar to that which exists around the Earth. But while the boundary of the "corotating region" is a result of electrostatic instabilities at 7 - 8 R_J, the boundary of the plasmasphere is derived from the interaction between the corotating electric field of the planet and the convective field induced by the solar wind inside the magnetosphere. Consequently this boundary must be very sensitive to the solar wind velocity variations, as has been demonstrated for the Earth, from experimental (Carpenter 1966, 1970) and theoretical papers (Nishida 1966, Grebowsky 1970).

OBSERVATIONS

To test this hypothesis, we have looked for a relationship between the solar wind velocity near Jupiter and the geocentric phase of Io during periods of decametric emission. We have chosen a period for which both solar wind data and Jupiter observations are available, that is from August 1964 to November 1964, from September 1966 to March 1967 and from November 1967 to April 1968. We have analyzed the Jupiter observations from Boulder and Florida observatories.

For these observations, we have distinguished the emission related to the 90° Io phase ($70° < \Phi < 110°$), that is the "Early Io" storms, and the emission related to the 240 arc deg Io phase ($150° < \Phi < 350°$), that is "Late-Io" storms ; this terminology is the same as defined by Duncan (1970).

The solar wind data are provided by Pioneers VI and VII, Explorers 33 and 35, and Vela 2 and 3.

EXTRAPOLATION PROCEDURE TO DERIVE THE SOLAR WIND VELOCITY NEAR JUPITER

To obtain the solar wind velocity near Jupiter, we calculated the time delay Δt of the solar corotating plasma flux from the satellite to Jupiter using the formula

$$\Delta t = \alpha / W + (1 + V_s)(r_J - r_E)$$

where α is the angle between the satellite-Sun line and Jupiter-Sun line, W the angular velocity of the Sun seen from Jupiter, V_s the radial velocity of the solar wind, assumed to be constant from the space-craft to Jupiter, and r_J and r_E are the radial distances. This formula was applied by using for V_s the value of each maximum of solar wind velocity, then translating from the spacecraft to Jupiter the whole profile around this maximum using the same value of V_s.

This procedure is reasonable

a - if the propagation effects between the Earth orbit and Jupiter (4 A.U.) are not very important. Until now this hypothesis has not been verified;

b - if the same maximum velocity is observed during many rotations, because in this case the temporal variations are not too high (Couturier and Leblanc, 1970);

c - if the latitudes of the probes are not very different from that of Jupiter ; indeed the solar orbit of the probes is 7° tilted on the solar equatorial plane. It has been shown

(Conseil, 1971) that long lived active regions can produce a latitude-dependence in the solar wind properties;

d - if the heliocentric longitude of the probes are not very different from that of Jupiter : a systematic study made by Gosling (1971) shows that exists a good agreement between the data obtained by two probes if the corotation between the two probes is less than 4 days (50° in longitude). In this case the deviations between the measured velocities are less than 50 Km/s.

For all our statistical results, we have taken into account the last three conditions.

RESULTS

A) Emission related to the 90° Io phase

We have found no significant correlation with the solar wind velocity. We have also looked for a relationship between the rate of change of the solar wind velocity and the phase of Io : Although this correlation is not very significant, it seems that when the rate of change is positive, the Io phase is less than 90°.

B) Emission related to the 240° Io phase

Fig.1 shows the solar wind velocity profile near Jupiter and the Io phase during the emission. In spite of the uncertainties, we can see that each maximum of the solar wind is correlated with a maximum of the phase of Io during emission. This latter maximum occurs generally one or two days later. Moreover before this maximum there is a well pronounced minimum of Io phase which corresponds to the beginning of the wind velocity increase. In other words, there appears to be a relationship between a maximum of velocity and a maximum of Io phase (Io phase is minimum when $\Delta V/\Delta t$ is positive, and Io phase is maximum when $\Delta V/\Delta t$ is negative).

In Fig 2 we have represented the histograms of the number of occurrences of emission as a function of the solar wind velocity and the phase of Io during the emission.

When V_s is less than 500Km/s, there appears a well pronounced maximum for the Io phase near 230° $\pm$ 30°. When V_s is higher than 500 Km/s the diagram is nearly flat with a weak maximum at about 250° of Io phase.

When the velocity is very high ($V_s \geq 600$ Km/s)

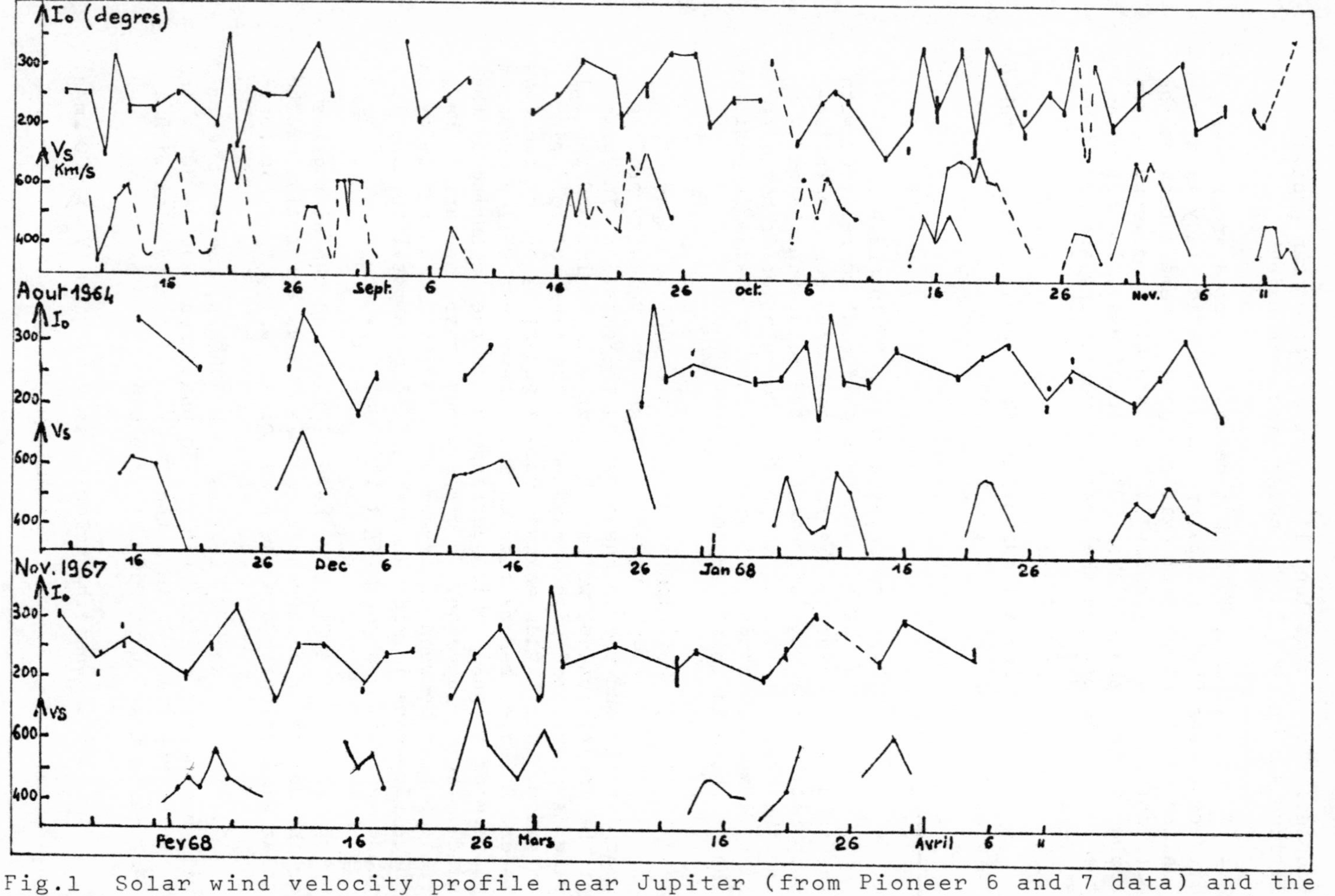

Fig.1 Solar wind velocity profile near Jupiter (from Pioneer 6 and 7 data) and the Io phase during emission-(between the maximum the velocity is less than 400 Km/s and for these low values the profile cannot be extrapolated.)

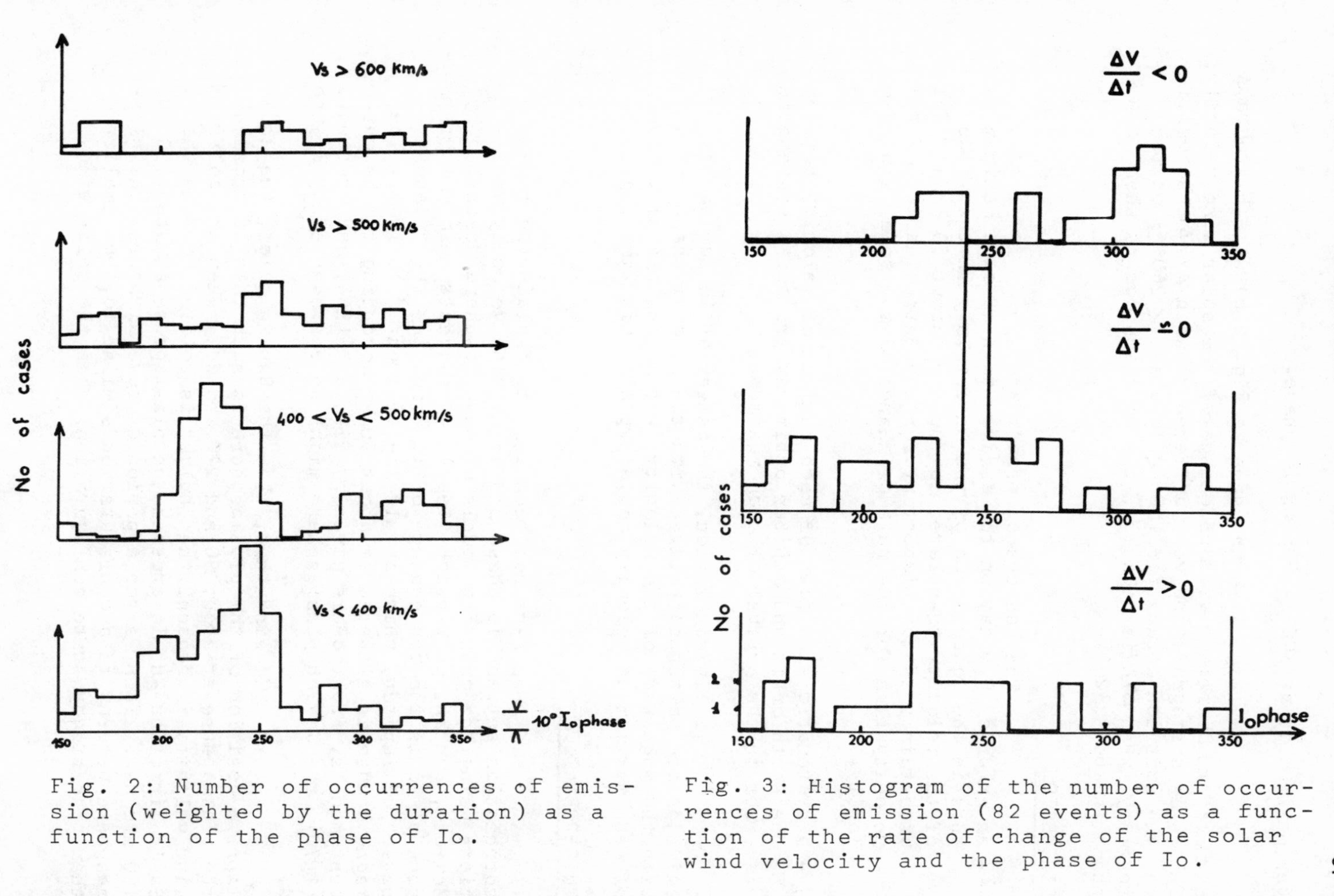

Fig. 2: Number of occurrences of emission (weighted by the duration) as a function of the phase of Io.

Fig. 3: Histogram of the number of occurrences of emission (82 events) as a function of the rate of change of the solar wind velocity and the phase of Io.

the maximum is at 300° ± 50° of Io phase.

Fig. 3 is the same as Fig. 2 but the histograms reflect different rates of change of the solar wind velocity. The three histograms are very different : when $\Delta V/\Delta t$ is positive, the emission occurs for 150° < Io phase < 260° ; on the contrary, when $\Delta V/\Delta t$ is negative, the emission occurs when Io phase is about 320°.

DISCUSSION

The sources of errors are :

1 - an error of 0.9 day on the arrival date of each maximum ; this is the consequence of the uncertainty in the value of the velocity translated to Jupiter (about 50 Km/s). This uncertainty is more important for the determination of the velocity than for the rate of change of the solar wind velocity.

2 - a systematic slow down of each maximum of velocity between the Earth orbit and Jupiter orbit may produce a systematic error in the arrival of each maximum.

In conclusion, the phases of Io during the emission are better correlated with the rate of change than with the magnitude of the velocity, but a possible slow down of each maximum of velocity could explain this time lag.

INTERPRETATION

We have not been able to explain the observed correlation between Io phase and solar wind velocity variations using the existing models of magnetosphere described by Melrose, Duncan, or Warwick (1967). Wu suggests that the influence of Io is the consequence of its crossing the "plasmapause" of Jupiter, which should lie at about 6 to 10 R_J. The magnetopause is situated at about 50 R_J according to the current models. From models given by Nishida (1966) and Grebowski (1970), the Jovian plasmasphere might be represented by Figure 4.

We find that, when the Sun is quiet, the equilibrium position of the plasmasphere is such that Io crosses the plasmapause at about 90 and 240 arc deg. When the solar wind velocity increases, the "bulge" is rotated towards the Sun's direction and Io crosses the plasmapause at about 90 and 180 arc deg. Then, after a period of several hours, the plasmasphere returns to a configuration similar to, but smaller than, the steady state configuration. When the solar wind

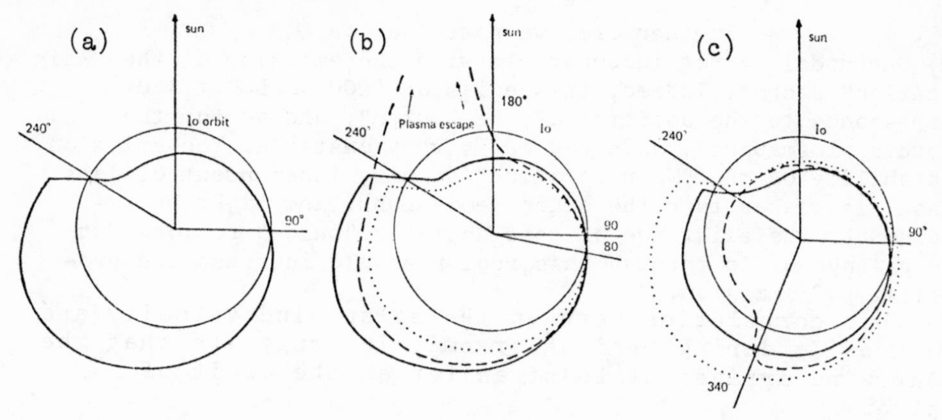

Fig. 4 Plasmapause location in the equatorial plane after Grebowsky (1970). The thick solid line depicts the initial state, the dotted line depicts the final state, and the thin solid line circle the Io orbit. (a) Steady state model of the plasmapause. (b) The evolution of the plasmapause when the solar wind velc-city increases (dashed curve). (c) The evolution of the plasmapause when the solar wind velocity decrea-ses (dashed curve).

decreases the bulge moves in the direction of rotation of Jupiter and Io crosses the plasmapause at abaut 90 and 340 arc deg. Afterwards the plasmasphere tends slowly to an enlarged steady state configuration.

CONCLUSIONS

The above hypothesis enables us to explain the following.

(a) The two positions 90 and 240 arc deg referred to the Earth-Jupiter line are not symmetrical, because of the asymmetric shape of the plasmasphere.

(b) The position 90 arc deg ("Early-Io" storms) is not very sensitive to the solar wind velocity.

(c) The position about 240 arc deg ("Late-Io" storms) is very widely scattered from 160 to 340 arc deg, and depends upon the wind velocity variations.

(d) The lack of influence of the other Jovian Galilean satellites on the decametric emissions (Dulk 1967) can be explained very simply by their trajectories being well beyond the plasmasphere. Such a simple explanation cannot apply to any magnetospheric model because of the magnetotail.

Furthermore, we note that :

(e) Our model is not incompatible with the emission of the "Main rotation" storms. Indeed, this emission ($200^\circ < \mathrm{LCM} < 360^\circ$) corresponds to the position of the "bulge", and as for the Earth's plasmapause, this region is very unstable. The emission probability of the "Main rotation" storms, independent of Io's phase, is weaker than the other emissions, and might be explained by the existence of this unstable "bulge" region. But the passage of Io through this region should increase the probability of emission.

(f) This correlation between the solar wind velocity and phase of Io during periods of emission suggests that the solar wind arrives at least as far as the orbit of Jupiter.

JUPITER PLASMASPHERE MODELS

Up to now, we have not a good theory and available data to predict the exact position of the plasmapause of Jupiter. We may consider theoretical models to see how the plasmapause can be set at 6 Jovian radii.

The boundary of the plasmasphere is derived from the interaction between the corotating electric field of the planet and the convective field induced by the solar wind inside the magnetosphere. We shall assume that the magnetic axis of the planet coïncides with its axis of rotation. The current lines will then represent the equipotential lines in the equatorial plane.

By supposing that the plasmapause lies at $6\ R_J$, it is possible to deduce the convective field; this latter is supposed to be uniform near the planet. The calculations have been performed by using a method described by Nishida (1966).

Two models of co-roration may be considered :

1) If the conductivity of the magnetospheric plasma is infinite, then the hydromagnetic forces tend to carry away the plasma at a V equal to $1.7\ 10^{-4}$ rd/s. It is the isorotation model of Ferraro. In this case the convective field is found to be about 200 mv/m when the plasmapause lies at $6\ R_J$.

2) If the conductivity of the magnetospheric plasma has a finite value, that is the equilibrium state is different from the isorotation, then the plasma may be stabilized in a partial corotation (Alfven, 1968).

In this case the plasma is carried away at the velocity

$$V = (2/3\ K m / r)^{1/2}$$

where K is the gravitational constant, m the Jupiter mass, and r the distance. Then we find that the plasmasphere would lie at 6 R_J, if the convective field is of the order of 15 mv/m.

The value obtained in this last case is still much higher than the estimated value of the convective field of the Earth (0.5 mv/m). Brice and Ioannidis (1970) have estimated the convective field of the Jovian magnetosphere by using Axford and Dungey models. They find a value of 0.05 mv/m. But the authors do not exclude the possibility that the transfer between the solar wind and the magnetospheric plasma is much more effective than in the Earth magnetosphere.

REFERENCES

Alfven, H., 1968, Icarus, 7, 387.
Bigg, E. K., 1964, Nature, 203, 1008.
Brice, N.M., and Ioannidis, G.A., 1970, Icarus 13, 173.
Carpenter, D.L., 1966, J. Geophys. Res., 71, 693.
Carpenter, D. L., 1970, J. Geophys. Res., 75, 3837.
Conseil, L., 1971, Submitted to Astron. Astrophys.
Couturier, P., and Leblanc, Y., 1970, Astron. Astrophys., 7, 254.
Dulk, G. A., 1967, Astrophys. J., 148, 239.
Duncan, R. A., 1970, Planet. Space Sci., 18, 217.
Gosling, J. T. 1971, Sol. Phys. 17, 499.
Grebowsky, J.M., 1970, J. Geophys. Res., 75, 4329.
Melrose, D.B., 1967, Planet. Space Sci., 15, 381.
Nishida, A., 1966, J. Geophys. Res., 71, 5669.
Warwick, J.W., 1967, Space Sci. Rev., 6, 841.

EFFECT OF A LARGE AMPLITUDE WAVE PACKET AND SECOND ORDER RESONANCE ON THE STIMULATION OF VLF EMISSIONS

A.C. Das

Physical Research Laboratory

Ahmedabad, India

This is an extension of a model of VLF emissions triggered by whistlers or man-made signals discussed by Das (1968), and a short introduction will be needed. Following the work of Kennel and Petschek (1966) a background noise level was assumed in the model;then the effect of a whistler mode wave packet on the particles was studied, which can be compared to pitch angle diffusion. The concept is equivalent to that of the quasilinear theory by Vedenov et al (1961, 1962), Engel (1965), Andronov and Trakhtengerts (1964), and Lutomirskiand Sudan (1966), but the problem was treated in a slightly different way which is applicable to the physical situation.

A loss cone in the distribution is introduced and the noise background in the model is maintained by the growth rate of the whistler mode wave because of the steep gradient at the loss cone. A wave packet is then assumed to propagate along the magnetic field line in this medium, and the effect on the particles at the boundary of the loss cone is studied. This forms a localised disturbed region. The particles inside the disturbed region are redistributed and form a fine structure in the distribution function. As time passes, the fine structure in the distribution is smeared out, a new distribution function is assumed and the growth rate is calculated. Recently a detailed computation of trajectories of the particles in the whistler mode wave packet near the loss cone by

Ashour-Abdalla (1970) has shown how the particles behave when the wave packet has gone through, and has confirmed the validity of the assumption of the distribution function in the model.

The results obtained in the model are interesting. The growth rate is reduced at the central frequency while enhancements occur at frequencies slightly above and below the central frequency. Further, the growth rate is found to be larger than the background noise by a factor of two and would be enough for emissions to be seen.

It is seen that there are two main limitations in the model; one is that there is a restriction in the amplitude of the Gaussian wave packet even in the linear theory, and the other is that the magnetic field is assumed to be uniform, which is not the case in reality. In this research note, the effects of the wave packet of amplitude larger than the critical value of the previous model and the non-uniformity of the ambient magnetic field have been studied separately.

MODIFIED MODEL

The modified model is seen in Fig.1. The disturbed region for the restricted model discussed earlier

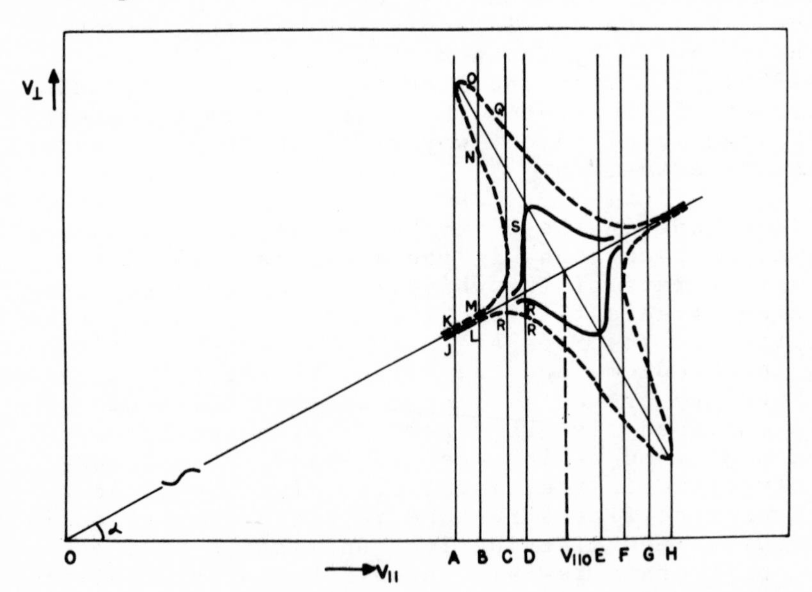

Fig. 1. Disturbed region in velocity space.

is represented by a continuous solid curve while for the present model it is described by the broken curve. The growth rate (δ) computation is done for both cases using the formula given by (Vedenov et al 1962, Das 1968)

$$\delta = \int 1/2\ f_o v_{\shortparallel}'\ v_{\perp}''\ \sec\alpha_o (2/\pi\ \frac{v_{\perp}/\delta v}{(1-(\varepsilon/\delta v)^2)^{1/2}}$$

$$-(1 + \frac{2}{\pi}\sin^{-1}\frac{\varepsilon}{\delta v})\cos\alpha_o)\ dv_{\perp}$$

where ($v_{\shortparallel}$, $v_{\perp}$) represent the position of a particle in the velocity space.

α_o = the loss cone angle

$v_{\shortparallel}' = v_{\shortparallel} + \delta v \sin\alpha_o$

$v_{\perp}'' = v_{\shortparallel}' \tan\alpha_o + \delta v \cos\alpha_o$

$\varepsilon = (v_{\perp} - v_{\shortparallel}\tan\alpha_o)\ v_{\shortparallel}' \sec\alpha_o/(v_{\shortparallel} + v_{\perp}\tan\alpha_o)$

f_o = the undisturbed distribution function such that f_o = constant when $\alpha > \alpha_o$
= 0 when $\alpha < \alpha_o$;

$\delta v = Ae^{-(v_{\shortparallel}' - v_{\shortparallel o})^2 \sec\alpha_o/d^2}$ determines the form of the envelope of the wave packet, A and d representing its amplitude and bandwidth.

The integration, however, becomes complicated in the latter case because the line along which it has to be carried out intersects the curve at four points. A suitable computational method has been used to find the points of intersection which would give the limits of integration. Obviously the limitation on the amplitude in the model has been removed and the growth rate is then calculated.

The results obtained are shown in Fig.2. The growth rate for different frequencies in the range of the wave packet is represented by the broken curve. The growth rate for the restricted model (continuous curve) is also shown for comparison. In both cases, it is seen that the growth rate is reduced at the central frequency while the enhancements occur at frequencies slightly above and below the central frequency,

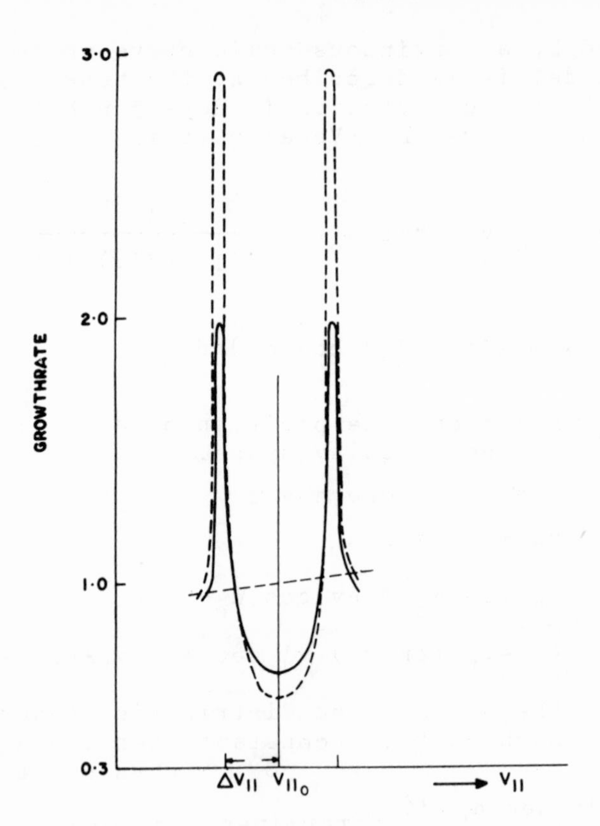

Fig.2. Computed growth rates.

but there is an important change to be noted: the peak in the growth rate for the unrestricted model is almost three times the background level while in the previous case it is only twice the undisturbed growth rate.

This is a clear indication that the nonlinear effects will be important for the large amplification in the model. Ashour-Abdalla (1970) has also obtained the current for the nonlinear case and has shown how the nonlinear effects are important.

In a recent report on occurrence of low latitude whistlers by Tantry (1970), it is noted that the duplicate traces of whistlers at a time difference of 15 milliseconds have been observed frequently. One of the explanations -- that these twin whistlers may be produced by lightning stroke occurring at an interval of 15 milliseconds -- frequently has been ruled out because the probability of occurrence of such lightning stroke systematically is very low. We have an interesting result in connection with these observations. The computation of time lag of these two peaks in the growth rate to be observed on the ground station for a suitable model shows a lag of the order of 20 ms. If the traces observed are treated as emissions the model described here is quite adequate to explain these observations.

TRAPPING TIME AND THE FREQUENCY OF THE TRIGGERING EMISSIONS

Trapping time T is given by $2\pi/T = e/mc(Bb \tan\alpha)^{\frac{1}{2}}$, where b is the amplitude of the wave packet and B and α are the magnetic field and the pitch angle of a particle. For b = 1 milligamma and L = 3, the trapping time T = 0.1 sec. This can be further reduced by increasing the amplitude b of the wave. The time taken by the resonant particle to pass through the wave packet of duration Θ is approximately $t = v_g \Theta / v_{\shortparallel} = 2\frac{w}{w_e}\Theta$ (writing $v_g = 2v_{ph}(w_e - w)/w_e$ and using resonant condition $kv_{\shortparallel} = w_e - w$ where w and w_e represent the frequency of the wave and gyrofrequency of the particle respectively).

For a dash with Θ = 150 ms, and if we assume that $w/w_e = 1/2$, t = 0.15 sec which is slightly greater than the trapping time. Hence the model of VLF emissions based on the trapping mechanism suggested earlier by Das (1968) will be very efficient. This seems to explain why most of the emissions have the tendency to occur at frequencies around one-half the minimum gyrofrequency (Helliwell (1969), Carpenter (1968)).

EFFECT OF NONUNIFORMITY OF THE MAGNETIC FIELD ON THE DISTURBED REGION

For $w \ll w_e$, the width $\delta v_{\shortparallel}$ of the disturbed region is related to the bandwidth and the change of magnetic field by $\delta v_{\shortparallel}/v_{\shortparallel} = -\delta k/k + \delta w_e/w_e$. Writing t_1= duration of the pulse and v_g= the group velocity and considering $v_g = 2v_{ph} = 2w/k$, $\delta k/k = \pi/wt_1 = 2\times10^{-5}$ for t_1=150 ms (dash) and

$w/2\pi=150$ kHz. Thus the distrubed region is very small indeed in a uniform field and the nonuniformity tends to enlarge this region if $\delta w_e/w_e > 2\times10^{-4}$.

SECOND ORDER RESONANCE

The effect of the wave on the trajectory of the particle in presence of a non-uniform magnetic field is discussed. The calculation of the effect will be made using the trajectory the particle would have had, if there were no waves. The quantities to be calculated here are the perturbations in the velocity components, and these give immediately the displacement of the particle in velocity space.

The unperturbed trajectory of the particle is given by

$$v_x = v_o \beta \sin\alpha_{eq} \cos(w_e(z)t + \phi_o)$$
$$v_y = v_o \beta \sin\alpha_{eq} \sin(w_e(z)t + \phi_o) \qquad (1)$$

where $\beta = (1+3\sin^2 x)^{1/2}/\cos^6 x$, x represents the geomagnetic latitude, and v_o the velocity in the equatorial plane assuming the earth's field as a perfect dipole. The magnetic field direction is taken along z-direction and ϕ_o is the initial phase.

Following Trevskoy (1967), we write the electric field vector of the wave travelling along the magnetic dipole field

$$E(z,t) = E(z) \exp i(wt - k_o \int_o^z q(z')dz') \qquad (2)$$

where $E(z) = E_o \beta q^{1/2}$, E_o is the field at z=0 and k_o is the wave number corresponding to the frequency w in the equatorial plane.

Using $\partial b/\partial t = -c$ curl E(z) the change in $v_{\parallel}$ of the particle becomes

$$\delta v_{\parallel} = \mathrm{Im}\,\left(\frac{v_o L E_o e}{mc} \int_{x1}^{x2} S(x) \exp(iw\phi(x))\, dx\right) \qquad (3)$$

where $\phi(x) = (1+w_e/w)\, t - (k_o/w) \int_o^z q(z')\, dz' + \phi_o/w$

and $S(x) = \beta v_{\parallel}^{-1} (1+2\sin^2 x)^{1/2} \cos x \sin \alpha_{eq}$.

The resonance occurs when $\phi(x) = 0$ at $x = x_r$; the stationary phase method is used to integrate.

$$\delta v_{\parallel} = \frac{v_o L E_o e}{2mc} \left(\frac{2\pi}{\phi''(x_r)}\right)^{1/2} \sin\left((w+w_e)t + k_o \int_o^z q(z')dz' + \phi_o - \pi/4\right) S(x_r) \quad (4)$$

The condition $\phi'(x) = 0$ gives

$$\partial w_e/\partial z \; v_{\parallel} \, t + w_e + w - k_o v_{\parallel} q = 0 \quad (5)$$

If $\partial w_e/\partial z \; v_{\parallel} \, t \ll w_e$, (5) is the same as the resonance condition we used before, $kv_{\parallel} - w = w_e$, which together with the dispersion relation becomes

$$w + a\beta - b\beta^{1/2} v_{\parallel} = 0, \quad \text{where } a = eB_o/mcL^3, \quad b = w^{1/2} w_p \, L^{3/2}/c \, w_{eo}^{1/2}.$$

This determines the latitude x_r and is approximately equal to the value obtained by the stationary phase method.

If now $\phi''(x_r) = 0$ also, the next higher order term of the expansion is included and the change in $v_{\parallel}$ is given by

$$\delta v_{\parallel} = \frac{v_o L E_o e}{mc} S(x_r)(6/\phi'''(x_r))^{1/3} \times 1/3 \, \Gamma(1/3) \sin(\pi/6 + \phi_o). \quad (6)$$

If $\phi''(x) = 0$ at x_r and $\phi'(x) = 0$ (x_o can be taken as zero for convenience), then second order resonance occurs and this gives

$$(t\delta w_e - k_o q)\delta v_{\parallel} = -v_{\parallel}(k_o v_{\parallel} q/2w_e + 2)\delta w_e \quad (7)$$

In general, it is seen from (4) that the resonance effects become stronger in this case. Using (6), (7) becomes quadratic in $v_{\parallel}$. A particle seems to have two velocities as a consequence of second order resonance and physically it means that there are particles in the disturbed region moving in opposite directions in velocity space as if they were trapped. Thus the effect of second order resonance seems to be nonlinear in the sense that the trapping occurs; the importance of these trapped particles has already been discussed in the model of VLF emissions. This calculation does not include the change in frequency with time. If the frequency is allowed to change with time then from $\phi''(x_r) = 0$, a shift in emission frequency due to this resonance is observed.

ACKNOWLEDGMENT

I would like to thank Mr. N.K. Vyas and Mr. V.H. Kulkarni for computations and useful discussions.

REFERENCES

Andronov A.A. and V. Yu. Trakhtengerts (1964) Geomagnetism and Aeronomy 4, 181.

Ashour-Abdalla M. (1970) Planet. Space. Sci. 18,1799.

Carpenter D.L. (1968) J. Geophys. Res. 73, 2919.

Das A.C. (1968) J. Geophys. Res. 73, 7457.

Engel R.D. (1965) Phys. Fluids, 8, 939.

Helliwell R.A. (1969) Reviews of Geophys, 7, 281.

Kennel C.F. and H.E. Petschek (1966) J.Geophys. Res. 71, 1.

Lutomirski R.E. and R.N. Sudan (1966) Phys. Rev. 147, 156.

Tantry B.A.P. (1970) Annual Technical Report 1, Occurrence of Low Latitude Whistlers, PL480 Research Project, ESSA Research Laboratories, Boulder, Colorado.

Tverskoy B.A. (1967) Geomagnetism and Aeronomy 7, 226.

Vedenov A.A., E.P. Velikhov and R.Z. Sagdeev (1961) Soviet Phys. Usp. 4, 332.

Vedenov A.A., E.P. Velikhov and R.Z. Sagdeev (1962) Nucl. Fusion Suppl. 2, 465.

UNIVERSAL INSTABILITY ASSOCIATED WITH THE PLASMAPAUSE AND ITS ROLE IN GEOMAGNETIC MICROPULSATIONS

H. KIKUCHI

NASA Goddard Space Flight Center

Greenbelt, Maryland

ABSTRACT

A steep plasma density gradient at the plasmapause is likely to be an origin of the 'universal' instability in the magnetosphere, as inferred from the theory and laboratory experiments of a non-homogeneous magnetoplasma. Drift waves excited at the plasmapause may be unstable in the direction of the electron drift and propagate eastwards nearly perpendicular to the magnetic field. The drift waves, however, tend to convert very soon to ion sound or Alfvén waves with a much larger phase velocity parallel to the magnetic field. This may be a possible source mechanism for rather regular geomagnetic micropulsations. In a very low β plasma (β = plasma pressure/magnetic pressure, $\beta < m_-/m_+$, m_- = electron mass, m_+ = ion mass) which is likely to exist just beyond the equatorial plasmapuse, the ion sound waves are purely electrostatic and are unstable for $\omega \lesssim \omega^*$ (ω = angular frequency, $\omega^* = k_y v_D$, k_y = wave number in the y-direction, v_D = electron drift velocity). When the perpendicular wavelength decreases and becomes comparable with or less than the ion Larmor radius ($Z = k_y^2 \rho_+^2 \gtrsim 1$), the accelerated ion sound waves tend to degenerate into the drift waves whose frequency decreases with increasing Z. In a not too low β plasma ($\beta > m_-/m_+$) which is likely to exist just below the equatorial plasmapause, the accelerated sound wave tends to couple to the Alfvén wave in an oblique direction in the case of nearly zero ion Larmor radius or at very low frequencies. This may be a possible excitation mechanism for long-period micropulsations. For the finite ion Larmor radius or higher frequencies, the drift wave with $\omega = \omega^*$ is unstable for perturbations which shift the wave frequency upwards, and tends to convert very soon to the Alfvén wave. The possibility of field-aligned plasma irregularities or ducts near the plasmapause may facilitate the mode coupling between drift waves, slow and fast Alfvén waves when the parallel wavelength of these waves is of the order of the scale of irregularities. This may be a possible excitation mechanism for short-period micropulsations (Pc-1,2) and indicates the role of the universal instability originated at the plasmapause in these geomagnetic micropulsations. These arguments are examined on the basis of a theory of the universal instability and experimental results obtained from satellite plasma and ground micropulsation data.

1. EXPERIMENTAL BACKGROUND

A steep plasma density gradient and fluctuating plasma distributions at and near the plasmapause observed from OGO-1, -3, and -5 are linked in some cases with rather regular micropulsations observed on the ground (Kikuchi and Taylor, 1969; Kikuchi, 1970, 1971). While the existence of the plasmapause is a rather permanent feature in the magnetosphere, plasmapause-associated irregularities are most likely formed along a magnetic field line during the recovery phase of a storm. These regular and irregular features may be important in two ways. First, the steep plasma density gradient and plasma irregularities at and near the plasmapause may excite drift waves and lead to the universal instability. Second, the wave-like fine structure (temporal) within the larger plasma fluctuations (spatial) has in some cases a periodicity comparable with the micropulsation wavelength in the medium near the equator. Thus, a duct that will provide some guidance for hydromagnetic waves can be formed between the trough of a plasmasphere fluctuation and the plasmapause.

A correlative study of satellite plasma measurements from OGO-3 and ground-based Pc-1 observations reveals that:

(a) substantial agreement exists between the plasmapause crossings identified from the satellite and the Pc-1 occurrence positions observed on the ground at midlatitudes during the nighttime (including dawn and dusk);

(b) comparison of closely spaced plasmapause crossings and Pc-1 events indicates that the Pc-1 propagation paths tend to fall within the region of plasmapause-associated irregularities observed from the satellite; (c) the propagation paths of nighttime Pc-1 events at midlatitudes appear to possess a maximum (bulge) in the dusk-side region and a minimum in the predawn hours, consistent with the local time variation of the plasmapause boundary.

These observations indicate a good correlation between the plasmapause and micropulsations and are discussed in this paper on the basis of a 'drift' wave model.

2. THEORETICAL BACKGROUND

The theory of universal instability has so far been developed primarily in connection with laboratory experiments on the so-called 'drift' waves (D'Angelo and Motley, 1963; Buchelnikova, 1964; Lashinsky, 1964). These experiments have been explained in terms of the collisionless 'drift' instability or the 'universal' instability. The collisionless theory of Rosenbluth (1965) applies to the case of very low β plasmas in which a pure electrostatic treatment is adequate. A more general case covering not too low β plasmas has been treated by Mikhailovskii (1967).

While applications of this model to magnetospheric plasmas are still very few, the relevance of the 'universal' instability theory to the observations of pulsating (optical) auroras and X-ray pulsations has been discussed by D'Angelo (1969).

In view of the application to tenuous magnetospheric plasmas, specifically to the region of the plasmapause, a very brief survey is given here of the universal instability in a non-homogeneous plasma with emphasis on physical intuition.

In a homogeneous Maxwellian plasma, particles absorb energy from the waves in the neighbourhood of the wave phase velocity, leading to Landau damping. In a non-homogeneous plasma, however, particles whose velocities are infinitesimally faster than the wave phase velocity may prevail owing to the existence of a plasma density gradient, leading to the universal instability. We take the magnetic field B_o (steady and uniform) in the positive z-direction and the density gradient ∇n in the negative x-direction. Assuming that perturbations vary as $\exp[-i\omega t + i(k_y y + k_z z)]$ and employing an electrostatic treatment, the work done by the electric field on the particles is written as

$$\delta W = -eE_z \left[f_1(x,v_z)v_z \delta v_z\right]_{v_z = \omega/k_z}$$
$$\simeq \frac{\pi e\omega}{k_z^2} E_z \left[-\frac{e}{m_-} E_z \frac{\partial f_o}{\partial v_z} + \frac{E_y}{B_o}\frac{\partial f_o}{\partial x}\right]_{v_z = \omega/k_z} \tag{1}$$
$$\simeq \frac{\pi e^2\omega^2}{KT_- k_z^3} E_z^2 (1 - \frac{\omega^*}{\omega})\, f_o(x, \frac{\omega}{k_z})$$

where

$$f_o(x,v_z) = \frac{n(x)}{\sqrt{\pi} v_{T_-}} \exp(-v_z^2/v_{T_-}^2),\ v_{T_-} = \sqrt{\frac{2KT_-}{m_-}}$$
$$\omega^* = k_y v_D,\ v_D = \kappa KT_-/m_+\omega_{c+} = \kappa\rho_s v_s = (\rho_s/a)v_s,\ \kappa = \frac{1}{a} = -\frac{1}{n}\frac{dn}{dx} \tag{2}$$
$$v_s = \omega_{c+}\rho_s = \sqrt{KT_-/m_+} = \text{velocity of sound}$$
$$\rho_s = \text{ion Larmor radius at the electron temperature}$$

The first term in Eq.(1) represents Landau damping. When the second term exceeds Landau damping, i.e. for $\omega < \omega^*$ or $v_p < v_D \sin\theta$, the plasma will be unstable. In the presence of a temperature gradient, ω^* may be replaced by $\tilde{\omega} = \omega^*(1-\eta/2)$, where $\eta = d\ln T_-/d\ln n$. In the magnetosphere, we normally have $\eta < 0$, because the electron temperature increases with increasing altitude. Then the instability boundary is extended more or less up to $\omega = \tilde{\omega} > \omega^*$. Furthermore, in a not too low β plasma, i.e. $\beta > m_-/m_+$, for instance, a hydromagnetic correction must be made in Eq.(1), yielding a new expression

$$\delta W \approx \frac{\pi e^2\omega^2}{KT_- k_z^3} E_z^2 \left[1 - \frac{\omega^*}{\omega} + \frac{\omega^*(\omega+\omega^*)}{\omega(\omega+\omega^*) - k_z^2 v_A^2}\right] \tag{3}$$

where the third term is a hydromagnetic correction. Then, the instability develops just downward of the Alfvén wave branch ($\omega \gtrsim 0$) of the dispersion curves, and accompanies a strong coupling between the drift and Alfvén waves when $k_z = \pm\sqrt{2}\,\omega^*/v_A$ (see Fig.1).

In order to investigate the drift instabilities in detail, the dispersion relation must be discussed on the basis of kinetic considerations. Combining Maxwell's equations for the ELF range (hydromagnetic equations) and the collisionless Boltzmann equation, the dispersion equation of a non-homogeneous plasma may be written, for the range $v_{T+} < |\omega/k_z| < v_{T-}$, as

$$\left\{(\omega-\omega^*)(1+i\frac{\sqrt{\pi}\omega}{k_z v_{T-}}) - \frac{k_z^2 v_s^2}{\omega^2} I_o e^{-Z}(\omega+\omega^*)\right\}(\omega^2+\omega^*\omega - \frac{Zk_z^2 v_A^2}{1-I_o e^{-Z}}) =$$
$$= Zk_z^2 v_A^2(\omega+\omega^*) , \tag{4}$$

where $Z = k_y{}^2\rho_+^2$, and $I_o(Z)$ is the Bessel function of imaginary argument. The first and the second factor on the left hand represent the ion sound and the Alfvén wave branch, and both branches are coupled to each other. We now consider long wave $(Z<<1)$ and short wave $(Z\gtrsim 1)$ perturbations separately.

(a) Case when $Z<<1$. Referring to the relation $I_o(Z)_e^{-Z} = 1-Z$, Eq.(4) may be separated into two nearly independent dispersion equations:

$$(\omega-\omega^*)(1+i\frac{\sqrt{\pi}\omega}{k_z v_{T-}}) - (\omega+\omega^*)\frac{k_z^2 v_s^2}{\omega^2} = 0 \tag{5}$$

and

$$\omega^2 + \omega^*\omega - k_z^2 v_A^2 = 0 \tag{6}$$

Eq.(5) indicates that the ion sound waves are unstable for $\omega<\omega^*$, as was shown in Eq.(1). When $|\omega/k_z|>>v_s$, the waves degenerate into the drift wave

$$\omega = \omega^* \tag{7}$$

In the presence of a temperature gradient, $\omega_i = \mathrm{Im}\omega$ may be written as

$$\omega_i = -\frac{\sqrt{\pi}}{2}\frac{\omega^{*2}}{k_z v_{T-}}\eta . \tag{8}$$

Then the drift wave is unstable for $\eta<0$.

The Alfvén waves in Eq.(6) have two branches. One is the decelerated branch $(\omega>0)$ in the direction of the electron drift, and the other is the accelerated branch $(\omega<0)$ in the direction of the ion drift. The dependence of ω on k_z determined by these dispersion equations is shown in Fig.1.

The conversion of the drift waves into the ion sound or Alfvén waves indicates a gradual change of the direction of wave propagation from y- to z-direction via oblique propagation. Furthermore, actual instability or coupling between these waves is most pronounced in some oblique directions when the wave phase velocities are close together. In this sense, the use of phase velocity surfaces seems most adequate for clarification of the physical picture. Fig.2 shows phase velocity surfaces for a variety of waves in a non-homogeneous plasma.

(b) Case when $Z \gtrsim 1$. For low frequency longitudinal waves ($|\omega| << |k_z| v_A$) in a low β plasma ($\beta < m_-/m_+$), the ion sound waves may be separated from Eq. (4) and written as

$$2\omega + i\frac{\sqrt{\pi}\omega}{k_z v_{T_-}}(\omega-\omega^*) - I_o e^{-Z}\left(1+\frac{k_z^2 v_s^2}{\omega^2}\right)(\omega+\omega^*) = 0 \; . \qquad (9)$$

For nearly perpendicular propagation ($|\omega| >> |k_z| v_s$), the accelerated ion sound wave tends to degenerate into the drift wave whose frequency and growth rate are expressed as

$$\omega_r = \omega^* \frac{\lambda}{2-\lambda} \; , \quad \omega_i = 2\sqrt{\pi}\,\frac{\omega^{*2}}{k_z v_{T_-}}\,\frac{\lambda(1-\lambda)}{(2-\lambda)^3} \; , \qquad (10)$$

where $\lambda = I_o(Z)e^{-Z}$. The drift wave frequency is now

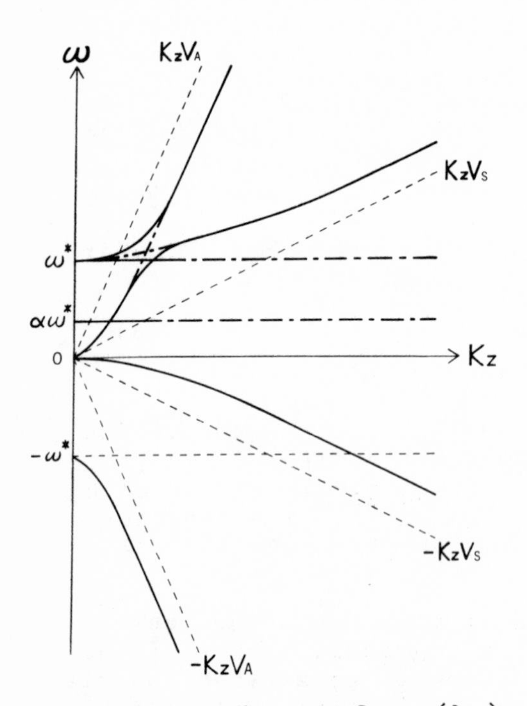

Fig.1. Dependence of $\omega=\omega(k_z)$ for waves in a non-homogeneous plasma.

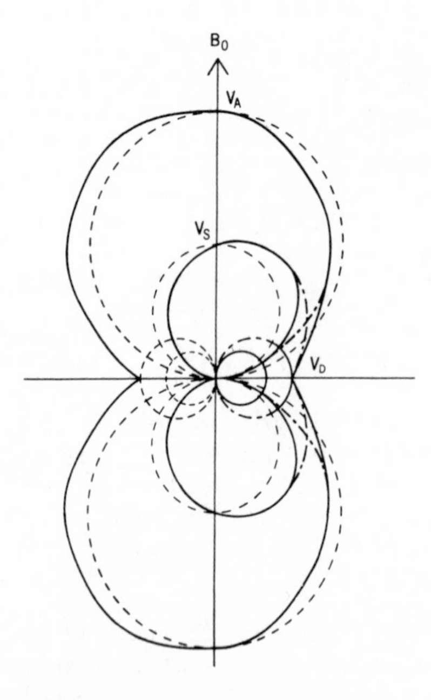

Fig.2. Phase velocity surfaces for waves in a non-homogeneous plasma.

shifted downwards because of the effect of the finite Larmor radius.

For the Alfvén wave branch in a not too low β plasma ($\beta \gtrsim m_-/m_+$), Eq.(4) leads to

$$(\omega-\omega^*)(1+i\frac{\sqrt{\pi}\,\omega}{k_z v_{T_-}})(\omega^2+\omega^*\omega-\frac{Zk_z^2v_A^2}{1-I_o e^{-Z}})=Zk_z^2v_A^2(\omega+\omega^*). \qquad (11)$$

For nearly perpendicular propagation ($|\omega|>|k_z|v_A>>\sqrt{2/Z}\omega^*$), the decelerated Alfvén wave tends to degenerate into the drift wave whose frequency and growth rate are expressed as

$$\omega \approx \omega^* + i\frac{2v_{T_-}}{\sqrt{\pi}}k_z \sim \omega^*(1+i\frac{2}{\sqrt{\pi}}) \quad . \qquad (12)$$

3. THE PLASMAPAUSE AND UNIVERSAL INSTABILITY

In order to apply the universal instability theory to the region of the plasmapause, we must fix the relevant plasmapause parameters. This has been done by utilizing satellite plasma and magnetic field data obtained from a series of OGO satellites. Fig.3 shows a model of equatorial Alfvén and thermal (electron and ion) velocity profiles based on satellite measurements. The plasmapause is identified at the position of L=5.5 by a sharp increase in the Alfvén velocity, indicating a decrease in plasma density. In fact, the magnetic field

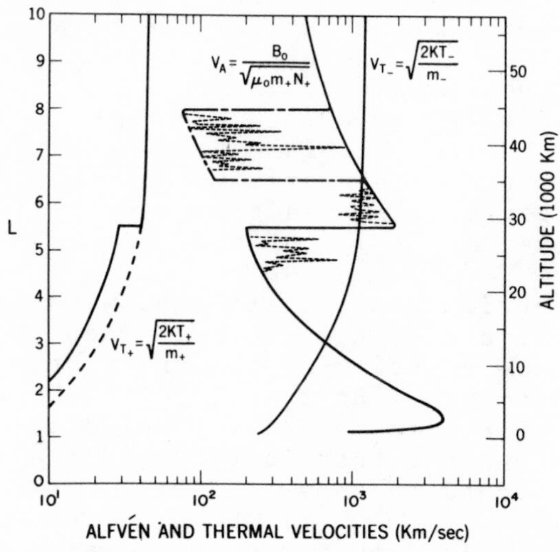

Fig. 3. Alfvén and thermal velocity profiles.

does not change appreciably across the plasmapause. The chain line in the Alfvén profile indicates the existence of the double plasmapause due to the plasma recovery in the dusk local time sector. The saw-tooth-like dashed lines near the plasmapause indicate plasma fine structure irregularities which seem to correlate in some cases with micropulsations and which are most likely formed during the recovery phase of a storm. The solid and dashed curves in the ion thermal velocity profile represent a nightside and dayside profile respectively and indicate an increase in the nightside ion temperature beyond the plasmapause. While the electron thermal velocity increases monotonically with increasing height, the Alfvén velocity displays a minimum and a maximum at the plasmapause in addition to another maximum at the height of about 3,000 km. The Alfvén velocity profile intersects the electron thermal velocity profile at a position of the outer plasmasphere, at the plasmapause, and again at a position of the inner plasma trough, and then tends to approach the ion thermal velocity near the magnetopause. Consequently, we have a magnetohydrodynamic region ($\beta > m_-/m_+$) just below the plasmapause.

We are now in a position to estimate the velocity of a drift wave. Using values, $\omega_{c+} \gtrsim 20$ rad/sec, $v_s \gtrsim 20$ km/sec, a = plasmapause thickness $\approx$1-20 km, we obtain $v_D = \rho_s v_s / a = v_s^2/\omega_{c+} a \approx 1\text{-}20$ km/sec. The ratio between the perpendicular and the parallel wavelengths can be roughly estimated from the relation $v_{T+} < |\omega/k_z| < v_{T-}$ as $k_z/k_y = \lambda_y/\lambda_z \approx 10^{-2}\text{-}10^{-3}$. If for λ_z we assume the overall length of a field line at the plasmapause ($\lambda_z \approx 10^5$km) as an example of long-period micropulsations, we obtain $\lambda_y \gtrsim$ 100-1,000 km. Then the frequency and phase velocity of the drift wave are roughly estimated as $f^* = v_D/\lambda_y \approx 10^{-2}\text{-}10^{-3}$ Hz and $v_z = \lambda_z f^* \approx 100\text{-}1{,}000$ km/sec. This indicates that just beyond the plasmapause the parallel phase velocity is less than the Alfvén velocity 2,000-4,000 km/sec, and an estimate of the growth rate of the wave from Eq.(10) gives a τ_{growth} of the order of 1-10 minutes.

For short-period micropulsations, we assume $\lambda_z \gtrsim$ 1,000 km, thus obtaining $\lambda_y \approx 1\text{-}10$ km. Then the frequency and parallel phase velocity of the drift wave are roughly estimated as $f^* \gtrsim 0.1\text{-}1$ Hz and v_z 200-1,000 km/sec, which is comparable with the Alfvén velocity. An estimate of the growth rate from Eq.(12) gives a τ_{growth} of the order of a second or somewhat less.

We have thus found that the drift waves excited at the plasmapause may grow east-north-wards and tend to convert very soon to the ion sound or Alfvén waves.

4. MICROPULSATIONS AND THE PLASMAPAUSE

We now discuss the observations which exhibit a close correlation between the plasmapause and micropulsations. Fig.4 shows an example of closely spaced OGO-3 plasmapause crossings and Pc-1 micropulsation events in terms of the local time-L coordinate system on the left panel, and in terms of the plasma density profile on the right panel. The boundary curve represents the average of the plasmapause locations obtained from all identifiable OGO-3 plasmapause crossings. The positions of the plasmapause and three Pc-1 events on June 27 are very close to the average plasmapause boundary, and three closely spaced Pc-1 events tend to fall within plasmapause-associated irregularities during the post-storm recovery. Plasma fluctuations just below the plasmapause within these irregularities possess a periodicity of 0.1-0.2 in L which is comparable with the Pc-1 wavelength in the medium (approximately 700-800 km). A survey of the literature indicates that longer-period micropulsations tend to occur at higher latitudes, while our preliminary observations indicate that short-period micropulsations occur most likely just below the plasmapause within plasma irregularities which appear to be elongated along field lines, forming a 'duct' or 'hydromagnetic' waveguide. The universal instability theory indicates that drift waves are likely to occur eastwards at the nightside plasmapause and tend to convert to either ion sound waves or Alfvén waves. For short-period micropulsations, these waves tend to propagate along field-aligned plasma irregularities or a 'duct' and part of them will couple to the fast Alfvén wave with right-hand polarization. This hypothesis might explain occasional observations of repetitive falling tones superimposed on rising tones in Pc-1 sonagrams. This is also supported by temporal displacements of Pc-1 propagation paths similar to the local time variation of the plasmapause boundary.

The nighttime enhancements of Pc-1 events particularly in the midnight to dawn local time sector seem to be supported by the universal instability model if one recalls that drift waves will grow in the region of the equatorial nightside plasmapause in the direction of the electron drift.

In summary, it can be stated that the universal

instability model may provide a most plausible explanation for a close correlation between micropulsations and the plasmapause which has been obtained from satellite and ground observations.

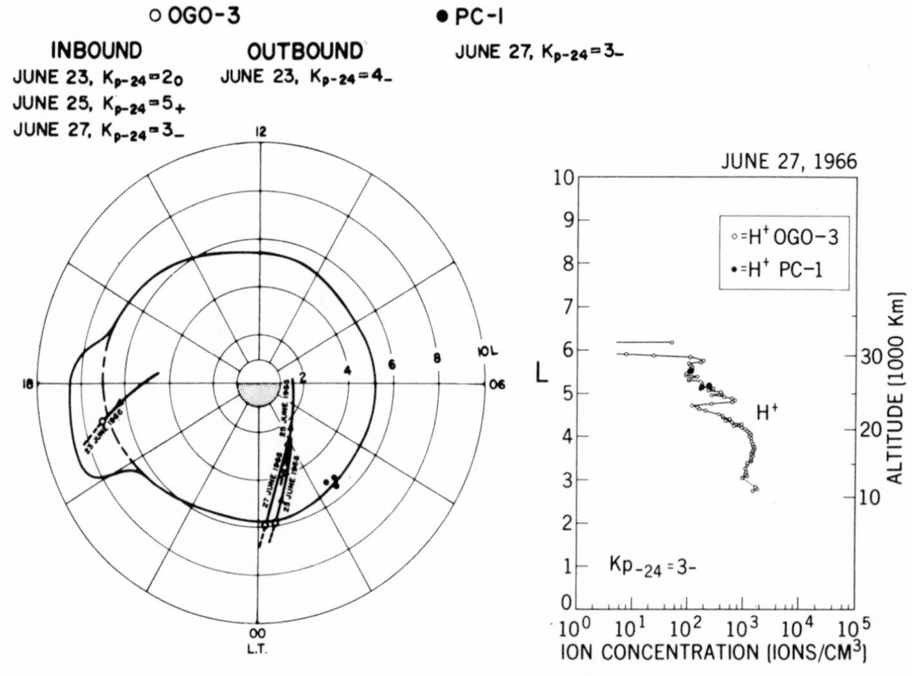

Fig. 4. OGO-3 plasmapause crossings and Pc-1 events in June, 1966.

REFERENCES

Buchelnikova, N.S., Nuclear Fusion 4, 165 (1964).

D'Angelo, N. and R.W. Motley, Phys. Fluids 6, 422 (1963).

D'Angelo, N., J. Geophys. Res. 74, 909 (1969).

Kikuchi, H. and H.A. Taylor, NASA/GSFC Document X-621-69-507 (1969); J. Geophys. Res. in press, 1971.

Kikuchi, H., Nature Phys. Sci. 229, 79 (1971).

Lashinsky, H., Phys. Rev. Letters 12, 121 (1964).

Mikhailovskii, A.B., Review of Plasma Phys., 1967.

Rosenbluth, M.N., in Plasma Physics, Intern. Atomic Energy Agency (IAEA), Vienna, 1965, p. 485.

DEFORMATION AND STRIATION OF BARIUM CLOUDS IN THE IONOSPHERE

F. W. Perkins*, N. J. Zabusky, and J. H. Doles

Bell Telephone Laboratories, Whippany, N. J.

Barium releases provide a means for testing our understanding of ionospheric plasma dynamics by introducing controlled perturbations.[1-4] This work is devoted to the theory of plasma clouds which are sufficiently large to dominate the Pederson conductivity on lines of force which pass through the cloud and sufficiently high so that $K_i = \Omega_i / \nu_{iA} \gg 1$. Figure 1 shows the model.

In these circumstances, the equations which describe the motion of Barium clouds reduce to

$$\nabla_\perp N \nabla_\perp \phi = \nabla_\perp N \cdot \underset{\sim}{E}_o \ , \tag{1}$$

$$\frac{\partial N}{\partial t} - \nabla_\perp N \cdot \nabla_\perp \phi \times \hat{b} \frac{c}{B} = 0 \ , \tag{2}$$

where $N = \int_{-L}^{L} n dz$ corresponds to the Pederson conductivity.

$\underset{\sim}{E}_o$ denotes the ambient electric field in the ionosphere. The boundary condition is that ϕ vanish as the distance from the cloud increases.

---

* Permanent address: Plasma Physics Laboratory, Princeton University, Princeton, N. J.

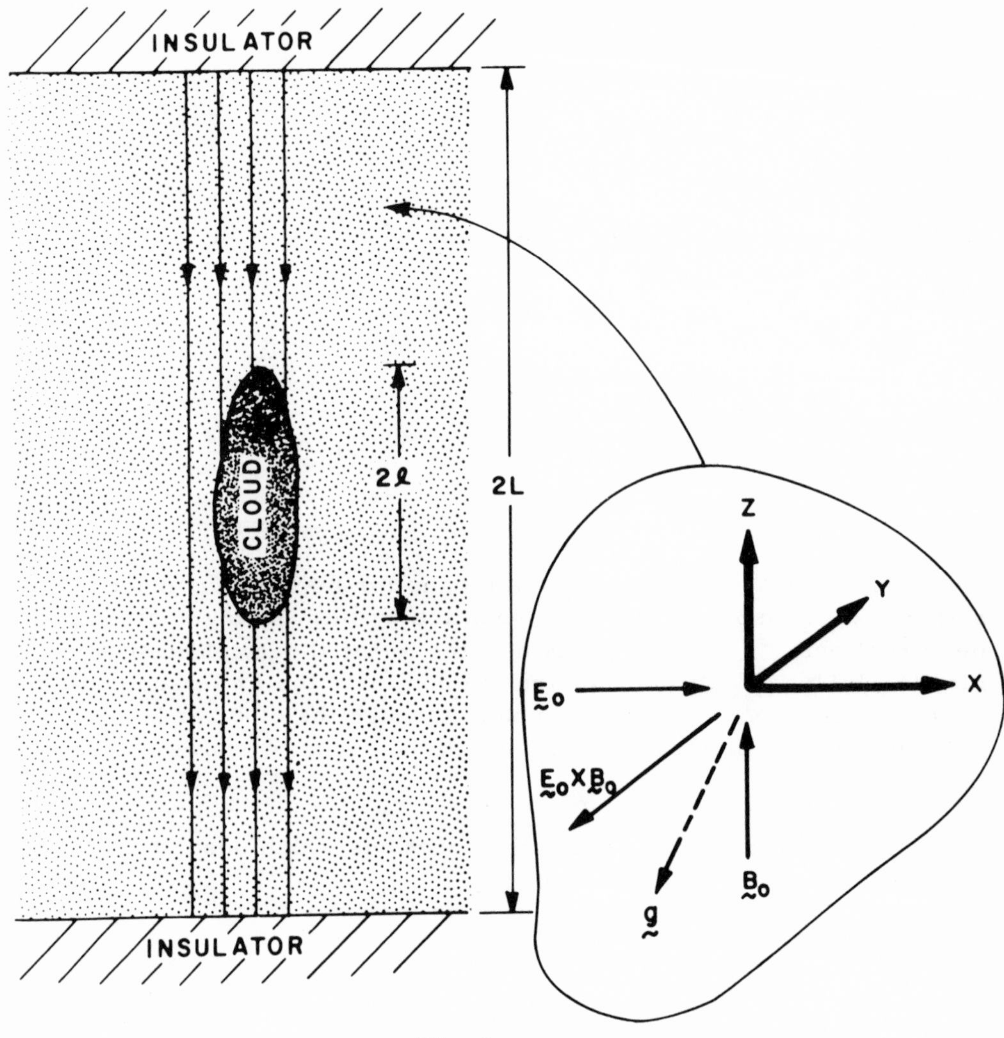

Fig. 1

Plasma cloud - ionosphere model. The ambient plasma has a uniform density n_A and ν_{in} is constant.

The set of Eqs. (1) and (2) has been integrated numerically, and Fig. 2 shows how an initially cylindrically symmetric cloud deforms.

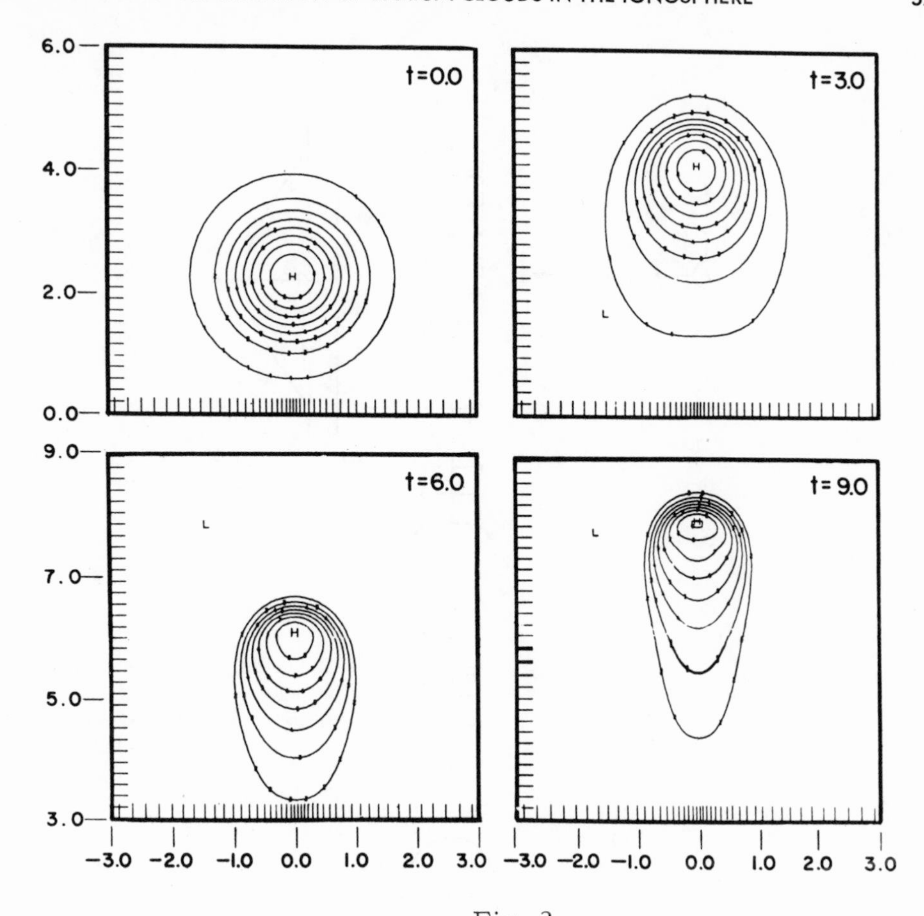

Fig. 2

Deformation of a cylindrical cloud.

The density gradient becomes steep on the side near the neutral cloud and the isodensity contours become elongated in the direction of motion. It should be emphasized that this deformation is the result of convective motions and not diffusion.

An approximate analytic solution to Eqs. (1) and (2) for a thin-bar model of a plasma cloud is shown in Fig. 3. This figure is useful in understanding the nonlinear development of Barium cloud striations. Once a striation contains a density maximum, the striation itself developes a steep density gradient and tends to "pinch off".

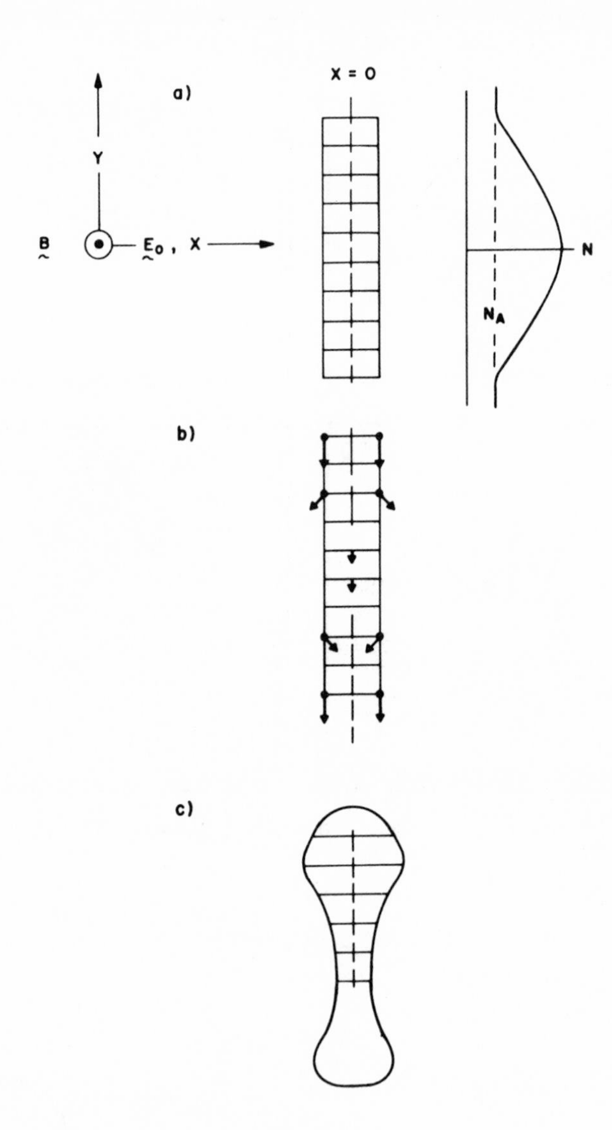

Fig. 3

a) Plasma cloud modeled as a thin bar. b) $\underset{\sim}{E} \times \underset{\sim}{B}$ velocities. c) Deformation of bar.

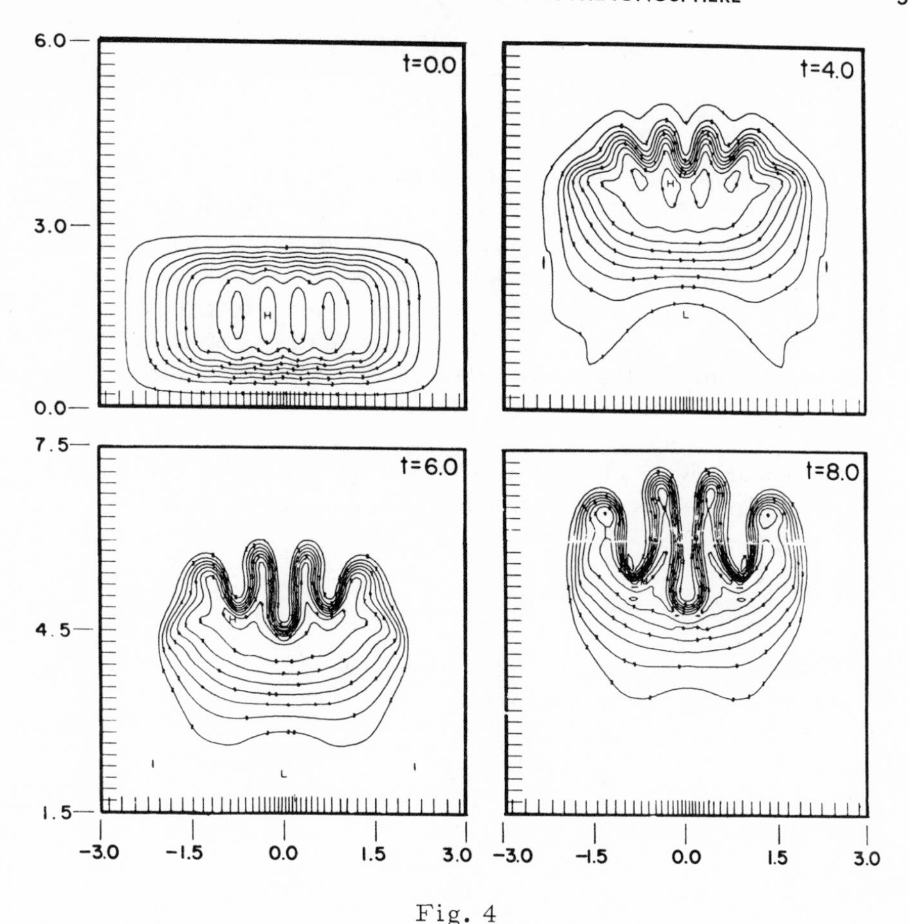

Fig. 4

Evolution of a perturbed cloud.

Figure 4 shows the evolution of a cloud with initial perturbations.

The pinching effect of Fig. 3 is evident in the later stages of evolution. The number of striations is closely tied with the initial conditions since Eqs. (1) and (2) do not permit reconnection of isodensity contours.

We have also carried out analytic investigations of the role of parallel resistivity on plasma cloud striations. The result of a quadratic form analysis is that the growth rate is given by

$$\gamma = \frac{cE_o}{R} \frac{\langle \nabla_o n \rangle}{\langle n \rangle} - \frac{2T_e}{m\nu_{eA}} \frac{\langle |\nabla_{\parallel} \phi|^2 \rangle}{\langle |\phi|^2 \rangle} . \tag{3}$$

Here the $\langle \; \rangle$ brackets denote an average along field lines, ν_{eA} is the electron collision frequency in the ambient ionosphere, and ϕ is that portion of potential which drives currents (i.e., not the part which forces diffusion to be ambipolar diffusion). The second term, which was not included in the work of Völk and Haerendel[2], determines the short wavelength cutoff for striation and requires threshold electric field to overcome the short-circuiting[2] effect.

This work was supported by the Department of the Army, Advanced Ballistic Missile Defense Agency under Contract DANC 60-69-C-0008.

REFERENCES

1. G. Haerendel, R. Lüst, and E. Rieger
Planet Space Sci. 15, 1 (1967)
2. H.J. Völk and G. Haerendel
J. Geophys. Res. 76, 4541 (1971)
3. A. Simon
J. Geophys. Res. 75, 6287 (1970)
4. L. M. Linson and J. B. Workman
J. Geophys. Res. 75, 3211 (1970).

THERMAL ENERGY TRANSPORT IN THE SOLAR WIND

Michael D. Montgomery

University of California

Los Alamos Scientific Laboratory, Los Alamos, New Mexico

INTRODUCTION

This paper is intended to summarize the present status of measurements of heat flux in the solar wind and to provide a comparison of these measurements with the theory for collision-dominated heat transport in a fully ionized medium developed by Spitzer and Härm [1953]. A short discussion of some recent models is included to illustrate the role of thermal conduction in the expansion. In addition, a brief description of the steady state solar wind is included in order to provide an observational background for this conference.

SOLAR WIND OBSERVATIONS

Observational study of the solar wind began with the flight of Mariner 2 to Venus in 1962 [Neugebauer and Snyder, 1966]. Since then, numerous spacecraft have carried instrumentation designed to measure the velocity distributions of solar wind ions and electrons and thereby define the important properties of the plasma. A number of excellent comprehensive reviews have been written on solar wind theory, observation, and the interpretation of the observations [Dessler, 1967; Lüst, 1967; Ness, 1967; Parker, 1967; Axford, 1968; Hundhausen, 1968; Wilcox, 1968; Parker, 1969; Holzer and Axford, 1970; and Hundhausen, 1970a, 1970b]. Here, we can only summarize the parameters obtained from measurements at 1 AU which describe the quiet and average solar wind during the period 1965-1968 in Table I. Quiet solar wind is defined in the usual way, i.e., periods of low (200-300 km/sec) bulk speed [Burlaga and Ogilvie, 1970; and Hundhausen, 1970b]. The information contained in this table was taken from long-time averages of Vela 3 proton measurements published by

Table I

Solar Wind Plasma Parameters

Quantities	All Data			Quiet Conditions ($V \leq 350$ km sec^{-1})	
	Average	σ	90% Range	Average	σ
$N(cm^{-3})$	7	3.3	3-14.7	8.3	3.6
V(km/sec)	410	72	305-550	--	--
T_p(Deg K)	8.1×10^4	4×10^4	$(2-24) \times 10^4$	4.6×10^4	2.6×10^4
T_E(Deg K)	1.4×10^5	$.32 \times 10^5$	$(.85-2.1) \times 10^5$	1.3×10^5	$.27 \times 10^5$
K_p	1.9	.47	1.1-3.7	2.0	1.0
K_e	1.1	.08	1.01-1.3	1.07	.57
F_E(erg/cm^{2}sec)	7×10^{-3}	6×10^{-3}	$(.6-20) \times 10^{-3}$	5×10^{-3}	4.2×10^{-3}
T_E/T_p	2.2	1.7	0.7-6.5	4.7	2.2
$B(\gamma)$	5.2	2.4	2.2-9.9	4.7	2.2
β_p	.95	.74	.09-2.5	.78	.69
V_A(km/sec)	43	17	18-88	36	16
M_A	10.7	4.8	4.4-20	10.7	5.0
α_p	.48	.40	.05-2.8	--	--

Hundhausen et al. [1970], combined Vela 3 proton and IMP 3 magnetic field data [Ness et al., 1972], and Vela 4 electron measurements during 1967-1968 [Montgomery, 1971 and Montgomery et al., 1971]. The variables are defined as follows: N is the proton density; V, the bulk speed; T_p, the proton temperature; T_E, the electron temperature; K_p, the ratio of maximum to minimum proton temperature (the temperature anisotropy); K_E, the electron temperature anisotropy; F_E, heat conduction flux density carried by electrons (parallel to the magnetic field); T_e/T_p, electron to proton temperature ratio; B, magnetic field magnitude ($\gamma = 10^{-5}$ gauss); β_p, the proton beta = $8\pi NkT_p/B^2$; V_A, the alfvén speed, M_A, the alfvén Mach number, and α_p the fire hose stability parameter = $4\pi NK(T_\parallel - T_\perp)_p/B^2$. This table provides a reasonably complete description of the solar wind near 1 AU during the periods specified above. A study of combined electron and magnetic field parameters is not yet available so quantities such as electron beta, and fire hose stability parameter (including the effects of the electron temperature anisotropy), have not been included. It can be seen, however, that the electron beta will be about twice the proton beta and the very small average electron temperature anisotropy will tend to further stabilize the plasma with respect to the fire hose instability. Since the solar wind is likely to be marginally stable with respect to one or another plasma instability that tends to limit the growth of thermal anisotropies and/or inhibit thermal conduction [Forslund, 1970], a careful multiparameter study of the thermal properties of the plasma along with plasma wave characteristics should lead to identification of those instabilities that are important.

SOLAR WIND MODELS

Numerous theoretical models have been proposed in an attempt to use the observational information to identify the dominant physical processes in the solar wind expansion. For an extensive list of references and a general description of many of the models see Hundhausen [1968, 1970b]. Here we intend to concentrate on some of the more recent models in order to provide an example of the present state of affairs concerning the comparison of theory with experiment and theory with theory, and to illustrate the role of heat conduction.

We begin the discussion by noting the difficulties of the basic 2-fluid model of Hartle and Sturrock [1968] (HS). Steady flow, spherical symmetry, collisional coupling between electrons and protons, and classical heat transport were assumed, while magnetic forces, magnetic modification of transport coefficients, viscosity, nonthermal energy sources, and enhanced collision rates between proton and electrons were neglected. As has been pointed out by others, this model predicted a proton temperature and bulk speed at earth that were too low $\sim 4 \times 10^3$ °K and ~ 250 km/sec while the electron temperature was more than a factor of 2 too high $\sim 3.5 \times 10^5$ °K. In addition, as Hundhausen [1970b] has pointed out, the heat flux density that resulted from the high classical thermal conductivity at this electron temperature was much too high $\sim 30 \times 10^{-2}$ erg cm^{-2} sec^{-1}. In fact, it is physically unrealistic.

It is clear that if energy were somehow added to the proton component of the plasma by an external source well beyond the base of the corona, but inside of $\sim$ 20-30 solar radii (R_s), the proton temperature and flow speed would be raised, and better agreement with observations could be obtained. However, it is also possible that agreement could be improved by inhibition of heat conduction beyond the critical radius by the tightening spiral of the solar magnetic field, or by plasma instabilities. In addition, an anomalous electron-proton collision rate provided by instabilities could reduce the electron to proton temperature ratio and the heat flux. These ideas in various combinations have been discussed at length by several authors [Parker,1964; Hundhausen,1970a,b; Forslund,1970;

Hartle and Barnes, 1970].

Hartle and Barnes [1970] (HB) have explored at length the effects of adding to HS an ad hoc external proton energy source having a radial dependence in the form of a density-weighted gaussian of variable width and position. After extensive variation of the width, position, and strength of this source function, they obtained the following results: 1. Energy deposition at low altitudes below the critical radius results mostly in an increase in bulk speed at 1 AU while deposition at higher altitudes primarily results in an increase in proton temperature. 2. In either case only a slight decrease in electron temperature and corresponding heat flux occurs. 3. Satisfactory agreement with observed quiet time proton parameters at 1 AU can be achieved, especially if the density at the inner boundary is reduced to a minimum acceptable value of 8.5×10^5 cm^{-3}; however, the electron temperature and heat conduction flux are still much too high. 4. Agreement can be obtained with the $T_p^{1/2} - v$ relation observed by Burlaga and Ogilvie [1970]. They argued that the electron temperature and heat flux excess can probably be reduced by inhibition of thermal conduction without much change in T_p or v.

Barnes et al. [1971] (BHB) have refined the above model by incorporating a heating function appropriate to the dissipation of fast-mode hydromagnetic waves. After definition of the form of the source function, the only free parameter that remained was the total input wave flux at the base. With this model and base boundary conditions of $N_o = 1.46 \times 10^6$ cm^{-3}, $T_{eo} = 1.3 \times 10^6$°K, $T_{po} = 1.7 \times 10^6$ °K, and $B_o = .18$ g at $R_o = 2R_s$, good agreement with proton observations was obtained except for a density that was a factor of ~ 2 too high. A wave flux density of 5.2×10^3 ergs $cm^{-2}sec^{-1}$ gave at 1 AU: n=14 cm^{-3}, v = 370 km sec^{-1}, $T_e = 2.2 \times 10^5$ °K, $T_p = 6.2 \times 10^4$ °K, and $B = 2.4\gamma$. In addition, for v < 420 km/sec, rough agreement was obtained with the $T_p^{1/2} - v$ relation by varying only the wave influx at R_o. The electron temperature was still somewhat high, and the heat flux was still much above the observed value. However, these quantities were lower than in the simple-two-fluid model and HB because of a different choice of inner boundary conditions.

For comparison, the results of the steady two-fluid model of Wolff et al. [1971] (WBS) are now considered. Classical proton viscosity, magnetic forces, solar rotation, and two fluids coupled only by classical binary collisions as well as inhibited electron viscosity and thermal conductivity were included in this model. An ad hoc radial power-law inhibition factor was used where an exponent of -.728 seemed to give best agreement with observations. This resulted in a strong reduction of about 1/50 the classical value at 1 AU. The inner boundary was located at 3 R_s where a rather low density of 1.7×10^5 cm^{-3} was used, and equal electron and proton temperature of 1.7×10^6 °K was assumed. It was thus implied that coronal heating maintains equal temperatures out to this distance. The parameters obtained at Earth were $n = 9$ cm^{-3}, $V_r = 303$ km sec^{-1}, $T_e = 2\times10^5$ °K, $T_p = 4\times10^4$ °K, and $B_r = 5\gamma$.

To check their numerical integration against HS, WBS modified their model by including classical thermal conductivity and assumed $T_e = T_p$ at an inner boundary of $2R_s$. The results were similar to those of HS with the protons too cool, electrons too hot, and bulk speed too low.

Their results may be summarized as follows: 1. The effect of the spiral field on the radial motion inside 1 AU was small. 2. Even with viscosity included, the protons cooled too rapidly when $T_e = T_p$ at an inner boundary of 2 R_s. 3. By inhibiting thermal conductivity and moving the inner boundary out to $3R_s$, reasonable agreement was obtained with the proton component of the quiet solar wind. The electron temperature of 2×10^5 °K was somewhat high, but unlike HB and BHB, the conduction energy flux density of 1.7×10^3 ergs cm^{-2}sec^{-1} was well below observed values. 4. Inhibition of thermal conductivity was necessary in order to achieve high enough bulk speeds without highly excessive electron temperature. 5. Nonthermal heating beyond $3R_s$ was not necessary to produce an adequate quiet solar wind model. 6. Viscous heating with the above boundary conditions was sufficient to hold the proton temperature up to adequate levels ($\sim 4\times10^4$ °K) at 1 AU.

It appears that either model can satisfactorily reproduce quiet time proton observations although both models have difficulty simultaneously matching observations at 1 AU and measured coronal densities [Newkirk, 1967]. By including classical viscosity and using a $3R_s$ inner boundary, WRS obtained a high enough proton temperature at 1 AU. However, as in the case with thermal conductivity, the proton viscosity is probably reduced by an enhanced collision rate, and the actual effect on proton temperature is therefore somewhat uncertain. A strong inhibition of the thermal conductivity, probably somewhat stronger than observations would indicate, provided adequate radial bulk speed along with a reasonable electron temperature and a small heat flux.

On the other hand, the freedom obtained by adding only a variable strength extended energy source to HS allows the HB and BHB models to considerably vary the flow speed at 1 AU. In fact, considerable emphasis was placed on the extent to which the $T_p^{1/2} - v$ relationship was obtained. A few words of caution are in order. HB and BHB both assumed that the $T_p^{1/2}$–v relation resulted from a continuum of steady macroscopic states and therefore could be used as a straightforward test for a steady-state model. This assumption may not be correct. Measurements show that the solar wind is not steady on the necessary time scale. Dynamic evolution of high speed streams arising from time variations in the corona, or the interaction of the steady fast streams with the slower ambient due to solar rotation could be the source of at least some of the observed $T^{1/2}$ -v relation. Examples of this effect can be seen in Burlaga et al. [1971] where both observed and calculated profiles of velocity and temperature in the interaction region between colliding streams are shown. Their Fig. 9 shows velocity, temperature, and density profiles calculated from the adiabatic nonsteady model of Hundhausen and Gentry [1969], where a colliding stream structure was produced by imposing a 100-hr long triangular-shaped temperature enhancement at the inner boundary. Disturbances such as these will show a strong T_p-v relationship [Hundhausen, personal communication]. It is important to note that considerable evolution in structure occurred even though the scale time for the change in temperature at the source was quite long ($\sim$ 1/2 - 1 times the solar wind transit time). Thus, the $T_p^{1/2}$-v relation measured by Burlaga and Ogilvie is likely to result from a mixture of nonsteady and source variation effects. It is possible that a time-dependent model could produce the observed $T_p^{1/2}$-v relation without extended nonthermal heating. Finally, from WBS it seems likely that inhibition of the heat flux within 1 AU increases the bulk speed as well as reducing the electron temperature. It is therefore not obvious that inclusion of heat conduction inhibition in the HB and BHB models would affect only the electron temperature and heat flux as they assumed.

As has been stated elsewhere [Parker, 1971], since the number of free parameters is so large, it is probably possible to match the quiet conditions at 1 AU with "reasonable" inner and outer boundary conditions in more than one way. In addition, a direct comparison between models that emphasizes the importance of the various physical processes is made difficult by a lack of uniformity of boundary conditions. A more accurate knowledge of coronal temperature and density profiles would be very helpful by imposing a smaller range of inner boundary conditions on the models. Space probe measurements of the radial dependence of solar wind quantities, particularly T_p, T_e, and heat flux should provide a means of distinguishing between extended nonthermal and viscous heating.

At present, it seems that extended nonthermal heating is a necessary ingredient of a steady solar wind model in order to produce flow speeds > 400 km/sec. However, if velocities this high are due only to time dependent or corotation effects, a nonsteady model might be able to satisfy observations without nonthermal heating. At this point, the answer is not clear. However, the need for inhibition of thermal conduction seems compelling and the observational evidence concerning this point will be explored in the following section.

HEAT CONDUCTION

Solar wind electron measurements comprehensive enough to

evaluate the heat flux have been carried out on the Vela 4 spacecraft [Montgomery et al. 1968; Montgomery et al., 1971; Montgomery, 1971] and were recently confirmed by the OGO-5 results of Ogilvie, et al. [1972]. In order to calculate the energy flux in the plasma rest frame it is necessary to measure the electron velocity distribution with enough resolution in velocity and angle to meaningfully evaluate the third velocity moment of the distribution. Neither of the above mentioned experiments was capable of a unique, unambiguous determination of the full three dimensional velocity distribution. The Vela spacecraft obtained a two-dimensional reduced (integrated over spacecraft polar angle) distribution by means of a hemispherical electrostatic analyzer mounted on a spinning spacecraft. (See Montgomery et al. [1970] for a description of their analysis.) The 2-dimensional nature of the Vela measurements amounts to the measurement of a projection of the true energy flux vector into the spacecraft equatorial plane. Ogilvie et al. used a triaxial electron spectrometer composed of 3 mutually perpendicular cylindrical electrostatic analyzers each with a 10° half angle of acceptance. Since the orientation of the spacecraft was fixed, they required that the interplanetary magnetic field smoothly change direction in order to rotate the velocity distribution with respect to the detectors. Thus, it was necessary to assume that the velocity distribution remained stationary during the time required for the rotation. Since the two measurement techniques were quite different, the OGO-5 results provide independent confirmation of the earlier Vela measurements. The energy flux parallel to the local magnetic field averaged over one year of Vela observations was 7×10^{-3} ergs $cm^{-2} sec^{-1}$. OGO-5 obtained 4×10^{-3}erg $cm^{-2} sec^{-1}$. It was pointed out [Montgomery et al., 1971; Montgomery, 1971] that the observed heat flux is significantly less than expected ($\lesssim 1/3$) assuming a conduction dominated radial electron temperature dependence $T \sim T_o(r/r_o)^{-2/7}$ and classical Spitzer-Harm thermal conductivity evaluated using simultaneously measured values of electron temperature. Evidence was thus provided for the failure of the collision dominated classical heat transport theory due to saturation effects or heat flux driven instabilities [Forslund, 1970]. An additional possibility mentioned was the validity of a quasi-collisionless transport theory proposed by Perkins [1971].

In order to provide observational information concerning the above alternatives, a description of the solar wind velocity distribution and its comparison with the calculated distributions of Spitzer and Härm [1953] will occupy the remainder of this presentation. Figure 1 shows a contour representation in the spacecraft equatorial plane of a reduced velocity distribution, $F(V,\emptyset)$, obtained by Vela 5A. The contours are logarithmically spaced--the ratio of the distribution function between adjacent contours is 4.3. An arrow has been included to show the flow direction of the heat flux, while the coordinates V_x and V_y are parallel to projections of the standard solar ecliptic x and y coordinates onto the spacecraft

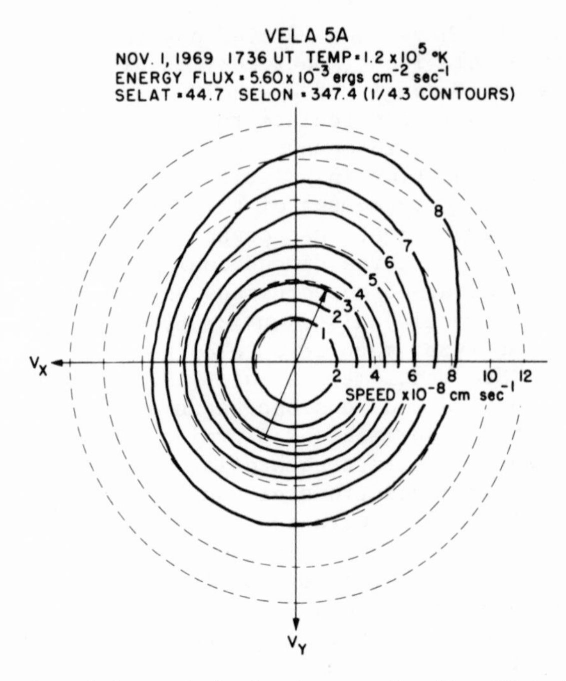

Figure 1. Solar wind electron velocity distribution.

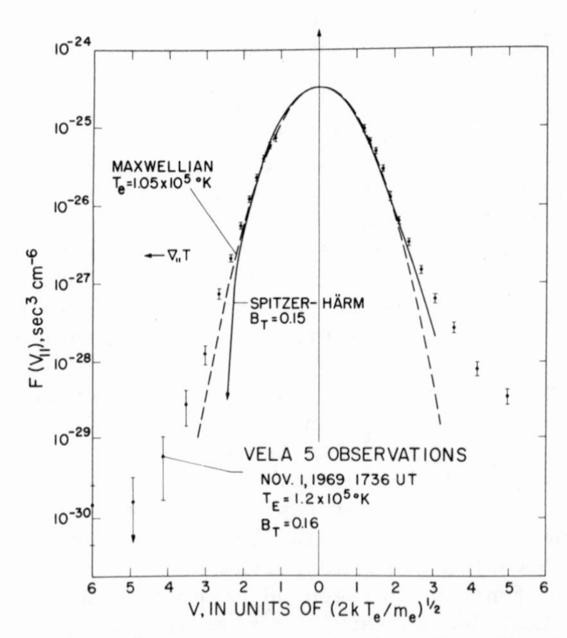

Figure 2. Plasma frame velocity distributions along the magnetic field.

equatorial plane. The skewing that results in a net energy flow is clearly visible and has been interpreted in terms of magnetic field aligned heat flux driven by a radial temperature gradient [Montgomery et al., 1968].

Figure 2 compares the measured velocity distribution with one derived from Spitzer and Härm [1953]. The curves represent cuts through the distributions along the direction of the magnetic field with heat flux flowing toward the right. The theoretical distributions have been integrated over a simulated spacecraft polar angle in order to make them directly comparable to the data. B_T is a dimensionless linear expansion parameter used by Spitzer and Härm to indicate the size of the heat conduction perturbation and it is also a useful parameter to express the relative magnitude of the heat flux [Forslund, 1970; Montgomery, 1971]. B_T can be defined by $B_T \equiv F_E/F_{ESAT}$ where F_E is the actual energy flux and F_{ESAT} is the saturated energy flux defined as the energy flux obtained if the internal energy of the distribution were convected at the thermal speed: $(3/2)nkT_e(2kT_e/M_e)^{1/2}$. T_e is the temperature of the Maxwellian part of the distribution while T_E, somewhat greater due to the enhanced high velocity tail, is the overall temperature of the observed plasma. At low speeds the experimental data fall nearly on the unperturbed Maxwellian instead of the Spitzer-Härm distribution, while at higher speeds the observed points indicate the usual nonequilibrium elevated tail.

In order to more clearly compare the deviations from Maxwellian that contribute to the heat flux, theoretical and observed differential energy flux curves are plotted in Figure 3. The left hand ordinate is defined by $\Delta F_{SH}(V) = F_{SH}(V_{11}) - F_{SH}(-V_{11})$ where F_{SH} is

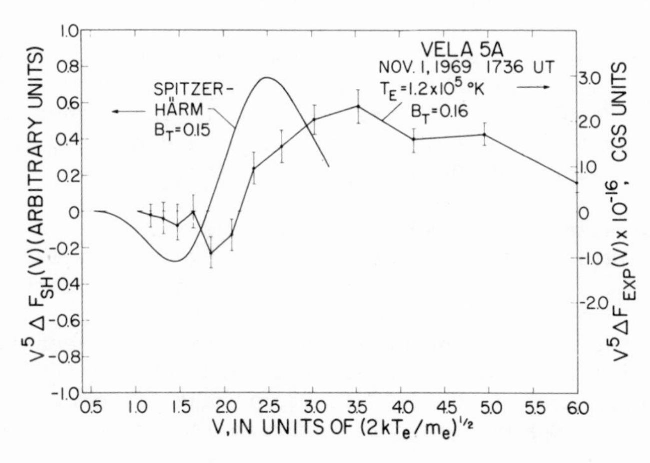

Figure 3. Differential heat flux comparison.

the reduced Spitzer-Härm distribution, and $V_{||}$ is velocity along the magnetic field in the direction of net energy transport ($-V_{||}$ is in the opposite direction). While there is similarity in shape between the predicted and measured curves, the heat flux in the solar wind is carried by much higher velocity electrons than predicted classically. The undershoot at low velocities is consistently present in the observations and results from the requirement for zero net current.

A hypothetical solar wind thermal conductivity model has been developed by Perkins [1971] in which the solar wind electrons bounce back and forth in a potential well formed by a combination of decreasing magnetic potential and growing electrostatic potential with increasing distance from the sun. The velocity distribution that results when the bounce frequency exceeds the collision frequency is essentially Maxwellian except for an asymetric high velocity tail due to escaping electrons at 1.5 - 3 times the thermal speed. The Maxwellian portion is shifted such that in the ion frame the peak of the electron distribution appears at a backward velocity equal to a significant fraction of the solar wind bulk speed. This shift is not observed. As with Spitzer and Härm, the electrons which carry the heat flux appear at lower velocity than observed.

In summary: 1. There is strong evidence that the thermal conductivity is $\sim$ 1/3 - 1/4 less than the classical Spitzer-Härm value. 2. The measured velocity distribution appears to be nearly in thermodynamic equilibrium near its center with $T_{e||}/T_{e\perp} \approx 1.1$ but has a nonthermal enhanced high energy tail. 3. the measured differential heat flux is similar in shape to the Spitzer-Härm model but peaks at higher velocities. 4. The shift in the body of the electron distribution by the solar wind bulk speed indicated by Perkins is not observed. 5. The pitch angle distribution of the heat carrying electrons is smooth and broad ($\sim$ 50°-60° wide)

The above results are consistent with an anomalously high electron-electron collision rate about 5 times the value expected on the basis of coulomb collisions near 1 AU. The higher energy electrons appear to be somewhat less collision dominated than those in the body of the distribution.

I wish to acknowledge useful discussions with Drs. A. J. Hundhausen, W. C. Feldman, and F. Perkins. Dr. S. J. Bame is the Principal Investigator for the Vela series of plasma detectors.

The Vela nuclear test detection satellites have been designed, developed and flown as part of a joint program of the Advanced Research Projects Agency of the U.S. Department of Defense and the U.S. Atomic Energy Commission. The program is managed by the U.S. Air Force.

References

Axford, W.I., Observations of the interplanetary plasma, Space Sci. Rev., 8, 331, 1968.

Barnes, A., R. E. Hartle, and J. H. Bredekamp, On the energy transport in stellar winds, Astrophys. J., 166, L53, 1971.

Burlaga, L.F. and K.W. Ogilvie, Heating of the solar wind, Astrophys. J., 159, 659, 1970.

Burlaga, L.F., K.W. Ogilvie, D.H. Fairfield, M.D. Montgomery, and S.J. Bame, Energy transport at colliding streams in the solar wind, Astrophys. J., 164, 137, 1971.

Dessler, A.J., Solar wind and interplanetary magnetic field, Rev. Geophys., 5, 1, 1967.

Forslund, D.W., Instabilities associated with heat conduction in the solar wind and their consequences, J. Geophys. Res., 75, 17, 1970.

Hartle, R.E. and P.A. Sturrock, Two-fluid model of the solar wind, Astrophys. J., 151, 1155,1968.

Hartle, R.E. and A. Barnes, Nonthermal heating in the two-fluid solar wind model, J. Geophys. Res., 75, 6915, 1970.

Holzer, T.E. and W.I. Axford, The theory of stellar winds and related flows, Annu. Rev. Astron. Astrophys., 8, 31, 1970.

Hundhausen, A. J., Direct observations of solar wind particles, Space Sci. Rev., 8, 690, 1968.

Hundhausen, A.J., Dynamics of the outer solar atmosphere, Proc. 4th Summer Inst. for Astron. and Astrophys, State University of New York, Stony Brook, 1970a.

Hundhausen, A.J., Composition and dynamics of the solar wind plasma, Rev. of Geophys. and Space Phys. 8, 729, 1970b.

Hundhausen, A.J. and R.A. Gentry, Numerical simulation of flare-generated disturbances in the solar wind, J. Geophys. Res., 74, 2908, 1969.

Hundhausen, A.J., S.J. Bame, J.R. Asbridge, and S.J. Sydoriak, Solar wind proton properties: Vela 3 observations from July 1965 to June 1967, J. Geophys. Res., 75, 4643, 1970.

Lüst, R., The properties of interplanetary space, in Solar-Terrestrial Physics, ed. by J.W.King and W.S. Newman, p.1, Academic, New York, 1967.

Montgomery, M.D., Average thermal characteristics of solar wind electrons, to be published in Proc. Solar Wind Conference, Asilomar, California, 1971.

Montgomery, M.D., S.J. Bame, and A.J. Hundhausen, Solar wind electrons: Vela 4 measurements, J. Geophys. Res., 73, 4999, 1968.

Montgomery, M.D., J.R. Asbridge, and S.J. Bame, Vela 4 plasma observations near the earth's bow shock, J. Geophys. Res., 75, 1217, 1970.

Montgomery, M.D., S.J. Bame, and J.A. Asbridge, Solar wind electrons--average properties (abstract) Trans. Amer. Geophys. Union, 52, 336, 1971.

Ness, N.F., Observed properties of the interplanetary plasma, Amer. Rev. Astron. Astrophys. 6, 79, 1967.

Ness, N.F., A.J. Hundhausen and S.J. Bame, Observations of the interplanetary medium Vela 3 - IMP 3 1965-1967, to be published in J. Geophys. Res., 1972.

Neugebauer, M. and C.W. Snyder, Mariner 2 observations of the solar wind, 1, Average properties, J. Geophys. Res., 71, 4469, 1966.

Newkirk, G.,Jr., Structure of the solar corona, in Annu. Rev. Astron. Astrophys., 5, 213, 1967.

Ogilvie, K.W., J.D. Scudder, and M. Sugiura, Electron energy flux in the solar wind, to be published in J. Geophys. Res., 1972.

Parker, E.N., Dynamical properties of stellar coronas and stellar winds, 2, Integration of the heat flow equation, Astrophys. J., 139, 93, 1964.

Parker, E.N., The dynamical theory of gasses and fields in interplanetary space, in Solar Terrestrial Physics, edited by J.W. King and W.S. Newman, Academic Press, New York, 1967.

Parker, E.N., Theoretical studies of the solar wind phenomenon, Space Sci. Rev., 9, 325, 1969.

Parker, E.N., Present developments in theory of the solar wind, to be published in Proc. Solar Wind Conference, Asilomar, Calif., 1971.

Perkins, F., Heat conducton in the solar wind, informal presentation to be published in Proc. Solar Wind Conf., Asilomar, Calif, 1971.

Spitzer, L.,Jr. and R. Härm, Transport phenomena in a completely ionized gas, Phys. Rev., 89, 977, 1953.

Wilcox, J.M., The interplanetary magnetic field: Solar origin and terrestrial effects, Space Sci. Rev., 8, 258, 1968.

Wolff, C.L., J.C. Brandt, and R.G. Southwick, a two-component model of the quiet solar wind with viscosity, magnetic field, and reduced heat conduction, Astrophys. J., 165, 181, 1971.

THE SOLAR WIND NEAR THE SUN: THE SOLAR ENVELOPE

Leonard F. Burlaga

Laboratory for Extraterrestrial Physics,

NASA Goddard Space Flight Center, Greenbelt, Maryland

This paper discusses the structure of the solar envelope and the physical processes which might occur in the envelope. The envelope is the region between $\approx 2R_\odot$ and $\approx (25\text{-}50)R_\odot$.

STRUCTURE OF THE ENVELOPE

To determine which physical processes can occur in the envelope, one must know the envelope's structure to zeroth order. This is given by steady, spherically-symmetric models for the "quiet" wind (the state corresponding to V=320 km/sec). Consistent speed, density, and magnetic field profiles are given by 1-fluid models (e.g. Whang, 1971a). Two-fluid models are needed to obtain proton and electron temperature profiles, T(r) and $T_e(r)$ (Sturrock and Hartle, 1966). The 2-fluid models are controversial because the thermal properties and heat sources of the solar wind are poorly understood. For example, one model (Hartle and Barnes, 1970) postulates Chapman heat transfer, negligible viscosity, and proton heating by an external heat source out to $25R_\odot$, while another model (Wolff et al., 1971) postulates non-Chapman heat transfer, non-negligible viscosity, and no external proton heat source beyond $3R_\odot$.

Despite uncertainties in the structure of the envelope, one can derive the following general results: 1) the ratio of the thermal energy density to the magnetic energy density increases from $\beta \approx 0.1$ to $2R_\odot$ to $\beta \approx 1$ at $50R_\odot$; 2) the flow energy is dominant above $\approx 15R_\odot$; 3) below $\approx 5R_\odot$, the Alfven speed V_A is much larger than the acoustic speed V_s, which is much larger than the bulk speed V; 4) the flow becomes supersonic at $\approx 5R_{\odot}$; 5) between $10R_{\odot}$ and $\approx 20R_{\odot}$, $V \approx V_A \approx V_s$; 6) beyond $\approx 20R_{\odot}$, $V >> V_A \approx V_s$.

PHYSICAL PROCESSES IN THE ENVELOPE

Relativistic solar-flare protons diffuse in the envelope (Lust and Simpson, 1957). An analysis (Burlaga, 1970) of the May 4, 1960, event (McCracken, 1962) gave $\approx 25R_{\odot}$ for the radius of the envelope and $\approx 10^{21} cm^2/sec$ for the diffusion coefficient. Similar results were found for the Nov. 18, 1968, event (Duggal et al., 1971) Such diffusion suggests the presence of MHD waves, and thus might be related to the heating and acceleration of the solar wind (Burlaga, 1970).

The mechanical, electrodynamical, and thermal processes in the envelope are inferred from studies of the solar wind states at 1 AU. These states can be described as relations n(V), T(V), $T_e(V)$, and B(V), together with the distribution of V. There is a simple relation between T and V (Burlaga and Ogilvie, 1970a) which is valid at all parts of the solar cycle. An inverse relation between n and V is observed (Neugebauer and Snyder, 1966 ; Burlaga and Ogilvie, 1970b), but the correlation between n and V is weak (Belcher and Davis, 1971). The electron temperature and magnetic field intensity are independent of V (Burlaga and Ogilvie, 1970a,b).

The quietest state ($V \approx 250$ km/sec) is approximately described by a 2-fluid model which postulates only Coulomb interactions in the envelope (Hartle and Sturrock, 1968); it gives T_e/T which is a factor of 6 too large, and it cannot explain the higher speeds that are usually observed. Thus, processes besides Coulomb interactions must occur in the envelope.

The range of solar wind speeds can be explained by postulating that T remains high in the envelope (Parker, 1963; Scarf and Noble, 1965). The T-V relation might be due to variations in the size of this high-temperature region (Burlaga and Ogilvie, 1970a). High T implies proton heating, which could be produced by an external energy source (Hartle and Barnes, 1970) or by heat transfer from electrons to protons (Hundhausen, 1970). These two mechanisms cannot be distinguished using only observations at 1 AU. The collisionless damping of a variable flux of hydromagnetic waves in the envelope can explain the T-V relation and gives speeds up to 400 km/sec (Barnes et al., 1971). The heat transfer mechanism implies some new plasma process (Hundhausen, 1970, Forslund, 1970), since Coulomb collisions are not sufficiently effective (Sturrock and Hartle, 1966).

The solar wind can also be mechanically accelerated to high

speeds in the envelope by Alfven waves moving away from the sun (Belcher, 1971) and by the Lorentz force (Whang, 1971b). In both cases the streaming energy is the result of a diminuition of the Poynting flux with distance from the sun. The relative and absolute importance of these two processes is yet to be determined.

The strong magnetic field in the envelope exerts a torque on the solar wind which causes the azimuthal speed to increase in the envelope where it may reach a maximum on the order of 5 km/sec (Weber and Davis, 1967; Wolff et al., 1971). Viscosity might cause a further increase in V_φ (Weber and Davis, 1970), but the importance of viscosity is debatable (Parker, 1965). Observations of V_φ at 1 AU (Brandt and Heise, 1970; Coon, 1968; Egidi et al., 1969) are not sufficient to allow experimental study the processes governing the angular momentum flux in the envelope because complicated interactions near 1 AU (Lazarus and Goldstein, 1971; Siscoe, 1971; Goldstein, 1971; Belcher and Davis, 1971) affect V_φ in ways that are difficult to evaluate.

In conclusion, present observations and models suggest the occurence of a variety of interesting physical processes in the envelope which are decisive in determining the solar wind states at 1 AU, but further measurements in or near the envelope will be needed to determine the relative importance of these processes.

REFERENCES

Barnes, A., R. E. Hartle, and J. H. Bredekamp, On the energy transport in stellar winds, Astrophys. J. Letters, 166, L53, 1971.

Belcher, J. W., and L. Davis, Jr., Large-amplitude Alfven waves in the interplanetary medium, 2, J. Geophys. Res., 76, 3534, 1971.

Belcher, J. W., Alfvenic wave pressures and the solar wind, Astrophys. J., (to be published) 1971.

Brandt, J. C., and J. Heise, Interplanetary gas XV. Nonradial plasma motions from the orientations of ionic comet tails, Astrophys. J., 159, 1057, 1970.

Brandt, J. C., and C. L. Wolff, On solar wind heating, (to be published) 1971.

Brandt, J. C., C. L. Wolff, and J. P. Cassinelli, Interplanetary gas XVI. A calculation of the angular momentum of the solar wind, Astrophys. J., 156, 1117, 1969.

Burlaga, L. F., Anisotropic cosmic ray propagation in an inhomogeneous medium, I. The solar envelope, in Proc. 11th Int. Conf. on Cosmic Rays, Budapest 1969, Acta Physica Academiae Scientiarum Hungaricae 29 Suppl. 2, 9, 1970.

Burlaga, L. F., and J. K. Chao, Reverse and forward slow shocks in the solar wind, NASA X-692-71-66, 1971.

Burlaga, L. F., and K. W. Ogilvie, Heating of the solar wind, Astrophys. J., 159, 659, 1970a.

Burlaga, L. F., and K. W. Ogilvie, Magnetic and thermal pressures in the solar wind, Solar Physics, 15, 61, 1970b.

Burlaga, L. F., K. W. Ogilvie, D. H. Fairfield, M. D. Montgomery, and S. J. Bame, Energy transfer at colliding streams in the solar wind, Astrophys. J., in press, 1971.

Carovillano, R. L., and G. L. Siscoe, Corotating structure in the solar wind, Solar Physics, 9, 1969.

Coon, J. H., in Earth's Particles and Fields, p.359, ed. by B. M. McCormac, Reinhold Book Corporation, New York, 1968.

Duggal, S. P., M. A. Pomerantz, and I. Guidi, The unusual anisotropic solar particle event of November 18, 1968, (to be published) 1971.

Egidi, A., V. Formisano, G. Moreno, F. Palmiotto, and P. Saraceno, Solar wind and location of shock front and magnetopause at the 1969 solar maximum, J. Geophys. Res., 75, 6999, 1970.

Egidi, A., G. Pizzella, and C. Signorini, Measurement of the solar wind direction with the IMP I satellite, J. Geophys. Res., 74, 2807, 1969.

Fisk, L. A., and K. H. Schatten, Transport of cosmic rays in the solar corona, NASA-GSFC X-661-71-313, to appear in Solar Physics, 1971.

Forslund, D. W., Instabilities associated with heat conduction in the solar wind and their consequences, J. Geophys. Res., 75, 17, 1970.

Goldstein, B. E., Non-linear corotating solar wind structure, (to be published) 1971.

Hartle, R. E., and A. Barnes, Nonthermal heating in the two-fluid solar wind model, J. Geophys. Res., 75, 6915, 1970.

Hartle, R. E., and P. S. Sturrock, Two-fluid model of the solar wind, Astrophys. J., 151, 1155, 1968.

Hundhausen, A. J., Direct observations of solar wind particles, Space Sci. Rev., 8, 690, 1968.

Hundhausen, A. J., Nonthermal heating in the solar wind, J. Geophys. Res., 74, 5810, 1969.

Hundhausen, A. J., Composition and dynamics of solar wind plasma, Rev. Geophys. and Space Phys., 8, 729, 1970.

Hundhausen, A. J., S. J. Bame, J. R. Asbridge, and S. J. Sydoriak, Solar wind proton properties: Vela 3 observations from July 1965 to June 1967, J. Geophys. Res., 75, 4643, 1970.

Hundhausen, A. J., and M. D. Montgomery, Heat conduction and non-steady phenomena in the solar wind, J. Geophys. Res., 76, 2236, 1971.

Jokipii, J. R., and L. Davis, Jr., Long-wavelength turbulence and heating of the solar wind, Astrophys. J., 156, 1101, 1969.

Jokipii, J. R., Propagation of solar flare cosmic rays in the solar wind, Rev. Geophys. Space Phys., 9, 27, 1971.

Lazarus, A. J., and B. E. Goldstein, Observation of the angular momentum flux carried by the solar wind, J. Geophys. Res., (to be published) 1971.

Lust, R., and J. A. Simpson, Initial stages in the propagation of cosmic rays produced by solar flares, Phys. Rev., 108, 1536, 1957.

McCracken, K. G., The cosmic ray flare effect 3. Deductions regarding the interplanetary magnetic field, J. Geophys. Res., 67, 447, 1962.

Meyer, P., E. N. Parker, and J. A. Simpson, Solar cosmic rays of February, 1956 and their propagation through interplanetary space, Phys. Rev., 104, 768, 1956.

Montgomery, M. D., S. J. Bame, and A. J. Hundhausen, Solar wind electrons: Vela 4 measurements, J. Geophys. Res., 73, 4999, 1968.

Ness, N. F., A. J. Hundhausen, and S. J. Bame, Correlated magnetic field and plasma observations, J. Geophys. Res., (in press, 1971).

Neugebauer, M. and C. W. Snyder, Mariner 2 observations of the solar wind 1. Average properties, J. Geophys. Res., 71, 4469, 1966.

Ogilvie, K. W., J. D. Scudder, and M. Sugiura, Electron energy flux in the solar wind, J. Geophys. Res., (in press) 1971.

Parker, E. N., Interplanetary dynamical processes, John Wiley, New York, 1963.

Parker, E. N., Dynamical theory of the solar wind, Space Sci. Rev., 4, 666, 1965.

Robbins, D. E., A. J. Hundhausen, and S. J. Bame, Helium in the solar wind, J. Geophys. Res., 75, 1178, 1970.

Scarf, F. L., and L. M. Noble, Conductive heating of the solar wind II. The inner corona, Astrophys. J., 141, 1479, 1965.

Schatten, K. H., Evidence for a coronal magnetic bottle at 10 solar radii, Solar Physics, 12, 484, 1970.

Serbu, G. P., Explorer 35 observations of solar wind electron density, temperature and anisotropy (to be published) 1971.

Siscoe, G. L., Structure and orientations of solar wind interaction fronts: Pioneer 6, to appear in J. Geophys. Res.,1971.

Strong, J. B., J. R. Asbridge, S. J. Bame, and A. Hundhausen, in the Zodiacal Light and the Interplanetary Medium, ed. J. L. Weinberg (Washington NASA SP-150) p.365, 1968.

Sturrock, P. A., and R. E. Hartle, Two-fluid model of the solar wind, Phys. Rev. Letters, 16, 628, 1966.

Tannenbaum, A. S., J. M. Wilcox, E. N. Frazier, and R. Howard, Solar velocity fields: 5-min. oscillations and supergranulation, Solar Physics, 9, 328, 1969.

Weber, E. J., and L. Davis, Jr., The angular momentum of the solar wind, Astrophys. J., 148, 217, 1967.

Weber, E. J., and L. Davis, Jr., The effect of viscosity and anisotropy in the pressure on the azimuthal motion of the solar wind, J. Geophys. Res., 75, 2419, 1970.

Whang, Y. C., Conversion of magnetic field energy into kinetic energy in the solar wind, Astrophys. J., (to appear), 1971a.

Whang, Y. C., Higher moment equations and the distribution function of the solar wind plasma, to be published in J. Geophys. Res., 1971b.

Wolfe, J., Large scale features of the solar wind plasma, presented at the second Solar Wind Conference, 1971.

Wolfe, J. H., and D. D. McKibben, Pioneer 6 observations of a steady-state magnetosheath, Planet. Space Sci., 16, 953, 1968.

Wolff, C. L., J. C. Brandt, and R. G. Southwick, A two-component model of the quiet solar wind with viscosity, magnetic field, and reduced heat conduction, Astrophys. J., 165, 181, 1971.

INFLUENCE OF NEUTRAL INTERSTELLAR MATTER ON THE EXPANSION OF THE SOLAR WIND

H.J. Fahr

Institut für Astrophysik und

extraterrestr. Forschung, Universität Bonn

I. INTRODUCTION

Parker's hydrodynamical solution of the solar wind expansion yields a supersonic plasma flow beyond a critical distance r_c from the sun with an essentially constant expansion velocity. This solution is derived from a hydrodynamical treatment of the coronal expansion that does not take into account any interaction of the solar wind plasma with neutral interstellar matter which enters the solar system. In that case the supersonic solar wind expands with a constant velocity v_p up to a specific distance r_s where the energy density of the solar wind plasma due to the geometrical divergence of the flow has decreased to the value $B_i^2/8\pi$, the energy density of the interstellar magnetic field.

Beyond this distance r_s the interstellar field B_i starts controlling the expansion of the solar wind, which means that the undisturbed hydrodynamical flow of the solar wind plasma in radial directions terminates at this region. At the distance r_s the solar wind is presumed to undergo a shock transition. Interstellar hydrogen and helium atoms which enter the solar system from the interstellar space become ionized while approaching the sun, due to charge transfer reactions with solar wind protons and due to the solar EUV radiation. These ionization processes which lead to the generation of secondary protons and helium ions in the solar wind give rise to the extraction of kinetic energy from the expanding solar wind plasma.

II. THEORY

If a solar wind proton of velocity v_p is lost from the plasma ensemble by a charge exchange process with an interstellar hydrogen atom, the solar wind plasma loses the momentum $m_p v_p$ and the kinetic energy $1/2 m_p v_p^2$. The simultaneous gain of energy due to the creation of a secondary proton can be neglected, since its initial velocity equals that of the former hydrogen atom and is at least about one order of magnitude smaller than v_p. Immediately after its creation the secondary proton is affected by the frozen-in magnetic field which moves with the solar wind. In general secondary protons take up energy from the moving magnetic field. This energy which is indirectly taken from the dominating kinetic energy of the plasma reappears in a kind of thermal and kinetic energy of these secondary protons. The average motion of secondary protons depends on the direction of the frozen-in magnetic field with respect to the bulk motion v_p of the plasma and is in general not synchronized to that motion. Only that part of the kinetic energy of secondary protons which corresponds to their average velocity component parallel to v_p, contributes to the kinetic energy of the solar wind plasma. Due to this fact it is found (Fahr, 1971a) that the following loss ΔE_{kin} occurs per secondary proton:

$$\Delta E_{kin} = -1/2 m_p v_p^2 \,(1+2\sin^2\alpha - \sin^4\alpha) \tag{1}$$

where α gives the angle between v_p and the frozen-in field B, if the secondary proton considered has been created by charge exchange. In case the secondary proton is generated by an EUV-ionization process, the energy loss ΔE_{kin} amounts only to:

$$\Delta E_{kin} = -1/2 m_p v_p^2 \,(2\sin^2\alpha - \sin^4\alpha) \tag{2}$$

Since the lines of the interplanetary magnetic field follow an Archimedean spiral, the angle α is a function of the distance r from the sun. However, in order to calculate numerically the extraction of kinetic energy from the solar wind plasma we shall assume that the inclination $\alpha \approx 90°$ in the whole interplanetary region of interest for this problem, since this produces only a 3% error in ΔE. Comparing the ionization frequencies it can be shown that only the energy loss ΔE_{kin} as given in (1), which is connected with secondary protons and helium ions originating from charge exchange collisions, has to be taken into account. We shall now calculate the rate of energy exhaustion which would be due to such secondary ions. Since the kinetic energy of the protons is the dominating energy in the solar wind plasma which therefore governs the wind dynamics, the energy extraction due to the interaction with neutral interstellar matter may be described as if the energy taken up by secondary ions during reacceleration, though primarily expended by the frozen-in field, is finally taken from the kinetic energy of the plasma. Therefore the following differential relation is valid:

$$d(1/2m_p v_p^2) = m_p v_p dv_p = dr\, \frac{v_{rel}}{v_p}\Big[q_{ex}(p,H)\Delta E_H n_H(r,\theta) + q_{ex}(p,\,He)\Delta E_{He} n_{He}(r,\theta)\Big] \qquad (3)$$

where v_{rel} is the relative velocity between solar wind protons and the neutral species, θ is the angle the velocity $\vec{v}_o$ of approach of interstellar matter makes with the direction to the sun. $q_{ex}(p,H)$, $q_{ex}(p,He)$; n_H, n_{He}; and ΔE_H,ΔE_{He} are the proton charge exchange cross sections, the particle densities and the specific energy losses of hydrogen and helium. According to the assumption $\alpha \approx 90°$ the specific energy losses ΔE can be taken not to be dependent on the distance r from the sun. In addition v_p can be regarded as being equal to v_{rel}, since the velocities of the neutrals can be neglected compared to v_p. Therefore (3) can be integrated to yield:

$$v_p(r,\theta) = v_{p,o} \exp\Big[-q_{ex}(p,H)\int_{r_o}^{r} n_H(r,\theta)dr - q_{ex}(p,He)\int_{r_o}^{r} 4n_{He}(r,\theta)dr\Big] = v_{p,o}W(r,\theta) \qquad (4)$$

This formula describes the decrease of the solar wind velocity v_p with increasing distance r in each direction θ. v_p is not sensitive to r_o, as far as values r_o smaller than 3AU are considered. Therefore we take r_o to be r_E. According to Fahr (1971a) the drift of secondary ions $\bar{v}_x$ in radial and $\bar{v}_y$ in azimuthal direction is given by:

$$\bar{v}_x = (B_y/B)^2 v_p \quad ; \quad \bar{v}_y = (B_x B_y/B^2) v_p \qquad (5)$$

which leads to $\bar{v}_x = v_p$, $\bar{v}_y = 0$, if α is close to 90° and $[(90°-\alpha)/90°]^2$ is negligible compared to 1. This means that the average motion of secondary ions is fully synchronized with the bulk motion v_p of the plasma, as far as distances r > 3AU are considered. In consequence there exists a reintegration of secondary ions into the bulk motion, and the equation of proton continuity reads:

$$n_{p,E}v_{p,E}r_E^2 = n_p(r,\theta)v_p(r,\theta)r^2 \quad (6)$$

where $n_{p,E}$ and $v_{p,E}$ are the density and velocity of the solar wind protons at $r = r_E = 1AU$. From this equation we obtain:

$$n_p(r,\theta) = n_{p,E}(r_E/r)^2 \frac{1}{W(r,\theta)} = \bar{n}_p/W(r,\theta) \quad (7)$$

where $\bar{n}_p$ represents the proton density of a $1/r^2$ density decrease which is connected with Parker's solution v_p = const. An interesting result can be obtained from (4) and (7), if we look at the interplanetary densities $n_{H,He}$ which are given by:

$$n(r,\theta) = n_o E_f(r,\theta)E_{ex}(r,\theta) \quad . \quad (8)$$

Here n_o is the density of interstellar hydrogen or helium outside the solar system, E_f is a function describing the focusing effect of the solar gravitational field (Blum and Fahr, 1970; Fahr, 1971c) and E_{ex} is a function describing the extinction due to ionization and charge exchange by the expression:

$$E_{ex}=\exp\left[-r_E^2 q_{euv} f_E \int_o^\theta \frac{d\theta}{\dot{\theta} r^2} - q_{ex} \int_\infty^{s(r)} \frac{ds}{v(r)} v_p(r,\theta)n_p(r,\theta)\right] \quad (9)$$

where q_{euv} is the average ionization cross section in the Lyman continuum, f_E the solar EUV photon flux at the orbit of the earth, $\dot{\theta}$ the angular velocity of the neutrals approaching the sun, s(r) the orbital length measured from the perihelion of the particle orbit which leads to the point $\{r,\theta\}$ and v(r) the velocity of the neutrals at this point. E_x contains both the velocity v_p and the density n_p of the solar wind protons which according to (4) and (7) are themselves functions of the density $n(r,\theta)$ of the neutrals. Fortunately, however, only the product of n_p and v_p enters the function E_{ex} as given in

(9). Since this product attains the following form

$$n_p(r,\theta)v_p(r,\theta) = n_{p,E}(r_E/r)^2 v_{p,E} = F_E(r_E/r)^2 \quad (10)$$

where F_E is the flux of solar wind protons at the orbit of the earth, the function E_{ex} can be found to be unchanged with respect to that derived from Parker's solar wind solution v_p=const.

III. NUMERICAL RESULTS

In order to calculate the density of the neutrals in the interplanetary space, we have to assume specific values for the ingredient parameters. The velocity v_o of approach of interstellar matter has been taken to be 20 km/sec. The density $n_o(H)$ of hydrogen and $n_o(He)$ of helium near the solar system has been assumed to be 0.1 and 0.01 cm^{-3}. The interplanetary density values have been calculated for a solar wind proton flux of F_E = 2×10^8 prot/cm^2sec, a solar EUV flux of 10^{11} phot/cm^2sec in connection with a mean weighted ionization cross section of $q_{euv}=1.2\times10^{-18}cm^2$ that has been determined by averaging the cross section $q_{euv,c}(\nu_c/\nu)^3$ with $q_{euv,c}$ given by Ambarzumjan (1957) over the solar EUV spectrum given by Hinteregger et al. (1963). The charge exchange cross section between protons and hydrogen atoms has been taken to be $2\times10^{-15}cm^2$ (Fite et al., 1960; Belyaev, 1964). The helium values have been calculated for a mean weighted ionization cross section of $q_{euv}(He)$ of $1.8\times10^{-18}cm^2$ with the critical cross section of $q_{euv,c}(He)=7.3\times10^{-18}cm^2$ at 504 Å taken from Po Lee and Weissler (1955) in connection with the solar helium continuum radiation below 504 Å given by Hinteregger et al. (1963). The main charge exchange processes of helium atoms in the solar wind are of the reaction type a) $He^{++}+He\rightarrow2He^+$; b) $p^++He\rightarrow H+He^+$. Reaction type a) is characterized by a cross section $q_{ex}(He)=5\times10^{-17}cm^2$ (Fite et al., 1962), and reaction type b) by $q_{ex}(He)=4.8\times10^{-17}$ (Mason and Vanderslice, 1957). Due to the resulting frequencies ν_i for radiative ionization and ν_{ex} for charge exchange, helium can penetrate much deeper than hydrogen into the solar system before it becomes ionized. This becomes obvious in the curves of fig. 1. It can be seen that the helium densities, though amounting only to 1/10 the density n_o of interstellar hydrogen at infinity ($r>10^3$AU), dominate over the hydrogen densities for distances smaller than $r\gtrsim3$AU.

Whereas the solar radiation pressure acting upon the hydrogen atoms compensates between 1/3 and 2/3 of the gravitational force during the 11-year solar cycle (Fahr,

1971c) the corresponding solar radiation pressure acting upon the helium atoms is negligible compared to the gravitational attraction. This is partly because the total energy contained in the solar helium α line at 584 Å (0.03-0.05 erg/cm^{2}sec) is about two orders of magnitude smaller than that contained in the solar Lyman-α line (3-6 erg/cm^{2}sec) (Hinteregger et al. 1963; Hall and Hinteregger, 1970; Tousey, 1963), and partly because the Einstein transition coefficient for the helium 584 Å transition $B_{1,2}=6.54\times10^{18}$cm^3/sec^{2}erg is about a factor of 3 smaller than $B_{1,2}=2.01\times10^{19}$cm^3/sec^{2}erg for the Lyman-α transition. This leads to a force K_r due to solar radiation pressure acting on the helium atoms which is about four orders of magnitude smaller than the gravitational force $K_g=3.96\times10^{-24}$erg/cm. Therefore the helium profiles do not change during the solar activity cycle, whereas the hydrogen profiles do. This is indicated by the dashed lines in figs. 1 and 2 which show the interplanetary density of helium and hydrogen in the downwind (θ=180°) and upwind (θ=0°) direction. It can be seen that in the downwind direction helium already becomes

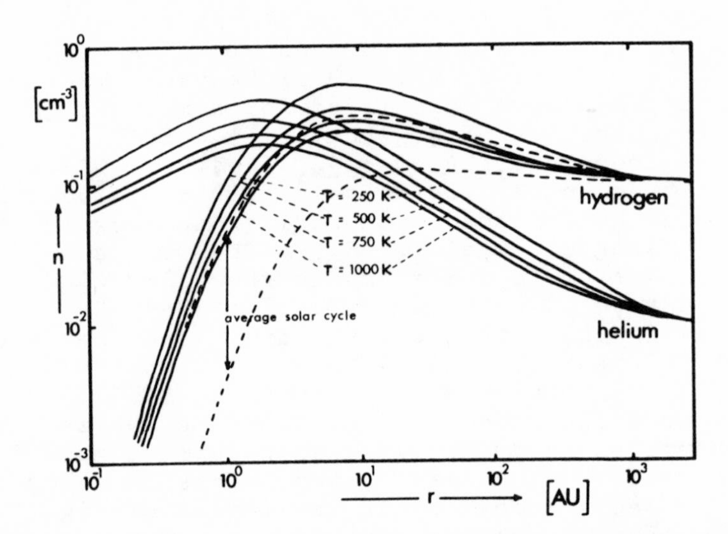

Figure 1: The densities of interstellar hydrogen and helium vs the distance r from the sun in downwind direction (θ=180°) for various interstellar temperatures T. The densities of the two interstellar gases near the solar system have been taken to be n_o(H)=0.1 cm^{-3} and n_o(He)= 0.01 cm^{-3}; V_o=20 km/sec. Dashed lines show the interplanetary density of He and H for θ=0° and θ=180°.

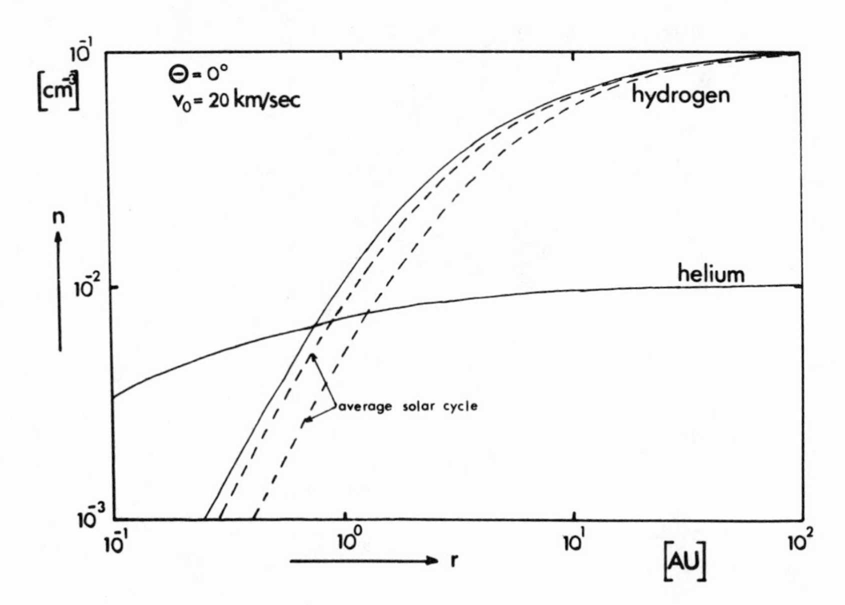

Figure 2: The densities of interstellar hydrogen and helium vs the distance r from the sun in upwind direction ($\theta=0°$). The densities of the two interstellar gases near the solar system have been taken to be $n_o(H)=0.1\ cm^{-3}$ and $n_o(H)= 0.01\ cm^{-3}$; $V_o=20$ km/sec. Dashed lines show the interplanetary density of He and H for $\theta=0°$ and $\theta=180°$.

dominant at about r=5-10AU depending on the actual radiation pressure on hydrogen, whereas in directions outside the density cone $165°<\theta<195°$ helium only dominates in a region much closer to the sun (r=1.1 - 0.5AU).

Fig. 3 shows how the solar wind velocity decreases due to the energy extraction by charge exchange reactions of interstellar neutrals with solar wind protons. Outside the density cone, the solar wind velocity profiles are shown for different Lyman-α radiation pressures characterized by a parameter u which is defined by $u=(K_g-K_r)/K_g$. This parameter vanishes if the radiation pressure K_r fully compensates the gravitational force K_g, and it attains the value 1 for vanishing radiation pressure. The strongest deceleration of the solar wind occurs within the density cone $\theta \approx 180°$, as is shown in the dashed-dotted line. Since the deceleration in the cone is mainly due to helium it is nearly independent of the solar cycle phase.

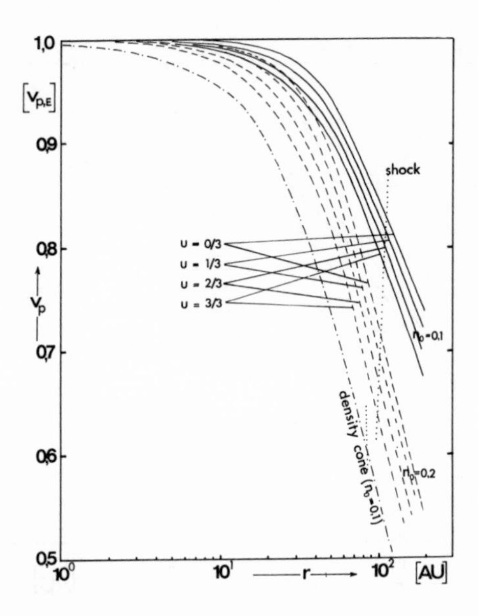

Figure 3: The solar wind velocity v_p vs the distance r from the sun. The full lines correspond to densities $n_o(H)=0.1\ cm^{-3}$; $n_o(He)=0.01\ cm^{-3}$, whereas the dashed lines correspond to $n_o(H)=0.2$; $n_o(He)=0.02\ cm^{-3}$. The dashed-dotted line gives the velocity profile within the downwind density cone ($\theta=180°$) for $n_o(H)=0.1\ cm^{-3}$; $n_o(He)=0.01\ cm^{-3}$. The dotted line in the nearly vertical direction gives the location of the heliospheric shock front.

In fig. 4 the profiles of the particle density n_p and the energy density E_p of the solar wind protons are shown. While the full lines show the $1/r^2$ profiles for $v_p=$ const., the dashed lines give the deviating profiles which are caused by the deceleration of the solar wind plasma for u=0 and u=1 which comprehends the variation to be expected during one solar activity cycle. The dashed-dotted lines give the profiles of n_p and E_p within the density cone ($\theta=180°$). The location r_s of the shock front is to be expected at that position where the energy density E_p of the solar wind plasma has dropped to the value of the energy density of the interstellar field B_i, because from here outwards B_i starts controlling the motion of the plasma. Fig. 4 shows where these points r_s are reached in case B_i has a value of 3,4,5,6×10^{-6} Gauss as is supported by Zeeman-splitting measurements of Verschuur (1968) and Davies et al. (1968). While at θ=150° (outside the cone) the shock front is expected at about 120AU, its location inside the cone is at r_s=90AU. The shock front location between 0°≤θ≤160° varies only by about Δr_s=5AU. The variation Δr_s due to varying Lyman-α radiation pressures is of about the same order of magnitude.

In fig. 5 the geometrical configuration of the heliospheric shock front resulting from these calculations can be seen. While the shock front is nearly spherical at angles between 0°≤θ≤160° with values of $r_s \gtrsim$120AU, it shows a pronounced cusp in the region of the density cone at θ=180° with values of r_s decreasing even to 90AU.

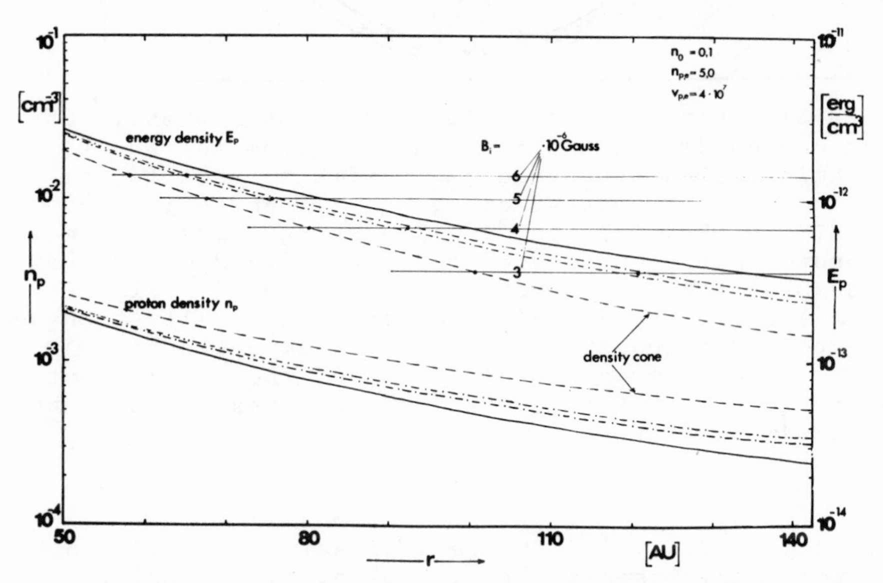

Figure 4: Particle density n_p and energy density E_p of the solar wind protons vs the distance r from the sun. The full lines give the $1/r^2$ profiles for v_p= const., the dashed lines indicate the range 0≤u≤1. The dashed-dotted lines give the profiles n_p and E_p for the density cone. The horizontal lines correspond to energy densities $B_i^2/8\pi$ of the interstellar field B_i=3,4,5,6×10^{-6} Gauss. The intersection of these lines with the E profiles shows the corresponding location r_s of the heliospheric shock front.

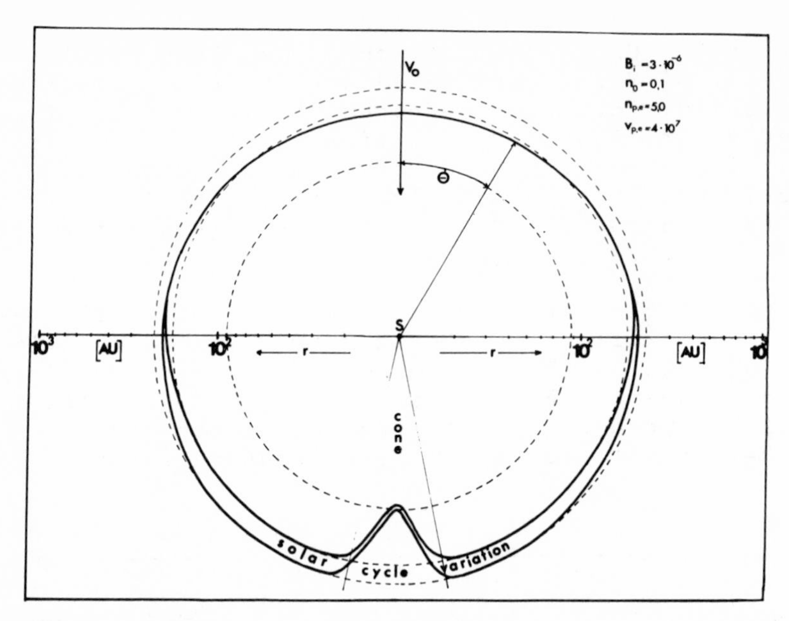

Figure 5: Shape of the heliospheric shock front. The polar coordinates are the distance r from the sun (logarithmic scale) and the angle θ with respect to the upwind direction. (n_E=5 cm^{-3}; $v_p = 4\times10^7$ cm/sec; n_o(H)= 0.1 cm^{-3}; n_o(He)=0.01 cm^{-3}). The energy loss per secondary proton has been assumed to be $\Delta E_{kin} = 2/2\ m_p v_p^2 \neq (\alpha \gtrsim 90°)$. The cusp of the shock front in the downwind region is due to the energy extraction of gravitationally focussed interstellar hydrogen and helium of temperatures T= 1000 K. The solar cycle variation of the shock front due to varying Lyman-α radiation pressures is shown by the two dashed circles. V_o=20 km/sec.

REFERENCES

Ambarzumjan, V.A., in "Theoretical Astrophysics" VEB Dt. Verlag d.Wissenschaften, Berlin, p. 45, 1957.

Axford, W.I. et al., Astrophys. J. 137, 1268, 1963.

Belyaev, V.A. et al., Soviet Physics JETP 25, 777, 1967.

Blum, P.W. and H.J. Fahr, Astronomy and Astrophys. 4, 280, 1970.

Dessler, A.J., Rev. Geophys. 5, 1, 1967.

Davies, R.D. et al., Nature 220, 1207, 1968.

Fahr, H.J., preprint 71-10, Institut für Astrophysik, Universität Bonn, 1971a.

Fahr, H.J., Planetary Space Science, in press, 1971b.

Fahr, H.J., preprint 71-11, Institut für Astrophysik, Universität Bonn. Astronomy and Astrophys., in press, 1971c.

Fite, W.L. et al., Proc. Roy. Soc. A 268, 527, 1962.

Hinteregger, H.E. and L.A. Hall, J. Geophys. Res. 75, 6959, 1970.

Hinteregger, H.E. et al., Space Research V, North-Holland Publishing Co., Amsterdam, p. 1175, 1965.

Mason, P. and D.L. Vanderslice, J. Chem. Phys. 40, 3552, 1955.

Po Lee and G.L. Weissler, Astrophys. J. 115, 570, 1952.

Tousey, R., Space Science Rev. 2, 3, 1963.

Verschuur, G.L., Phys. Rev. Letters 21, 775, 1968.

HYDROGEN - HELIUM EXPANSION FROM THE SUN

Edmund J. Weber

Kitt Peak National Observatory*

Tucson, Arizona

In general, solar wind models have treated a plasma consisting of hydrogen-ions and electrons. However, already the first measurements of the properties of the interplanetary medium taken in 1962 by Mariner 2 had indicated that α-particles were the most important minor ion constituent in the solar wind, accounting for approximately 4% of the total number of ions at 1 AU. The helium abundance in the solar wind has been measured on many subsequent spacecraft (Pioneer 6, Vela 3, Explorer 34) and even by the Apollo solar wind experiments on the moon. The equations required to represent the complete model of a steady-state hydrogen-helium-electron plasma are extremely complicated and difficult to solve. Thus to enable us to make some progress at all in trying to understand the motion of such a two-ion gas, let us use the simplest of all possible models, being fully aware of its shortcomings yet hoping that we might learn something from its, admittedly crude, quantitative and quali-

*Operated by the Association of Universities for Research in Astronomy, Inc. under contract with the National Science Foundation.

tative features.

We shall use a hydrodynamic description of the hydrogen-helium plasma in which all such effects as magnetic fields, finite conductivities, viscosities, etc. are neglected. Furthermore, we assume that the ion constituents are only weakly coupled by means of the electric field and that all plasma components have constant, but not necessarily equal, temperatures. We can then write the momentum equations for the two ions in the following form:

$$u_i \frac{du_i}{dr} = -\frac{kT_i}{n_i m_i}\frac{dn_i}{dv} - \frac{GM_\odot}{r^2} - \frac{kT_e}{m_i}\frac{Z_i}{n_e}\frac{dn_e}{dr} \tag{1}$$

where the subscripts i and e refer to a given ion and the electrons, respectively. All symbols have their usual meaning, with Z_i representing the charge on the ion. In the above equation we have neglected terms of the order $Z_i\ m_e/m_i$. Conservation of mass requires that

$$n_i u_i r^2 = J_i = \text{constant} . \tag{2}$$

Since the plasma is electrically neutral we require further that

$$n_e = \sum_i Z_i n_i \quad . \tag{3}$$

Thus, the term dn_e/dr in equation (1) can be rewritten as

$$\frac{dn_e}{dr} = -\sum_i Z_i \frac{dn_i}{dr} = -\sum_i Z_i n_i \left(\frac{1}{u_i}\frac{du_i}{dr} + \frac{2}{r}\right) \quad . \tag{4}$$

This equation shows clearly that the individual ion

momentum equations are coupled together due to the presence of the electric field. This can be exhibited more clearly by rewriting equation (1) using equation (4) as follows:

$$\frac{1}{u_i}\frac{du_i}{dr}\left[u_i^2 - \frac{kT_i}{m_i} - \frac{kT_e}{m_i}\left(\frac{n_i}{n_e}\right)Z_i^2\right] - \frac{kT_e}{m_i}\frac{Z_i}{n_e}\sum_{j\neq i} n_j Z_j \frac{1}{u_j}\frac{du_j}{dr}$$

$$= \frac{2}{r}\frac{k}{m_i}\left[T_i + Z_i T_e\right] - \frac{GM_\odot}{r^2} \tag{5}$$

This set of two coupled differential equations can be solved to give us two differential equations which can be written in general as

$$\frac{1}{u_i}\frac{du_i}{dr} = \frac{N_i}{D} \tag{6}$$

where

$$N_i = \left[u_j^2 - \frac{kT_j}{m_j} - \frac{kT_e}{m_j}\left(\frac{n_j}{n_e}\right)Z_j^2\right]\left[\frac{2}{r}\frac{k}{m_i}\left(T_i + Z_i T_e\right) - \left(\frac{GM_\odot}{r^2}\right)\right]$$

$$+ \frac{kT_e}{m_i}\frac{Z_i}{n_e} n_j Z_j \left[\frac{2}{r}\frac{k}{m_j}\left(T_j + Z_j T_e\right) - \frac{GM_\odot}{r^2}\right] \qquad i\neq j \tag{7}$$

and

$$D = \left[u_i^2 - \frac{kT_i}{m_i} - \frac{kT_e}{m_i}\left(\frac{n_i}{n_e}\right)Z_i^2\right]\left[u_j^2 - \frac{kT_j}{m_j} - \frac{kT_e}{m_j}\left(\frac{n_j}{n_e}\right)Z_j^2\right]$$

$$- \left(\frac{kT_e Z_i Z_j}{n_e}\right)^2 \frac{n_i}{m_i}\frac{n_j}{m_j} \qquad i\neq j \tag{8}$$

Thus we see immediately that wherever D equals zero a singularity occurs in the differential equations. We recall that for the single hydrogen plasma of the solar wind as discussed by Parker (1963) this occurs where $u_i^2 = 2kT_i/m_i$. The more general result for a single ion plasma can be obtained from the above equation (8) by letting n_j=0. We then find that for D to equal zero,

$$u_i^2 = \frac{k}{m_i}\left[T_i + Z_i^2 T_e \frac{n_i}{n_e}\right] = \frac{k}{m_i}\left[T_i + Z_i T_e\right] \tag{9}$$

which reduces, of course, identically to Parker's result for Z_i=1 and T_e=T_i. In the case of the two-ion plasma we find that there exists many more singular points which depend in general on the flow properties and especially the flow velocities. To illustrate this, it is convenient to introduce the Mach number

$$M_i^2 = u_i^2 \Big/ \left(kT_i/m_i\right) \tag{10}$$

in terms of which the condition that D equals zero reduces to

$$\left(M_i^2 - 1 - \alpha_i \frac{Z_i^2\, n_i}{n_e}\right)\left(M_j^2 - 1 - \alpha_j \frac{Z_j^2\, n_j}{n_e}\right) = \left(\frac{kT_e Z_i Z_j}{n_e}\right)^2 \frac{n_i n_j}{m_i m_j} \qquad i \neq j \tag{11}$$

where $\alpha_i = T_e/T_i$.

Since this expression is explicitly only a function of the ion Mach numbers, temperatures (assumed to be constant) and the hydrogen-helium mass flux ratio (J_1/J_2) in the solar wind, we can locate all the points where D=0.

The curves for all possible (M_1, M_2) combinations for which equation (11) is satisfied for a given hydrogen-helium ratio and a given temperature of the species are shown in Figure 1. As can be seen quite clearly, there are two distinct branches which do not connect for posi-

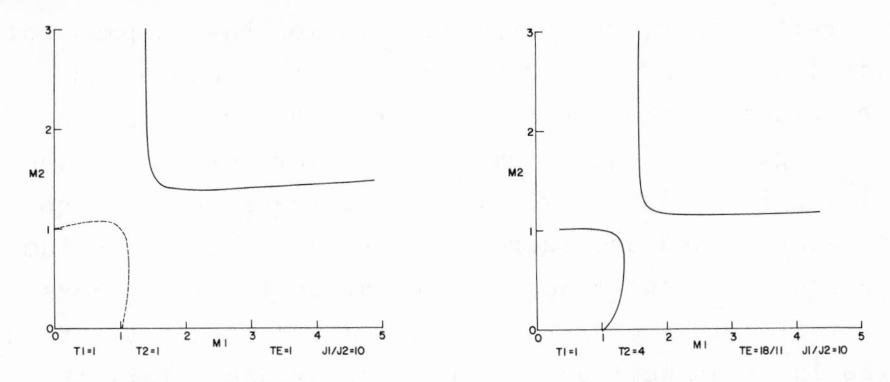

Fig.1 (left) & Fig.2 (right) - Loci of (M_1,M_2) for which D=0. Solid (dashed) lines indicate the loci for which the critical radius $r_c \geqslant r_\odot$ ($r_c \leqslant r_\odot$). Helium temperature T_2 is $1\times10^{6\circ}$ and $4\times10^{6\circ}$, respectively.

tive Mach numbers. This is a general property of D and exists for any combination of T_1, T_2 and T_e. In order to determine where the critical points are located we use now the standard argument that du_i/dv has to be finite everywhere for a physical solution. Thus we require that the numerator function $N_i(r)$ vanishes wherever the denominator D goes to zero. Since the numerator depends on the radial distance and since we know all pairs of (M_1,M_2) which make D=0, we can calculate the values of the critical radius r_c where then the numerator also vanishes. If this is done for the values of temperature shown in Figure 1, we find that values of $r_c > r_\odot$ can only be found for values of (M_1,M_2) on the upper branch. On the lower branch the values of (M_1,M_2) will result in $r_c < r_\odot$, which means that we have non-physical critical points. But in order to have supersonic expansion we require that the solution passes in the M_1-M_2 plane from a point close to the origin (since near the sun both hydrogen and helium are clearly subsonic) to a region where both hydrogen and helium are supersonic, as has been observed by the spacecraft. Thus, the physical solution has to pass both critical lines and this clearly can not be done with those values of the ion and electron temperatures. As we use higher but equal temperatures for both ions, the results still hold. However, if one keeps the hydrogen temperature fixed and increases the value of the helium temperature T_2, one finds a temperature for which physically real values of r_c are obtained for values of (M_1, M_2) on the lower as well as on the upper branch. This is shown in Figure 2. In particular, a helium temperature of 4×10^6 °K was used. To determine the actual solution we need additional information which can be obtained by observing that equation (1) can be integrated directly. This will give us two Bernoulli-type equations for the system, namely

$$\frac{u_i}{2} + \frac{kT_i}{m_i} \ln n_i - \frac{GM_\odot}{r} + \frac{kT_e Z_i}{m_i} \ln n_e = \text{constant} = E_i . \quad (12)$$

Both of these equations have to be satisfied at the two critical points and this allows us to determine uniquely the two critical points through which the solution has to pass for a given set of ion and electron temperatures and the H/He ratio at infinity (*i.e.* on J_1/J_2). This is indicated in Figure 3, which shows the values of M_1, M_2, E_1 and E_2 as functions of the position of the critical radius r_c. The lowest value for T_2 for which such a solution can be found will indicate the threshold temperature at which a solar wind type expansion is possible for a H/He gas. This temperature is near 3.5×10^6 °K. For an expanding plasma with a higher (lower) ratio of J_1/J_2 [*i.e.* a lower (higher) helium abundance] than that used to obtain the values shown in the last two figures, lower

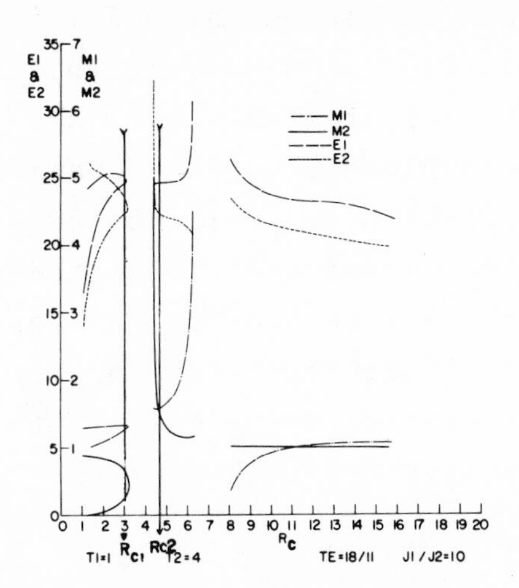

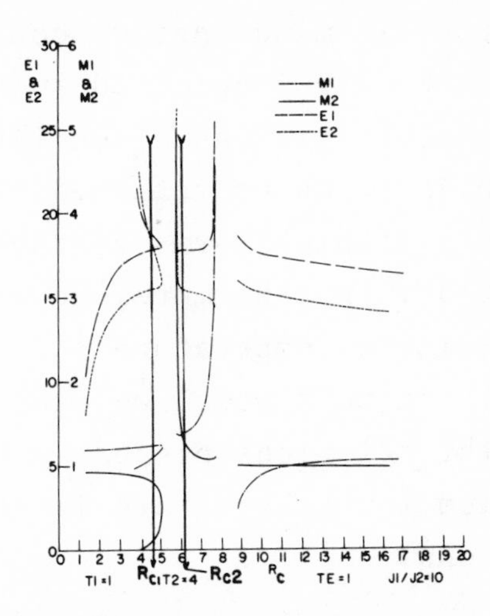

Fig.3 (left.) & Fig.4 (right) - Mach numbers and constants of Bernoulli's equation as functions of r_c. At the critical radii r_{c1} and r_{c2} both E_1 and E_2 [in units of 10^{14} cm^2 per sec^2] have the same values, respectively.

(higher) helium temperatures are required. The curves shown in this figure depend on the choice of the electron temperature T_e. However, the latter does not influence the existance or non-existance of critical points. This is indicated in Figure 4.

Conclusions. This greatly simplified model of the hydrogen-helium solar wind has provided us with the important result that no solar wind type expansion of the hydrogen-helium gas is possible when both gases have equal temperatures, and only when the helium ions are hotter than the hydrogen ions can a supersonic expansion from the sun occur. The lower the helium/hydrogen mass flux ratio is, the lower needs to be the helium temperature which will allow this two-ion solar wind flow. Thus we expect that this effect is very small for the other ions in the solar wind which have only a very small abundance. In obtaining this overall result, one has to keep in mind that we consider an isothermal model in which we have neglected such interactions between the ions as collisions. Yeh (1970) has treated a similar case in which he retained the complete electron equations (*i.e.* in which terms of the order $Z_i m_e/m_i$ were not neglected). Furthermore, he assumes immediately that the helium ion temperature is four times that of the hydrogen ion. In this model we exhibit clearly that the topology of the solutions present a physical solution only if the helium ion temperature is about 3.5 times that of the protons.

REFERENCES

Parker, E. N.: 1963, *Interplanetary Dynamical Processes*, (New York: Interscience Publ.).

Yeh, T.: 1970, *Planet. Space Sci.* 18, 199.

HEATING OF THE SOLAR WIND IONS

N. D'Angelo[†] and V.O. Jensen[††]

[†]European Space Research Institute, Frascati

[††]Research Establishment Risø, Roskilde

ABSTRACT

The suggestion is advanced that the high ion temperatures in the solar wind, observed near the Earth's orbit, may be associated with the presence of an influx of neutral hydrogen from the boundary of the heliosphere. Photoionisation and charge exchange transform neutral hydrogen atoms into protons. The ionised hydrogen component near the Earth's orbit is then double-peaked and unstable. The kinetic energy of the protons (originating from the neutral hydrogen component) in the rest frame of the solar wind ions may then be sufficient to produce the observed ion heating.

Full paper to be published in Cosmic Electrodynamics.

ON THE GENERATION OF SHOCK PAIRS IN THE SOLAR WIND

V. Formisano and J.K. Chao

Laboratorio Plasma Spaziale del CNR, Roma, and

European Space Research Institute, Frascati

Numerical studies (neglecting the magnetic fields) have been carried out by Hundhausen and Gentry (1969a, b) to develop shock pairs in the solar wind. They considered a step function increase of momentum exerted upon the ambient solar wind for longer than 5 hours approximately which would produce a double shock structure. Since shock pairs are rarely observed at 1 A.U., they suggested that a rarefaction wave behind the reverse shock weakens it and eventually causes it to disappear. The basic idea of our paper is that shock pairs are generated mainly within the solar wind at some distance (possibly large) from the solar corona, in connection with sufficiently large velocity gradients. As a consequence double shocks are related to slow changes in the temperature of the solar corona between hot regions and cold regions. A velocity gradient, indeed, will give rise to a pressure pulse given by

$$\Delta P(r) = \frac{P_o}{r^2}\left(\frac{\Delta\tau_o}{\Delta\tau} - \frac{1}{r^{1/2}}\right) + \frac{P_{mTo}}{r^2}\left(\left(\frac{\Delta\tau_o}{\Delta\tau}\right)^2 - 1\right)$$

where
P_o = pressure at 0.1 A.U.
r = radial distance measured in 0.1 A.U.
$\Delta\tau = \Delta\tau_o - (r-1)r_o\left(\frac{1}{v_1} - \frac{1}{v_2}\right)$
r_o = 0.1 A.U.
V_1, V_2 = plasma velocity before and after the gradient

$\Delta\tau_o$ = time separation between V_1 and V_2 at r_o

P_{mto} = transverse magnetic pressure at r_o.

When $\Delta\tau$ starts to become small, there will be within the gradient a pressure larger and larger than the ambient pressure, and it is this pressure pulse that can push the ambient plasma in both directions, away and toward the sun. If this pushing is strong enough it is reasonable to compare the gradient region with two expanding pistons.

It is possible to compute, on the basis of this model, the location of generation of the shock pair when V_1, V_2 and $\Delta\tau_o$ (or $\Delta\tau$ at 1 A.U.) are given. If at 1 A.U. the shock pair is not developed it is possible to estimate pressure, density and temperature ratios, between the gradient and the ambient plasma. The model agrees well with observations of Chao et al. 1971 and Burlaga et al. 1971.

REFERENCES

Hundhausen, A.J. and Gentry, R.A., 1969(a), J. Geophys. Res. 74, 2908.

Hundhausen, A.J. and Gentry, R.A., 1969(b), J. Geophys. Res. 74, 6229.

Chao, J.K., Formisano, V. and Hedgecock, H., Shock pair observation in Conference on the solar wind, Asilomar, California, 21-26 March 1971.

Burlaga, L.F., Ogilvie, K.W., Fairfield, D.H., Montgomery, M.D. and Bame, S.J., 1971, Astrophys. J. 164, 137.

SPECTRAL ANISOTROPY OF ALFVÉN-WAVES IN THE SOLAR WIND

Heinrich J. Völk and Werner Alpers
Max-Planck-Institut für Physik und Astrophysik
Inst.f.extraterr.Physik, Garching

According to the analysis of Mariner 5 data (Belcher and Davis, 1971, hereafter referred to as BD) large-amplitude Alfvén-waves of solar origin dominate the microscale fluctuations in the solar wind for at least 50% of the time. This claim is disputed by other observers and it may have to be revised towards a weaker statement regarding the dominant occurrence of these waves.

Here we will not have to add anything to this discussion from the observational side. Under the assumption, however, that Alfvén-waves constitute at least at times a well distinguishable part of the fluctuation spectrum we wish to emphasize some consequences of the hypothesis of their solar origin.

Thus, assuming the existence of a spectrum of Alfvén-waves in the outer corona at some radial distance r_0 from the sun, say $r_0 \geqslant 2\ R_\odot$, we discuss the propagation of these waves into the interplanetary medium. This is done in the linear approximation for the wave amplitudes and in the approximation of geometrical optics, using the model of a stationary, spherically symmetric solar wind. In this approximation which should be very good at least at the higher frequencies reported by BD, there is no coupling of Alfvén-waves to other waves, effects which undoubtedly play an important role in the solar wind as will be discussed later. The ray tracing method describes the change of wave amplitude $|\delta \underline{B}|$

as well as the direction of energy flow and the direction of the wave vector $\underline{k}$ as the wave propagates towards the orbit of earth. From the modifications of the $\underline{k}$-vectors in particular we shall derive a relation for the directional distribution of the magnetic field vectors associated with the waves.

The time rate of change of the position $\underline{x}$ of a point in a wave-front and of the wave-normal $\underline{k}$ is given by (Bazer and Hurley, 1963).

$$(1) \quad \frac{d\underline{x}}{dt} = \underline{v} + \frac{\underline{B}}{\sqrt{4\pi\rho}} \quad ; \quad \frac{d\underline{k}}{dt} = \left\{ -(\underline{k}\cdot\underline{\nabla}) - \underline{k}\times\underline{\nabla} \right\} \left(\underline{v} + \frac{\underline{B}}{\sqrt{4\pi\rho}} \right)$$

where $\underline{v}$, ρ and $\underline{B}$ are the solar wind velocity, mass density and magnetic field, respectively. $\underline{B}$ is assumed to point away from the sun. Only outgoing waves are considered. In terms of the frequency ω, $\underline{k}$ is given by the local dispersion relation

$$(3) \quad \omega = \underline{k}\cdot\left(\underline{v} + \frac{\underline{B}}{\sqrt{4\pi\rho}} \right) \quad ; \quad \frac{d\omega}{dt} = 0$$

ω is constant in a frame fixed in the sun. This is to a good approximation also the frequency seen by a space probe like Mariner 5.

The rays $\underline{x}(t)$ are mainly radial with a small tendency to spiral with the magnetic field. Using a special coordinate system (r, θ, ϕ) centered at the sun we solve eqs. (1)-(3) for the $\underline{k}$-vector in the solar equatorial plane ($\theta = \pi/2$) ignoring the small tilt of this plane against the ecliptic plane in which plane the observations were made. Thus we get

$$(4) \quad k_\phi = \frac{r_0}{r} k_\phi(r_0) \quad ; \quad k_\theta = \frac{r_0}{r} k_\theta(r_0)$$

while k_r then follows as

$$(5) \quad k_r = \frac{\omega - k_\phi c_\phi}{v_r + c_r} \quad ; \quad \underline{c} \equiv \frac{\underline{B}}{\sqrt{4\pi\rho}}$$

where we neglected the small azimuthal wind veloctiy v_ϕ.

The $\frac{1}{r}$ dependence of k_ϕ and k_Θ is due to the spherically symmetric expansion of the solar wind and does not depend on any other characteristic of the flow.
Neglecting for a moment the azimuthal component of $\underline{B}$ and observing that v_r+c_r is roughly constant, we have
$k_r \approx$ const.
Thus the wave normal is turned towards the radial direction as an Alfvén-wave propagates outward. Since the wave magnetic field vector $\delta\underline{B}$ is proportional to $\underline{k} \times \underline{B}$, $\delta\underline{B}$ is turned into a direction perpendicular to $\underline{e}_B$ and $\underline{e}_r$, where $\underline{e}_B$ and $\underline{e}_r$ are the unit vectors in the $\underline{B}$ and $\underline{r}$ directions; this produces a directional anisotropy of $\delta\underline{B}$.

The wave amplitude $|\delta\underline{B}|$ obeys the eq.

$$\text{(6)} \quad \frac{d}{dt} \ln\left(\frac{|\delta\underline{B}|^2}{\sqrt{\rho}}\right) = -\operatorname{div}\underline{v}$$

Since from eq. (1) $\frac{d\underline{x}}{dt}$ is independent of $\underline{k}$, the same is true for $\frac{d}{dt}|\delta\underline{B}|^2$. Therefore two amplitudes being equal for different $\frac{\underline{k}}{|k|}$ remain equal in course of time. Eq. (6) can be integrated to give:

$$\text{(7)} \quad \ln\left(\frac{\delta B^2(r)}{\delta B^2(r_o)}\sqrt{\frac{\rho(r_o)}{\rho(r)}}\right) = -\int_{r_o}^{r} dr \frac{1}{r^2(v_r+c_r)} \frac{d}{dr}\left(r^2 v_r\right)$$

Using an approximate representation for v_r and $|\underline{c}|$ as a function of r (Burlaga, 1971, private communication), $\left|\frac{\delta B}{B}\right|$ reaches a maximum of $\approx 20\left|\frac{\delta B_o}{B_o}\right|$ very near the orbit of earth and decreases again beyond (Figs. 1 and 2).

To calculate the directional anisotropy of the spectrum of $\delta\underline{B}$ as a function of r we decompose, as BD have done, $\delta\underline{B}$ into components parallel to $\underline{e}_B \times \underline{e}_r$, $\underline{e}_B \times (\underline{e}_B \times \underline{e}_r)$, $\underline{e}_B$. Assuming now at r_o a source of Alfvén-waves which for each ω provides a spectrum where all wave normal directions and wave amplitudes are equally probable, we calculate the average of

$|\delta B_{\underline{e}_B x(\underline{e}_B x \underline{e}_r)}|^2$ and $|\delta B_{\underline{e}_B x \underline{e}_r}|^2$ over the resulting spectrum at $r > r_o$. By the assumption of Alfvén-waves we have $\delta B_{\underline{e}_B} \equiv \underline{e}_B \cdot \delta \underline{B} = 0$ and the ratio

$$A = \left\langle |\delta B_{\underline{e}_B x(\underline{e}_B x \underline{e}_r)}|^2 \right\rangle \cdot \left\langle |\delta B_{\underline{e}_B x \underline{e}_r}|^2 \right\rangle^{-1}$$

is the power-anisotropy in the plane perpendicular to $\underline{B}$.

This anisotropy is displayed in Figs. 3 and 4. Starting with an isotropic spectrum at $r_o = 2 R_\odot$, the resulting value of A at 1 AU is roughly 90. Even with $r_o = 4 R_\odot$ we still get $A \approx 50$.

A qualitatively similar but quantitatively much smaller value has been observed by BD. For average conditions they obtained, independent of frequency:

$$\left\langle |\delta B_{\underline{e}_B \times \underline{e}_r}|^2 \right\rangle : \left\langle |\delta B_{\underline{e}_B \times (\underline{e}_B \times \underline{e}_r)}|^2 \right\rangle : \left\langle |\delta B_{\underline{e}_B}|^2 \right\rangle = 5:4:1$$

while in the compression regions of high velocity streams, where also the wave amplitudes are very high, the above ratio is quoted as 6:3:1.

BD seek to explain the anisotropy essentially by coupling of waves with $k_\theta \neq 0$ into damped magnetoacoustic waves through the spiralling of the magnetic field $\underline{B}$; correspondingly the observed small deviation from isotropy is thought of as a measure of the amount of heating of the solar wind plasma by the Alfvén-waves.
Our results show clearly that the average anisotropy is essentially a geometric effect due to the expansion of the solar corona into interplanetary space. Also the observation of an "instantaneous" anisotropy, i.e. relative to the "instantaneous" average field $\underline{B}$, is readily explained in this way.
Although it is easy to convince oneself that the measured anisotropy must be smaller than the one actually present this cannot explain a discrepancy of almost two orders of magnitude. We believe that scattering either due to convected irregularities (Valley, 1971) or due to interaction between the large-amplitude waves is responsible for keeping the anisotropy down to the low values observed. Nonlinear plane Alfvén-waves (Kantrowitz and Petschek, 1966) which are of a semi-circularly polarized

nature could explain a lower anisotropy, assuming that these nonlinear waves interact only weakly. This would, however, be inconsistent with the observation of both high amplitudes and (relatively) high anisotropy in compression regions because in this case higher amplitudes would imply lower anisotropy. Thus, tentatively we suggest the following picture: The Alfvén-waves emanating from the solar corona are strongly scattered in the interplanetary medium. This scattering is probably strongest near the sun, where all amplitudes are largest. Since this scattering is certainly not only elastic, i.e. produces only changes in the direction of k, but will generate other (damped) magnetohydrodynamic waves, also the relative amplitudes $\delta B/B$ of the Alfvén-waves will be prevented from growing in the way predicted by the WKB-theory (Parker, 1965) and given in equation (7). This dissipation determines both the amount of heating of the solar wind and the distribution of turbulent magnetic pressure (Belcher, 1971) as far as they are due to Alfvén-waves. In the specific case of the compression regions - following now the argument of BD - the "arriving" low anisotropy is enhanced by coupling with the background which undergoes fairly rapid changes there, leaving both higher amplitudes and higher anisotropies at the orbit of earth.

References

Bazer, J., and J. Hurley, J. Geophys. Res., 68, 147(1963)

Belcher, J., Ap. J. 168, 509 (1971)

Belcher, J., and L. Davis Jr., J. Geophys. Res., 76, 3534 (1971)

Kantrowitz, A.R., and H.E. Petschek in "Plasma Physics in Theory and Application" (Ed. W.B. Kunkel), pp. 148-206, McGraw Hill, New York (1966)

Parker, E.N., Space Sc. Rev., 4, 66 (1965)

Valley, G.C., Ap. J. 168, 251 (1971)

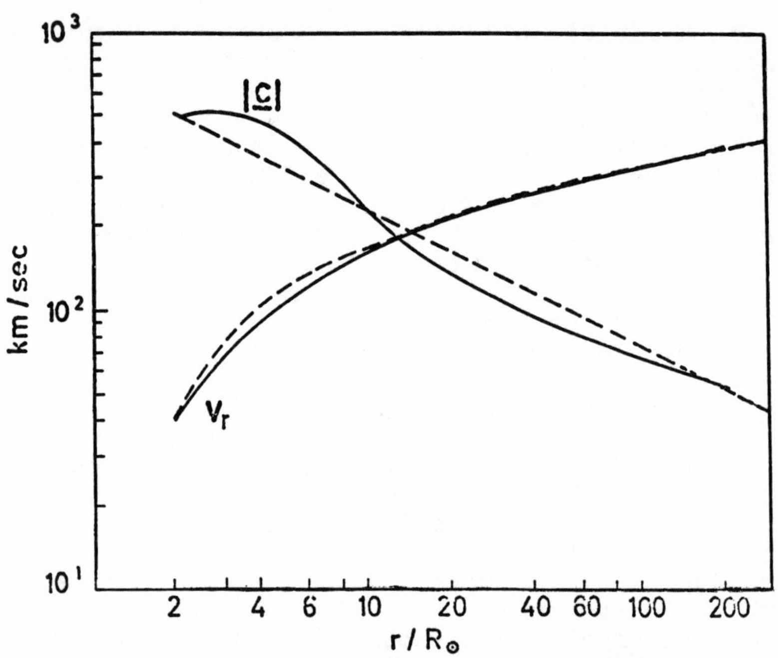

Fig 1. The r-dependence of the radial expansion velocity v_r and the Alfvén-velocity $|\underline{c}| = |\underline{B}| \cdot (4\pi\rho)^{-1/2}$ after Burlaga (1971). Dashed curves are approximations used to calculate the curves of Fig 2 from eq. (7).

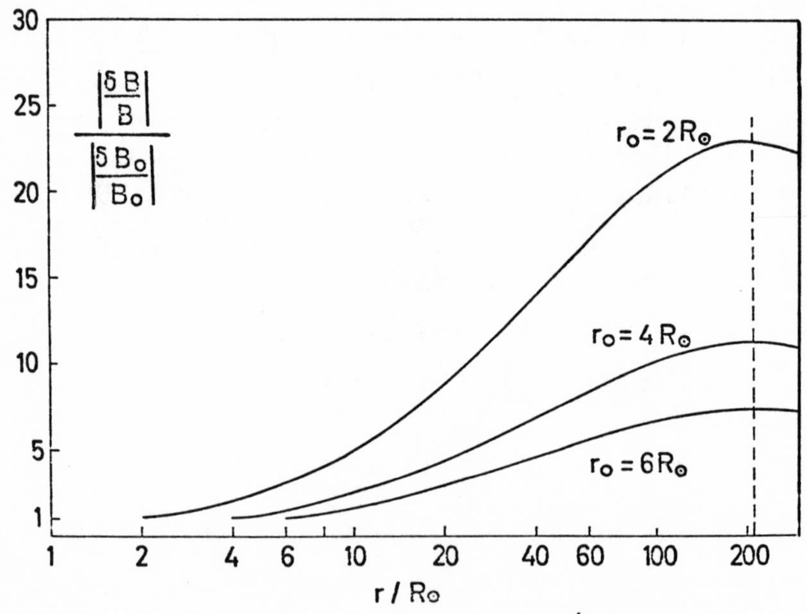

Fig 2. The amplification factor $\left|\frac{\delta B}{B}\right| \Big/ \left|\frac{\delta B_o}{B_o}\right|$ as a function of r. $\left|\frac{\delta B_o}{B_o}\right|$ is the relative amplitude at r_o.

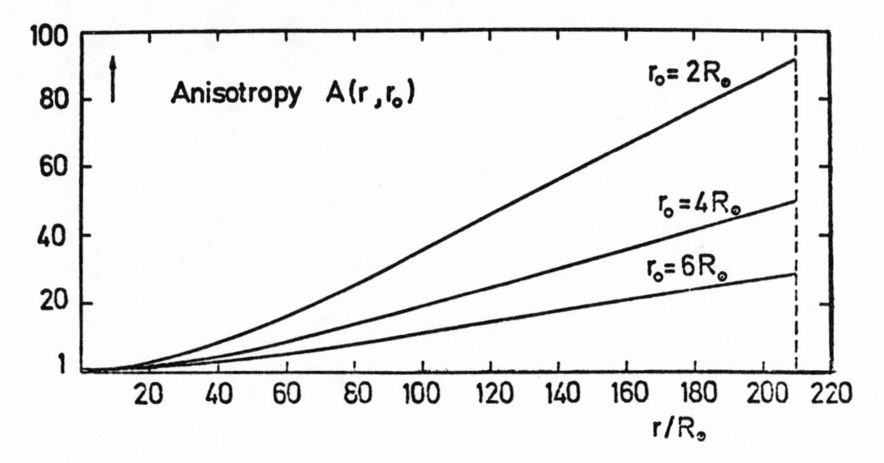

Fig 3. The anisotropy factor $A(r,r_o)$ as a function of r assuming isotropy (A = 1) at different source levels r_o.

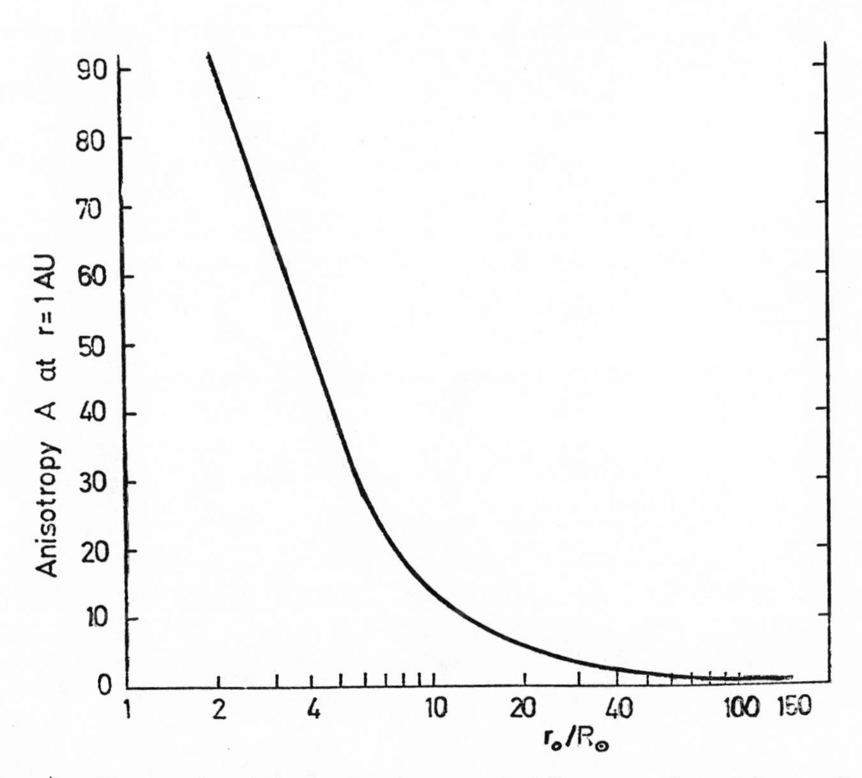

Fig 4. The anisotropy A at r = 1 AU as a function of the radial distance r_o of the source.

EVIDENCE FOR WAVES AND/OR TURBULENCE IN THE VICINITY OF SHOCKS IN SPACE

J.K. Chao

Laboratorio Plasma Spaziale del CNR, Roma, and

Istituto di Fisica, Università di Roma

The magnetohydrodynamic Rankine-Hugoniot relations have been tested for many shocks found in space (Sonett et al.,1964, Ogilvie and Burlaga, 1969, Mihalov et al., 1969, Chao, 1970, Lepping and Argentiero , 1971, and others). This was possible because more refined plasma and magnetic field data have been collected by various spacecraft during recent years. In the previous studies, plasma data were not complete enough, so that only the magnetic field data were used to deduce the shock normal direction (Taylor, 1968, Ness and Wilcox, 1967). Multiple spacecraft methods have also been used to obtain the shock speed and normal direction (Burlaga, 1970). However, it is not possible to check the MHD R-H relations unless both the plasma and the magnetic field data are available. Many fast shocks (Sonett et al., 1964, Chao, 1970, and others), a few slow shocks (Chao and Olbert, 1970, Burlaga and Chao, 1971), a reverse shock (Burlaga, 1970) and a shock pair (Chao et al., 1971) have been identified in interplanetary space by applying the MHD R-H relations. One case of the earth's bow shock has also been studied (Mihalov et al., 1969, Mariani et al., 1971). Many of these shocks have been independently checked for their transit time and shock speed using the multiple spacecraft method.

In most of the above-mentioned studies, the MHD R-H relations were not totally satisfied because only those relations which do not involve the plasma pressure were satisfied. The conservation equations for normal momentum and energy flux were not satisfied in many of the above-mentioned cases. It is the purpose of this study to re-examine critically the conservation equations for normal momentum and energy flux in all the possible cases for which complete plasma and magnetic field data are available. We have been able to find eighteen shocks which will be examined later.

There are eight shock relations, namely mass flux (1 eq.), transverse-momentum flux (2 eq.), transverse electric field (2 eq.), normal magnetic field (1 eq.), normal momentum flux (1 eq.) and energy flux (1 eq.). We let the subscript "1" refer to the direction normal to the shock front. Since the coplanarity theorem also holds for an anisotropic plasma, we may set $B_2 = B_2' = 0$, where the prime refers to parameters in the post-shock state. Brackets indicate the difference between the pre-shock and post-shock states. The shock equations become (adopting rationalized M.K.S. units):

$$[B_1] = 0 \tag{1}$$

$$[V_2] = 0 \tag{2}$$

$$[\rho V_1^{*}] = 0 \tag{3}$$

$$[V_3 B_1 - V_1^{*} B_3] = 0 \tag{4}$$

$$B_2 = B_2' = 0 \tag{5}$$

$$\left[\rho V_1^{*} V_3 - \xi \frac{B_1 B_3}{\mu_o}\right] = 0 \tag{6}$$

$$\left[\rho V_1^{*2} + P + \frac{1}{3}(\xi + \frac{1}{2}) \frac{B^2}{\mu_o} - \xi \frac{B_1}{\mu_o}\right] = 0 \tag{7}$$

$$\left[(\frac{1}{2} \rho V^{*2} + \frac{\gamma}{\gamma-1} P + \frac{1}{2}(\xi+2) \frac{B_2^{\,2}}{\mu_o} - \xi \frac{B_1^{\,2}}{\mu_o}) V_1^{*} - \xi \frac{B_1 B_3}{\mu_o} V_3\right] = 0 \tag{8}$$

The ratio of the specific heat γ in Eq. (8) should be 5/3 for a monatomic gas with 3 degrees of freedom if we derive the energy flux equation by taking the 2nd moment of the Vlasov equation. However, we will retain γ in Eq. (8) as a free parameter which may be varied later.

In deriving these equations, we assume that protons and electrons can be approximated by bi-Maxwellian distribution functions in a frame of reference moving with the mean plasma velocity $\underline{V}$.

Here P is the trace of the pressure tensor and ξ is a parameter measuring the anisotropy of the plasma:

$\xi = 1 - \frac{P_{\parallel} - P_{\perp}}{B^2/\mu_o}$. $V_1^* = V_1 - V_s$ is the component of velocity normal to the shock front in the frame of reference travelling with the shock, V_s being the shock speed.

The equations (1) to (5), if written in an arbitrary frame of reference, contain seventeen parameters, namely: $\underline{V}$, $\underline{V}'$, $\underline{B}$, $\underline{B}'$, $\underline{V}_s$, N and N'. However, the shock velocity $\underline{V}_s$ is not measured. Thus, formally, by using the five conservation equations we are able to predict any five of the seventeen parameters provided that the remaining 12 are known. Since we have 14 measured parameters, it implies that two of them are overdetermined. This overdetermination allows us to test the validity of the MHD shock relations and to solve for the unknown shock velocity $\underline{V}_s$.

We can find a set of values of $\underline{B}$, $\underline{V}$, N as a first approximation which satisfies equations (1) to (5); these equations do not depend on ξ and ξ'. We require this set of values to be as close as possible to their corresponding measured values (see Chao and Olbert, 1970). Equation (6) gives a linear relation between ξ and ξ'. Using equations (7) and (8), and substituting the ξ and ξ' relation from (6), we can determine the total particle pressure P and P' as a function of ξ or ξ'. Since the proton pressure of the solar wind is measured, we can solve for the electron pressure in terms of ξ or ξ', (assuming $\gamma = \gamma' = 5/3$).

We have been able to collect eighteen events which have been shown to be shocks by various authors. The measured parameters satisfy Eqs. (1) to (5), and their shock speeds and normal directions have also been computed. Figure 1 shows the normal of these eighteen shocks in solar ecliptic coordinates. Note that the distribution of θ_s and ϕ_s are symmetric with respect to the sun-earth line for both the fast and slow shocks. Some of these shocks have also been observed by more than one spacecraft, so that the shock velocity can be checked independently. Hence we are confident that

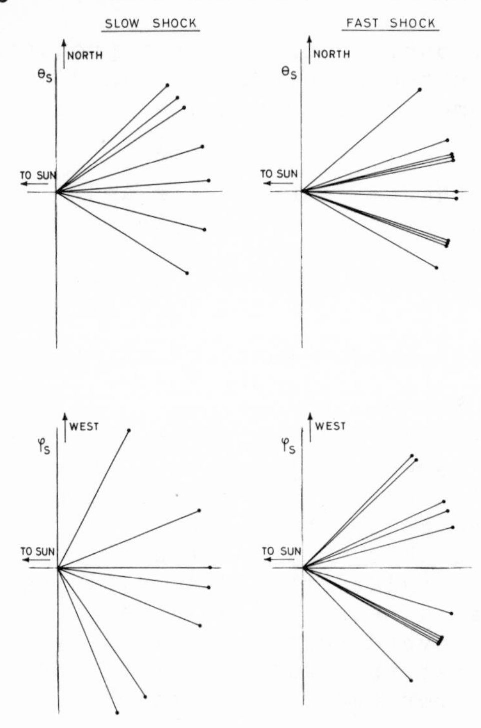

Fig. 1 The orientations of eighteen shock normals. The angle θ_S is solar ecliptic latitude and ϕ_S is solar ecliptic longitude.

Eqs. (1) to (5) are well satisfied by the plasma and magnetic field measurements. Note again that the computed shock velocity $\underline{V}_S$ is independent of pressure anisotropies.

The requirement that the plasma be stable against the firehose and mirror instabilities, restricts ξ and ξ' to the range between 0 and 1. Then the electron temperature can be computed in terms of ξ or ξ' for each of the eighteen shocks. It is well known that the electron temperature in the solar wind remains fairly constant in the range 0.8 to 1.5 × 10^5 °K (Montgomery et al., 1970, Burlaga, 1968). The proton temperature shows much larger fluctuations. However, it turns out that the computed electron temperatures from Eqs. (6) to (8) for some of the shocks are much less than 0.8 × 10^5 °K.

The computed electron temperature T_e in the upstream state can be related to the magnetosonic Mach number $M_{F \text{ or } S}$. In Figure 2, T_e is plotted versus $M_{F \text{ or } S}$ where $M_{F \text{ or } S}$ is the fast or slow magnetosonic Mach number. Those events which lie below the dashed line have T_e much less than 10^4 °K.

We can see that except for case 16, the electron temperature is "reasonable" for Mach numbers between 1 and 2, but the electron temperature is too low for those shocks with Mach numbers greater than 2. Adding any reasonable thermal anisotropy in our computation will not change this conclusion.

We will now investigate the possibility of modifying the normal momentum and energy flux equations (Eqs. 7 and 8) to obtain a reasonable T_e for all the shocks considered.

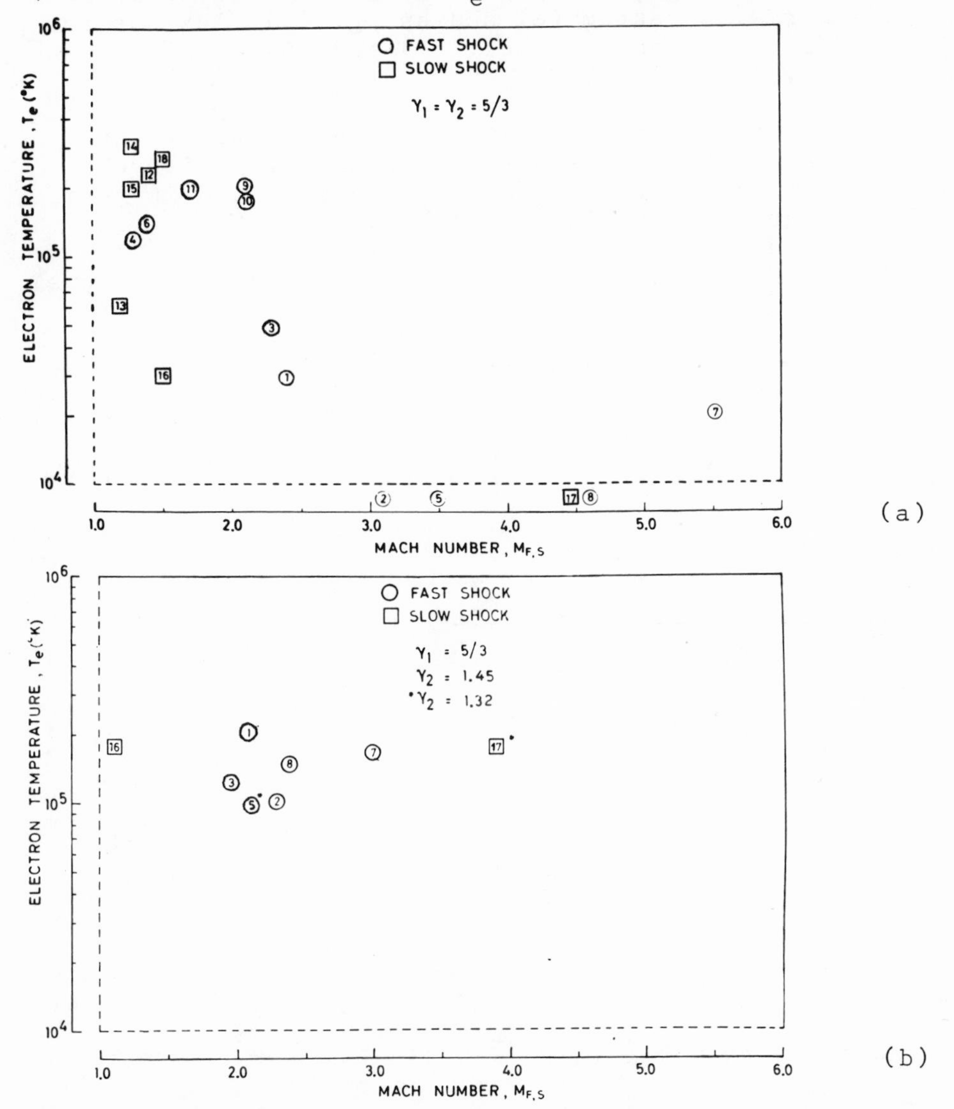

Fig. 2 Computed electron temperature, T_e, vs magnetosonic Mach number, $M_{F \text{ or } S}$, for the 18 shocks (see Table 1), assuming $\gamma_1 = \gamma_2 = 5/3$. (b) Same plot as (a) for those shocks having T_e's less than 0.8×10^5 K, assuming $\gamma_2 = 1.45$ or 1.32 and $\gamma_1 = 5/3$.

It is well known that fluctuations in the vicinity of shocks in space are very pronounced, especially in the downstream state. These fluctuations may manifest the existence of waves and/or turbulence associated with shocks. In particular, the fluctuations for the (high Mach number) strong shocks are very much enhanced. Let us add a normal momentum (G) and energy (F) flux due to waves and/or turbulence in Eqs. (7) and (8) and let $\Delta G = G' - G$, $\Delta F = F' - F$.

Since the downstream state is more turbulent than the upstream state, we assume ΔG and ΔF are positive. The electron temperature T_e and T_e' can be found in terms of the measured quantities $\underline{B}$, $\underline{B}'$, $\underline{V}$, $\underline{V}'$, N, N', T_p and T_p', and the unknown parameters of anisotropy ξ or ξ', and ΔG and ΔF, where γ and γ' are taken to be 5/3. Thus we have:

$$T_e = \frac{1}{Nk(V_1^{*} - V_1^{*\prime})} \left\{ \left(\Delta B \frac{\gamma - 1}{\gamma} - V_1^{*\prime} \Delta A\right) + \left(\Delta F \frac{\gamma - 1}{\gamma} - V_1^{*\prime} \Delta G\right) \right\} - T_p \quad (9)$$

and

$$T_e' = \frac{1}{N'k(V_1^{*} - V_1^{*\prime})} \left\{ \left(\Delta B \frac{\gamma - 1}{\gamma} - V_1^{*} \Delta A\right) + \left(\Delta F \frac{\gamma - 1}{\gamma} - V_1^{*} \Delta G\right) \right\} - T_p' \quad (10)$$

where

$$\Delta A = \left[\rho V_1^{*2} + \frac{1}{3}\left(\xi + \frac{1}{2}\right) \frac{B^2}{\mu_o} - \xi \frac{B_1^2}{\mu_o} \right] \quad (11)$$

and

$$\Delta B = \left[\left(\frac{1}{2} \rho V^{*2} + \frac{1}{3}(\xi + 2) \frac{B^2}{\mu_o} - \xi \frac{B_1^2}{\mu_o}\right) V_1^{*} - \xi \frac{B_1 B_3}{\mu_o} V_3 \right]. \quad (12)$$

Since we do not have an explicit expression for ΔG and ΔF without a specific model of waves and/or turbulence, we cannot use these equations to compute T_e and T_e'. Assuming that only Alfvén waves contribute to G and F, and using the formulas given by Scholer and Belcher (1971) for ΔG and ΔF, we found the computed T_e's are even lower than those shown in Fig. 2(a). Hence, the presence of Alfvén waves will not resolve these difficulties. Other types of waves and/or turbulence are needed. If we define P_2 and γ_2 as follows:

$$P_2 = P' + \Delta G \tag{13}$$

and

$$\frac{\gamma_2}{\gamma_2 - 1} P_2 V_1^{*\prime} = \frac{\gamma}{\gamma - 1} P' V_1^{*\prime} + \Delta F \tag{14}$$

where $\gamma = 5/3$, then we can write the electron temperature T_e and T_e' in the following form:

$$T_e \simeq \frac{\left(\Delta B - \frac{\gamma_2}{\gamma_2 - 1} V_1^{*\prime} \Delta A\right)}{Nk \left(\frac{\gamma}{\gamma - 1} V_1^{*} - \frac{\gamma_2}{\gamma_2 - 1} V_1^{*\prime}\right)} - T_p \tag{15}$$

and

$$T_{e2}' = \frac{\left(\Delta B - \frac{\gamma}{\gamma - 1} V_1^{*} \Delta A\right)}{N'k \left(\frac{\gamma}{\gamma - 1} V_1^{*} - \frac{\gamma_2}{\gamma_2 - 1} V_1^{*\prime}\right)} - T_p' \tag{16}$$

Now, we have written T_e and T_{e2}' in terms of an unknown parameter γ_2 which represents the effects of waves and/or turbulence. For given values of the measured plasma and magnetic field and with a "reasonable" assumed value of ξ or ξ', we can adjust γ_2 such that T_e lies between 0.8 to 1.5×10^5 °K. It turns out that the events (1), (2), (3), (7), (8) and (16) will have a "reasonable" value of electron temperature if γ_2 is taken to be 1.45. However, we have to take $\gamma_2 = 1.32$ for events (5) and (17). Fig.2(b) shows the electron temperature vs. Mach number plot for $\gamma_2 = 1.45$ and $\gamma_2 = 1.32$ for those events which previously had too low electron temperatures. Now we have all shocks with "reasonable" electron temperatures.

The amount of wave and/or turbulence energy flux can be estimated from Eq. (14). If we assume $P_2 \simeq P'$, and

$R = \frac{\Delta F}{(1/\gamma-1)P'V_1^{*\prime}}$, then R can be readily computed and can be interpreted as the ratio of the energy flux due to waves and/or turbulence to the thermal energy flux in the downstream state. For γ_2 equals 1.45 and 1.32, the corresponding R equals 0.5 and 1.2, respectively. That is, the amount of energy flux for waves and/or turbulence is comparable to the thermal energy flux.

Table 1 gives a list of the shocks under investigation and some computed parameters. M_A is the Alfvén Mach number defined by V_1^*/C_A, where C_A is the Alfvén speed in the pre-shock state based on the magnetic field component normal to the shock front. $\theta_{B,n}$ is the angle between the shock normal and the pre-shock magnetic field vector. Both M_A and $\theta_{B,n}$ are obtained independently of Eqs. (6) to (8). $M_{F\text{ or }S}$ is the Mach number based on the fast or slow magnetosonic waves and β_T is the ratio of total thermal pressure to the magnetic pressure. Both $M_{F\text{ or }S}$ and β_T depend on Eqs. (6) to (8). β_p, which is a measured quantity, is the ratio of proton thermal pressure to the magnetic pressure. T_e is the computed electron temperature of the pre-shock state.

We would like to point out that by varying ξ or ξ' between 0 and 1, we do not change our results for T_e by more than 10 per cent (also see Lepping, 1971). Therefore, we will neglect the ξ or ξ' effect for our estimation of energy flux due to waves and/or turbulence.

It is interesting to note that Dryer (1970) and Shen (1971) have suggested that the specific heat ratio γ_2 should be lower than 5/3 when computing the bow shock shape. Shen (1971) suggested that the presence of hydromagnetic waves would lower the γ_2 value. However he did not consider the modification of the normal momentum flux. In the present study, the local shock conservation equations were examined using both plasma and magnetic field data recorded by instrument on board various spacecraft. We found it necessary to modify the energy and momentum flux equations in order to satisfy the shock conservations. We suggest the presence of waves and/or turbulence in the vicinity of shock waves could explain the discremancy shown in Fig. 2(a). Let us, instead, add a term due to heat-flow in the energy flux equation (Eq. (8)). We replace ΔF by $\Delta q = q'-q$ and set $\Delta G = 0$ in Eqs. (9) and (10), where q and q' are the heat flow in the pre- and post-shock states. We can obtain an equivalent result if we accept an Δq which is greater than zero. That is a heat-flow toward the downstream side is needed.

Acknowledgements: I would like to thank Drs. F. Mariani and K. Schindler for their comments and suggestions.

REFERENCES

Burlaga, L.F., Solar Physics 4, 67, 1968.

Burlaga, L.F., Cosmic Electrodynamics 1, 233, 1970.

Burlaga, L.F. and J.K. Chao, J. Geophys. Res. (1971). To be published.

Chao, J.K., Interplanetary Collisionless Shock Waves, MIT Center for Space Research, CSR TR-70-3, 1970.

Chao, J.K. and S. Olbert, J. Geophys. Res. 75, 6394, 1970.

Chao, J.K., V. Formisano and P.C. Hedgecock, Shock Pair Observation. Submitted to the Proceedings of the Solar Wind Conference, Asilomar, Pacific Grove, California, March 21-26, 1971.

Lepping, R.P. and P.D. Argentiero, J. Geophys. Res. 76, 4349, 1971.

Lepping, R.P., in preparation for publication, 1971.

Mariani, F., N.F. Ness and J.K. Chao. Submitted for publication, 1971.

Mihalov, J.D., C.P. Sonett and J.H. Wolfe, J. Plasma Phys. 3, 449, 1969.

Montgomery, M.D., J.R. Asbridge and S.J. Bame, J. Geophys. Res. 75, 1217, 1970.

Ness, N.F. and J.M. Wilcox, Solar Phys. 2, 351, 1967.

Ogilvie, K.W. and L.F. Burlaga, Solar Phys. 8, 422, 1969.

Scholer, M. and J.W. Belcher, Solar Phys. 16, 472, 1971.

Sonett, C.P., D.S. Colburn, L. Davis Jr., E.J. Smith and P.J. Coleman, Phys. Rev. Letters 13, 153, 1964.

Taylor, H.E., Sudden Commencement Associated Discontinuities in the Interplanetary Magnetic Field Observed by IMP-3, GSFC Report X-616-68-239, 1968.

TABLE 1: SOME SHOCK PARAMETERS

	TIME		M_A	$\theta_{B,n}$	$M_{F,S}$		β_T		β_p	$T_e(10^5\ °K)$		TYPE
	DAY YEAR				$\gamma_2=5/3$	$\gamma_2=1.45$	$\gamma_2=5/3$	$\gamma_2=1.45$		$\gamma_2=5/3$	$\gamma_2=1.45$	
1	82 1966	P 6	4.7	51°	2.4	2.1	0.9	2.1	0.7	.3±.2	2.±.5	FAST
2	241 1966	P 7	4.8	47°	3.1	2.3	0.5	1.7	0.4	.005	1.±.3	FAST
3	177 1967	M 5	11.2	45°	2.3	1.9	14.0	35.0	6.6	.5±.2	1.±.4	FAST
4	241 1967	M 5	1.3	33°	1.4	1.0	0.8	1.9	0.4	1.2±.2	1.±.4	FAST
5	11 1968	E 35	2.5	21°	3.5	1.9★	-	2.1	0.6	<0	1.±.3	FAST (in magnetosheath)
6	84 1969	H 1	2.0	45°	1.4	1.3	0.3	0.5	0.4	1.4±.2	3.7±.5	FAST REVERSE
7	350 1965	P 6	10.0+	90°	5.5	3.0	1.6	6.2	0.5	.2±.1	1.7±.3	FAST NORMAL BOW SHOCK
8	266 1966	P 7	4.4+	90°	4.6	2.4	1.1	2.7	0.9	<0	1.5±.2	FAST NORMAL
9	26 1968	E 33	3.0+	90°	2.1	1.6	1.1	3.2	0.2	2.±.5	5.±1.	FAST NORMAL
10	26 1968	E 35	3.0+	90°	2.1	1.6	1.7	4.5	0.2	1.8±2.	4.5±1.	FAST NORMAL
11	84 1969	H 1	2.1+	90°	1.7	-	0.5	0.8	0.2	2.0±1.	-	FAST NORMAL
12	360 1965	P 6	0.8	64°	1.4	1.0	0.5	1.6	0.2	2.3±.3	10.±1.	SLOW REVERSE
13	19 1966	P 6	0.9	42°	1.2	0.9	0.7	8.7	0.6	.6±.3	9.±1.	SLOW REVERSE
14	20 1966	P 6	0.9	60°	1.3	1.0	0.9	13.3	0.2	3.±1.	50.±10.	SLOW
15	26 1966	P 6	0.9	42°	1.3	1.0	0.9	6.5	0.4	2.±1.	30.±10.	SLOW
16	201 1967	M 5	0.9	46°	1.5	1.1	0.6	1.9	0.3	.3±.1	1.8±.5	SLOW
17	242 1967	M 5	0.9	81°	3.9	1.5★	0.1	0.7	0.2	<0	1.8±.5	SLOW
18	299 1967	E 35	0.9	62°	1.5	1.0	0.5	2.9	0.2	2.8±1.	25.±10.	SLOW

P - Pioneer M - Mariner E - Explorer H - HEOS

★ $\gamma_2 = 1.32$

\+ Based on the total magnetic field

COMETS IN THE SOLAR WIND

(Dedicated to W. Heisenberg
on the occasion of his 70th birthday)

L. Biermann

Max-Planck-Institut für Physik und Astrophysik

München

Two bright comets, Tago-Sato-Kosaka and Bennett, appeared in winter and spring of last year. They were the first ones to be observed by means of satellite-borne instruments in the UV down to approximately 900 Å. The most important discovery made during these observations was the existence of a huge atmosphere of atomic hydrogen visible in the resonance line Lyman α 1216 Å; the probability for excitation of this line by the flux of solar Lyman α quanta is once every few minutes, at the solar distance in question.

The first two pictures show first the normal appearance of comet Bennett (Fig. 1) as seen from the ground around April 1, and then the isophotes in Lyman α (Fig. 2) with the first picture copied into it. The scale as indicated at the lower left is approximately 2 million km per 1°. While the head of the comet had the usual diameter of several 100 000 km, the size of the hydrogen atmosphere is seen to be $\gtrsim$15 000 000 km, or 1/10 a.u. The total number of hydrogen atoms present at the time was of the order of 10^{36}. From the probable lifetime before ionization, which as we shall see should have been of the order of 10 days, the production rate is found to be $\gtrsim 10^{30}$ atoms/sec. With the usual assumptions on the chemical composition of comets, this means at least some 10^7 gr/sec.

These results are of obvious importance for the understanding of the constitution and the origin of comets as well as of their interaction with the solar

Fig. 1: Comet Bennett 1969i on April 1, 1970 (Hamburg-Bergedorf Observatory).

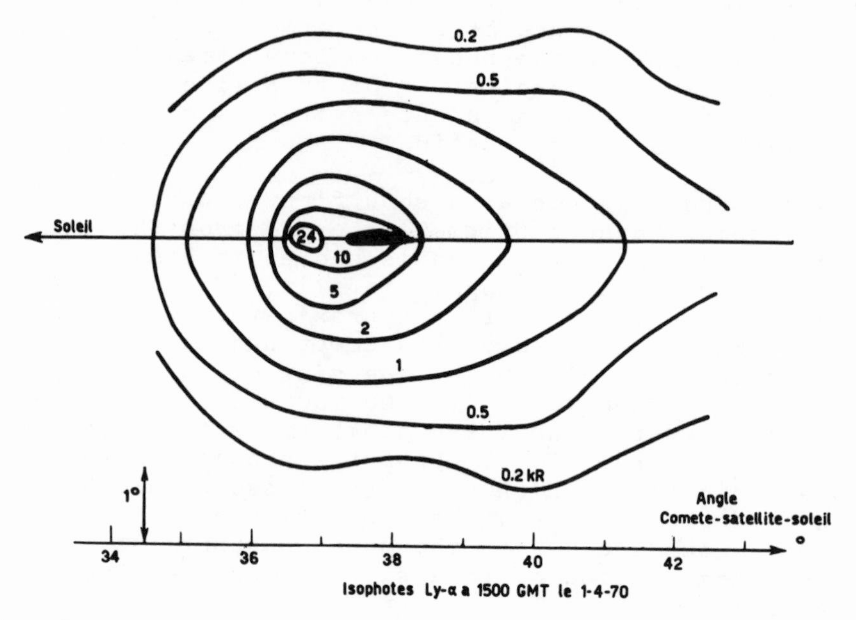

Fig. 2: Comet Bennett as seen in the light of the line Lyman α 1216 Å with the photographic picture taken on the same date copied into the diagram.

wind, which is our main concern at this conference. The existence, the brightness and the size of the escaping hydrogen atmosphere had actually been anticipated in theoretical work done since approximately 1964. This will be described first. Next we turn to the UV observations made with three different instruments on board the satellites OAO 2 and OGO 5. Their interpretation will be discussed along the lines of recent work by H.U. Keller and related prior theoretical studies, on the basis of which it is expected that it will become possible before too long to derive the solar wind flux outside the ecliptic plane from such observations. In conclusion some remarks will be made on the main changes since 1951 in our overall understanding of the interaction of the solar wind with comets.

The central subject of the investigations to which I turn first was the total gas production of comets*. The roundish coma or head of a comet consists of neutral molecules like C_2 and CN, which stream outwards with velocities of around 1/2 to 1 km/sec to the distance given by their lifetime before dissociation or ionization of 1 or several 10^5 sec. From the number of spectral transitions observed from the ground the true number density and the production rate of the molecules in question can be derived; for moderately bright comets this is found to be of the order of 10^{27} mol/sec. The gas tails, on the other hand, which extend in the anti-solar direction over several up to tens of million km, consist of molecular ions like CO^+ and N_2^+, most of which seem to originate fairly close to the nucleus. These ions move with typical velocities of the order of several 10 or even 100 km/sec, due to accelerating forces which are evidently much higher than solar gravity. After it had been realized, just 20 years ago, that their momentum could not be derived from solar light pressure, the concept of a continuous component of the sun's corpuscular radiation -- now called the solar wind -- was formulated, the velocity of which, from the finite lag angle of these tails against the solar radius vector, had to be of the order of 10 times the usual orbital velocity of comets (30 to 50 km/sec), more exactly its azimuthal component.

Under the conditions given in interplanetary space, with the sun's light diluted roughly in the proportion $1:10^5$, all excited states of molecules are heavily underpopulated; as a result only resonance lines are absorbed,

* cf. the report of IAU Commission 15 to the IAU General Assembly 1970; also an article by the present author (Biermann, 1971)

which in most cases lead to downward transitions in the same or neighboring frequencies. These circumstances determine the selection of those molecules which can be observed from the ground, to which unfortunately all those molecules do not belong which would on general grounds be expected to be abundant, as for instance H_2O and hydrocarbons. For this reason the total gas content of cometary atmospheres cannot be estimated from the observed numbers of those molecules which happen to be visible from the ground; an upper limit can of course be written down easily from the solar heat flux and the evaporation heat of each molecular species, provided the radius of the comet's nucleus is known (see below).

The first important change in this state of affairs began with the discovery, by P. Swings and J. Greenstein (1962), of the forbidden red oxygen doublet near 6300 Å in cometary spectra. As a matter of fact this line had often been observed, but was always ascribed to the night sky background; a separation of a cometary emission from that of the night sky required a spectral resolution which was not available before approximately 1950. The unambiguous identification of the cometary emission from its Doppler displacement suggested a re-discussion of older spectra of moderate dispersion, which led to the conclusion that the presence of the forbidden red oxygen line is in fact a typical feature in the spectra of bright comets. The surface brightness due to cometary oxygen atoms is usually of a similar order as that of the night sky emission.

The subsequent attempts to interpret these cometary emissions led actually to the first clue to the actual gas output of comets. The forbidden character of the transition in question has the consequence that excitation by solar continuum light is very unlikely and would therefore require enormous quantities of atomic oxygen. I cannot go into the detailed description of the several attempts to account for the observed intensities. I would just like to mention that according to a study made in Munich (L. Biermann and E. Trefftz, 1964) one very probable mode of origin is part of the process of dissociation by which the oxygen atoms come into existence. In this interpretation the total number of quanta per sec, of the order of 10^{30} for a moderately bright comet, gives also the order of magnitude of the production rate of atomic oxygen. Taking into account the relative abundance of oxygen and the relative probability of this process, the total number of molecules produced per sec, by order of magnitude, was thus estimated to be of the order of 10^{30} to 10^{31} for such comets.

Between 1965 and 1970 these estimates, which at first sight appeared to be rather high, were confirmed in three different ways. First, it was shown that using the values of the radii of cometary nuclei given by E. Roemer in 1965* and the relations of physical chemistry for the vapor pressure, a consistent picture was arrived at (W. Huebner, 1965; A. Delsemme, 1966). Second, the theory of dust tails as developed by Finson and Probstein in 1968-69 led again to an estimate of the gas output by mass, which for comet Arend-Roland was found to be of the same order of magnitude (Probstein 1968, Finson and Probstein 1968). Finally, D. Malaise succeeded in finding pressure effects in cometary spectra of high resolution, which again demanded the same order of magnitude (D. Malaise, 1970).

In the aforementioned paper of 1964, E. Trefftz and the present author had already concluded that comets should be very bright in the line Lyman α, since hydrogen atoms would be expected to be equally (if not several times more) abundant as the molecules producing atomic oxygen. A more detailed study (based on Whipple's (1951) icy-conglomerate model) made three years later (L. Biermann, 1968) led to the conclusion that comets should possess a hydrogen atmosphere extending to quite a number of million km, from the expected lifetime of the atoms of 10^6 sec -- mainly given by charge transfer in the solar wind -- and the expected velocity of between 5 and 10 km/sec. This atmosphere was predicted to be optically thick in the central parts (some 10^{10} cm radius), with the projected density and the surface brightness decreasing outwards approximately with the inverse first power of the distance from the nucleus.

The first observations of comet Tago-Sato-Kosaka made in mid-January 1970, using the UV spectrophotometer on OAO 2, confirmed at once the existence of a huge hydrogen atmosphere and its great integral brightness in Lyman α, the central brightness being approximately 30 kR. These spectrophotometric observations indicated also the presence in large quantity of the OH radical, which is somewhat difficult to observe from the ground. No new molecules were found, with the possible exception

---

* for new comets -- most bright comets belong to this class -- usually of the order of a few to ten km if an albedo of the order 0.1 is assumed; cf. E. Roemer, Report to the 13th Liège Colloquium of 1965.

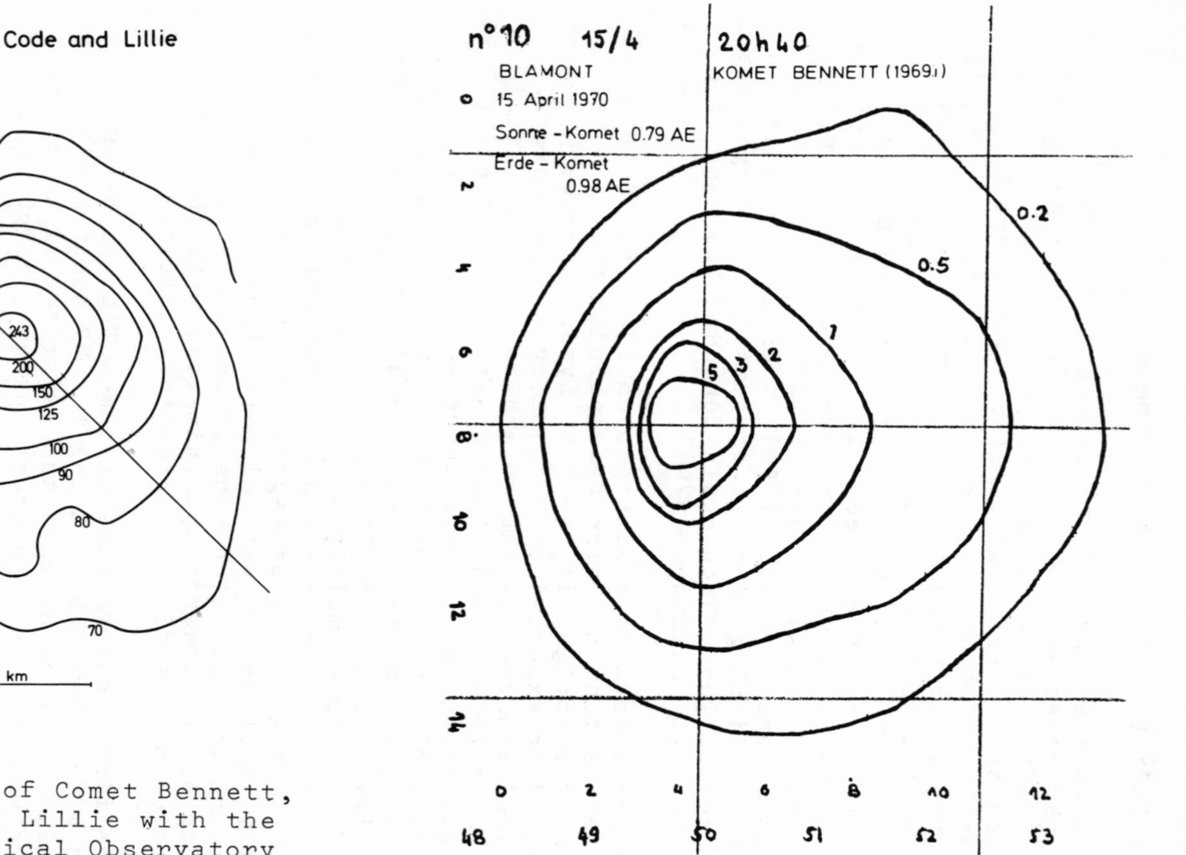

Fig.3: Isophotes of Comet Bennett, taken by Code and Lillie with the Orbiting Astronomical Observatory OAO 2. In order to reduce the figures to the brightness in kR they have to be multiplied by approx. 0.3.

Fig.4: Isophotes of Comet Bennett taken by Bertaux and Blamont with OGO 5 on April 15, 1970.

of molecular hydrogen, the presence of which has still to be confirmed. This comet was observed to about mid-February 1970 (Code and Lillie, 1970).

When in March 1970 another bright comet, Comet Bennett, had come into view, it was observed not only with the same instrument on board OAO 2 (Fig. 3), but also with two UV photometers, placed on board the eccentric orbit satellite OGO 5 by the French group under Bertaux and Blamont (Fig. 4), and the U.S. group at the University of Colorado under C. Barth and G. Thomas. These two instruments had a much poorer spatial resolution -- 40 arc minutes and 1 1/2° respectively against 1' -- but a much higher sensitivity (0.1 kR and a fraction of 0.1 kR, respectively), which permitted the measurement of the atomic hydrogen out to the ionization limit and, for the second instrument, even at distances of up to 40 million km in approximately the anti-solar direction. The measurements made with the instrument of Barth and Thomas were the only ones pertaining to the period of perihelion and the passage through the ecliptic plane in late March.

The measurements revealed that the innermost part of the hydrogen cloud was indeed optically thick in Lyman α, and showed the also expected influence of solar light pressure, which for hydrogen atoms compensates largely solar gravity. The actual velocity with which the hydrogen atoms stream out into interplanetary space, and their average lifetime, could, however, be found only by comparing theoretical models with the observed isophotes. This was done by H.U. Keller (1971) of Munich, who, from the isophotes obtained by Bertaux and Blamont (1970), found a velocity of around 7 or 8 km/sec and a lifetime before ionization close to 10^6 sec, at a solar distance of approximately 0.8 a.u. (cf. Figs. 5a and 5b). In particular, he concluded that the lifetime of the hydrogen atoms in directions away from the sun was considerably longer (Fig. 5b), obviously as a result of the interaction of the comet with the solar wind. Assuming for instance equality of the gas production (gr/sec) with the solar wind flow through a circle around the comet with 3×10^6 km radius, we would get $\approx 10^{32}$ atomic mass units/sec, or $1...2 \times 10^8$ gr/sec, which is slightly more than indicated by the Lyman α isophotes. The lifetime itself measures directly -- after due allowance had been made for collisional and photoionization -- the flux of the solar wind, which determines the rate of charge transfer reactions. Such determinations can obviously be made just as well when the comet is outside the plane of

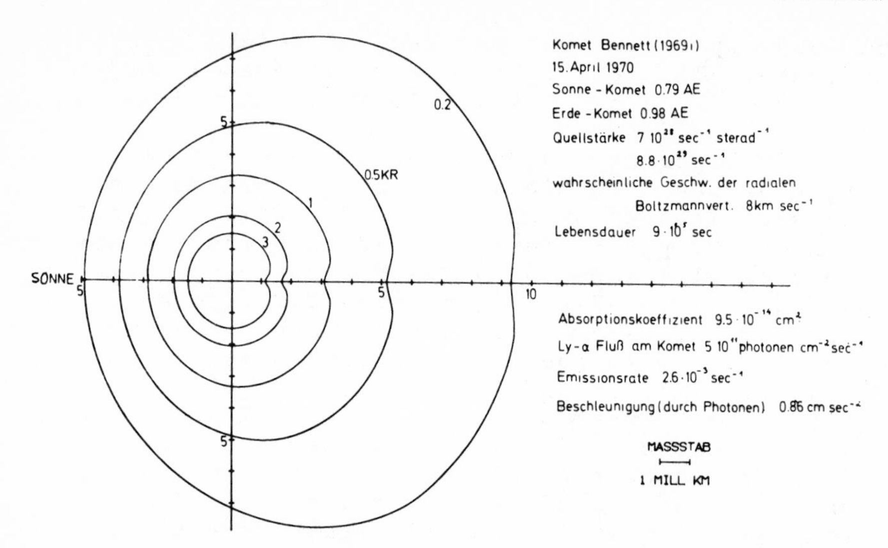

Fig. 5a: Keller's model of the hydrogen atmosphere of Comet Bennett on April 15.

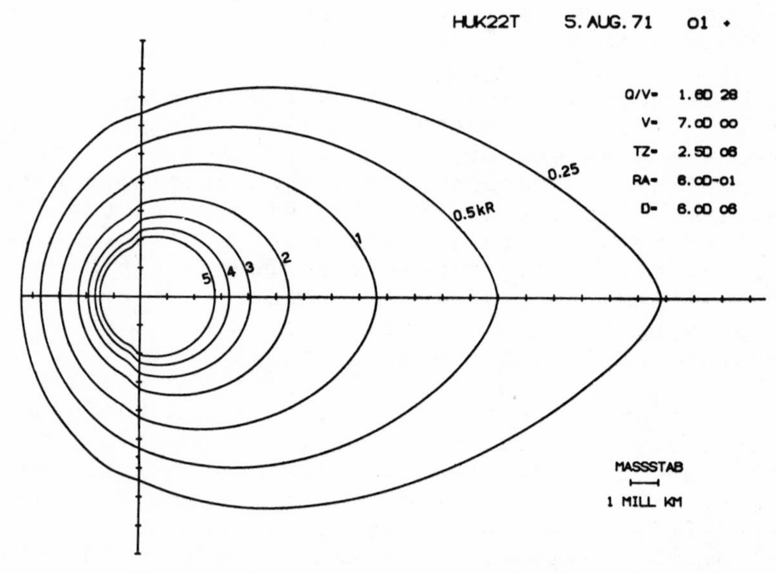

Fig. 5b: Model of the hydrogen atmosphere of Comet Bennett (1969i) referring to the measurement of Bertaux and Blamont on April 1, 1970.

Q/v production [H atoms × km^{-1} $sterad^{-1}$]

v mean velocity of H atoms [km × sec^{-1}]

TZ mean lifetime for comet-sun distance of 1 a.u.

RA comet-sun distance [a.u.]

D characteristic distance from the solar radius vector [km] for the lifetime dependence $\tau = \tau_o \times \frac{D}{d}$ ($\tau_o = RA^2 \times TZ$).

the ecliptic as when it is near to it.

This interaction had been theoretically investigated already some years earlier (L. Biermann, B. Brosowski, H.U. Schmidt, 1967; Brosowski and Schmidt, 1967) in a study which I can describe only very briefly (Fig. 6). Attention was fixed on the flow of the cometary plasma, which originates by the ionization of the neutral gas emitted from the nucleus. It can easily be seen that the comet is thus a source of plasma, which necessarily interacts with the magnetized solar wind plasma. Since the large majority of the ions of cometary origin must come into existence (by charge transfer or photo-ionization of the parent molecules or atoms) in the solar wind flow itself, one has to use the equations of fluid dynamics with (positive or negative) source terms for the mass, the numbers, the momentum, and the energy. This work refers to the vicinity of the solar radius vector; more recently R. Wegmann in Munich has begun work on the 3-dimensional situations.

It is obviously necessary to assume that there is some analogue to a contact surface in the sense of fluid dynamics, which should be, by order of magnitude, at a distance of 10^5 km upstream from the nucleus; since there must be a stagnation point, it furthermore seems necessary to conclude that there should also be a shock front upstream at a distance of the order of some 10^6 km, which

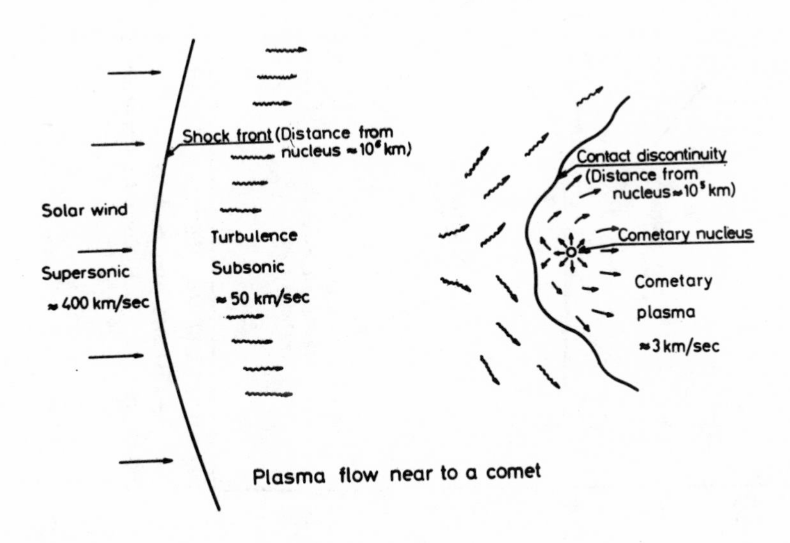

Fig. 6: Plasma flow near a comet.

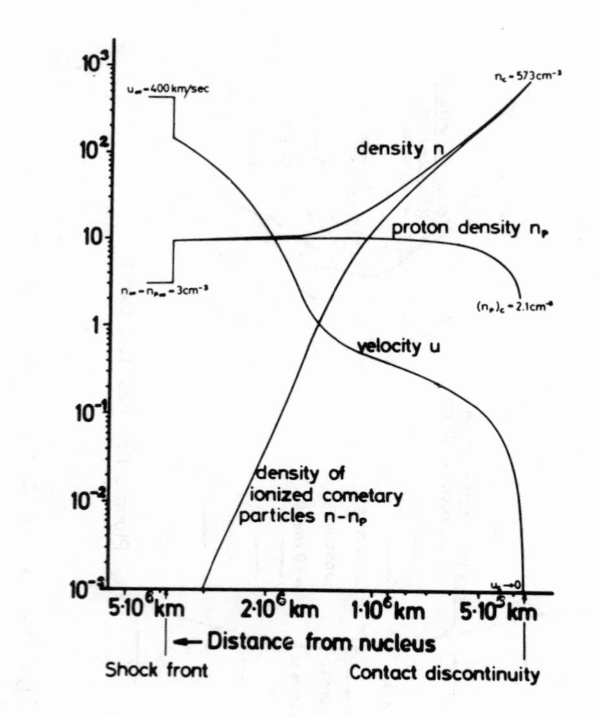

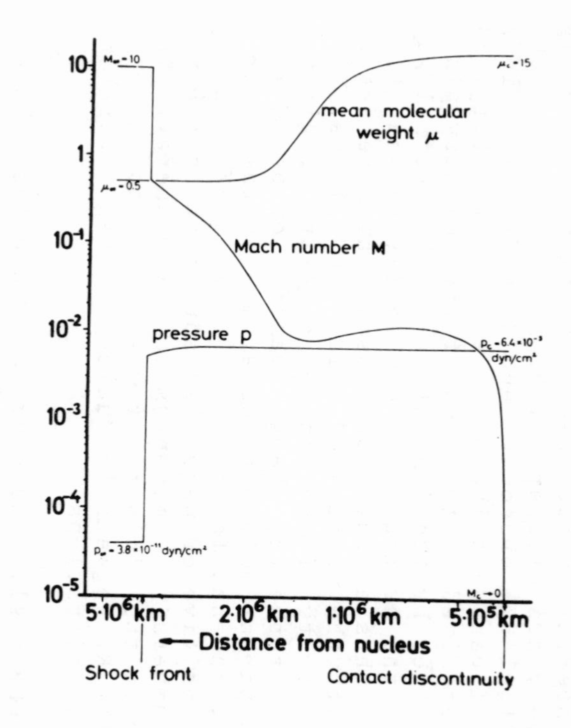

Figs. 7 and 8: Model solution for a production rate of $G=10^{30}$ molecules/sec.

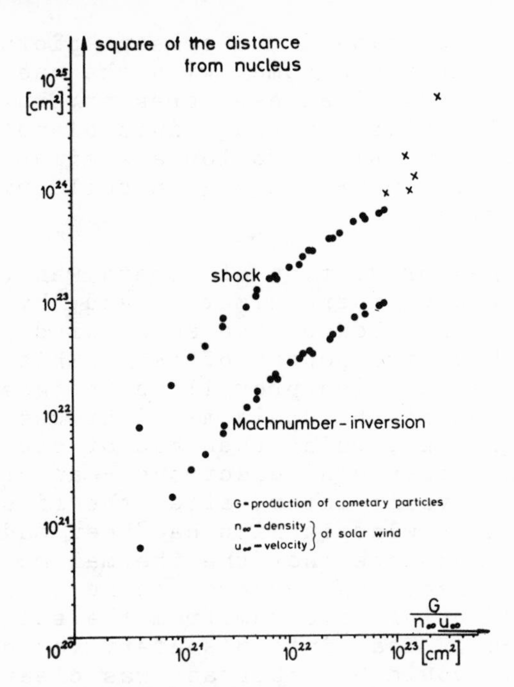

Fig. 9: Distance of the shock front as a function of $G/(n_\infty u_\infty)$ for all models.

separates the hypersonic solar wind from the subsonic transition region between the shock front and the contact surface (Figs. 6-9). This shock front introduces, at the present time, the main uncertainty in the relation between the lifetime of the hydrogen atoms and the flux of the solar wind.

In these investigations the magnetic fields are not taken into account explicitly; they make, however, fluid dynamics applicable in the same way as in the theory of the flow around the earth as discussed by Spreiter and others in about the last 10 years.

The importance of the UV observations described here is that they permit for the first time comparisons between theoretical models and observations for the large scale features of comets. Especially intriguing are the observations of hydrogen atoms of cometary origin at very large distances, in approximately the anti-solar direction. These observations obviously indicate again the large total amount of the gas production, though their detailed interpretation has not yet been arrived at.

In closing let me compare the present picture of the solar wind flow around a comet with the one given in 1951, when the existence of an ever present, but not steady, flow of solar plasma through interplanetary space was first proposed. In this I follow a similar exposition which I gave a few months ago at a colloquium held at Leeds University.

In 1951, the gas production of comets was (as we have seen) considerably underestimated, and, by an unfortunate coincidence, the flow of the solar wind plasma overestimated by about two powers of ten. This was a consequence of the assumption prevailing in those years, that the polarization of the zodiacal light was due to interplanetary electrons, which then had of course to be identified with the solar wind electrons - an error which was realized only 10 years later, after the first measurements of the solar wind in situ had been made. This had the further consequence that the thermal motion of the interplanetary electrons appeared to be an adequate mechanism of transferring momentum from the solar wind to the cometary ions. That interplanetary magnetic fields, if present, would be important was clear, as was later pointed out in more detail in Alfvén's well known paper of 1957, but their real properties also began to emerge only with the advent of space vehicles. The most striking change in the present picture as compared to the old one is, however, connected with the gas output. It is now obvious that the solar wind flow is affected over a region of interplanetary space of almost comparable extent as that indicated by the Lyman α picture, and that the visible plasma tail is just the central flow line in this very much bigger flow pattern. Future cometary missions should thus have no difficulty in measuring the influence of the comet on the interplanetary plasma flow, even if they do not succeed in coming close to the comet's nucleus itself.

REFERENCES

Alfvén, H., 1957, TELLUS 9, 92.

Bertaux, J., and J. Blamont, 1970, C.R. Acad. Sci., Paris, 270, 1581.

Biermann, L., and E. Trefftz, 1964, Z. Astrophys. 59, 1.

Biermann, L., B. Brosowski and H.U. Schmidt, 1967, Solar Phys. 1, 244.

Biermann, L., 1968, JILA Report No. 93, Univ. of Colorado.

Biermann, L., 1970, Sitzungsber. Bayer. Akad. Wiss., Sonderdr. 2, p. 11.

Biermann, L., 1971, Nature 230, 156.

Brosowski, B., and H.U. Schmidt, 1967, ZAMM 47, T140.

Code, A., and C. Lillie, 1970, Report at the meeting of IAU Commission 15 to the Brighton IAU General Assembly.

Delsemme, A., 1966, Nature et Origine des Comètes, Rpt., 13th Liège Colloquium of 1965.

Finson, M., and R.F. Probstein, 1968, Ap. J. 154, 327 and 353.

Hall, L., and H. Hinteregger, 1970, J. Geophys. Res. 75, 6959.

Huebner, W., 1965, Z. Astrophys. 63, 22.

Keller, H.U., 1971, to be published in Mitt. Astron. Ges.

Malaise, D., 1968, thesis Liège (1970, Astronomy and Astrophysics).

Probstein, R.F., 1968, Problems of Hydrodynamics and Continuum Mechanics, (Soc. Indust. Applied Mech., Philadelphia).

Swings, P., and J. Greenstein, 1962, Ann. Astrophys. 25, 165.

Whipple, F., 1951, Ap. J. 113, 464.

COMET-LIKE INTERACTION OF VENUS WITH THE SOLAR WIND

Max K. Wallis

Institut d'Astrophysique, Université de Liège;

Div. Plasma Phys., Royal Inst. Tech. Stockholm

The solar wind flow through the outer cometary atmosphere is presumably controlled by new ions /1/, picked up by the plasma by means of convected or fluctuating fields. Enhanced ionization via heated electrons — a critical velocity effect /2/ — can then underlie the space and time fluctuations in plasma structures /3/. Ionizing flows should thus differ appreciably from pressure-driven flows.

The data of Venera 4 and Mariner 5 show several comet-like features, explicable on the hypothesis that the solar wind is interacting with an extended upper atmosphere (of helium) /4/. The main features are

1. an extensive disturbance of the solar plasma, reaching too far upstream and too far laterally for simple pressure effects;
2. no high energy 'spikes' of electrons or ions, indicative of a shock Mach number $M \leq 3$ /cf. 5/;
3. a very broad upstream 'discontinuity', possibly with the absence of shock-like heating and dissipation;
4. absence of a distinct ionosphere;
5. presence of structures and discontinuities in a broad wake.

To explain the comet-like interaction of Venus requires a helium outer atmosphere of some $10^{4.5}$ atoms/cm^3 at 10,000 km radius /4/. Such a high value may be plausible

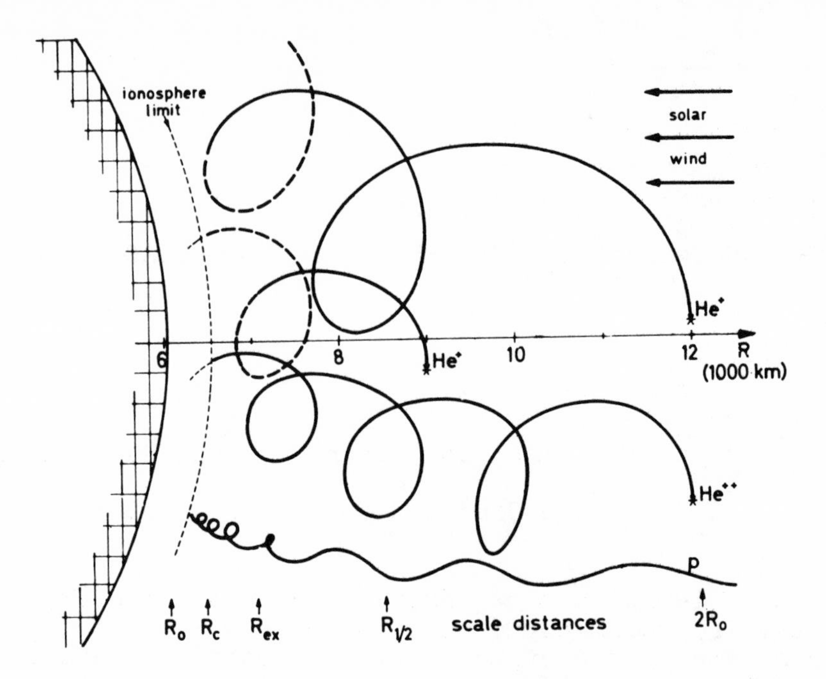

Fig. 1 Scale radii and sample orbits of new ions and a solar wind proton in the sub-solar exosphere. Note that the scale level for charge exchange R_{ex} occurs outside the critical level (exosphere base) R_c.

if solar wind heating of the upper atmosphere is significant. The 'two-temperature' picture of the H atmosphere /6/ would qualitatively support this idea. Fig. 1 illustrates how solar wind streaming energy would transfer preferentially to the helium atmosphere. The new He ions decelerate the solar wind by the scale radius $R_{1/2}$, and take up most of its energy into their gyration and streaming motions. Resonant charge exchange processes at around the position R_{ex} convert the energetic He ions to fast atoms and most of these atoms subsequently contribute to collisional heating around the critical height R_c.

Apart from the general heating effect, the incoming flux of energetic atoms can control the high energy tail of the velocity distribution and thereby the density at high altitudes. To examine this quantitatively, a model is set up in which the supra-thermal (non-Maxwellian) tail is relaxed via collisions with thermal particles but maintained by the incoming flux. In the Venus exosphere, charge exchange interactions happen to dominate the plasma deceleration, and then the energy spectrum of the incoming flux turns out to be little dependent on the flow model,

$$\phi(E) \sim E^{-\nu}, \; \nu \simeq 1{\cdot}2. \qquad (1)$$

This expression ignores a possibly important angular dependence. Assuming the concept of a sharp critical level is valid here /7/, the change in spectral flux ψ at $R=R_c+$ is described by

$$\partial\psi/\partial t = \Lambda\phi - (\psi-\Lambda\psi)/\tau, \qquad (2)$$

τ being the particle time of flight in the exosphere. The main problem is to choose suitable forms of the collisional operator Λ, such that $\Lambda\psi$ describes the upward-moving, supra-thermal products of an incoming spectral flux ψ.

A linear integral operator form of Λ can be used to describe the products of one or two binary collisions.

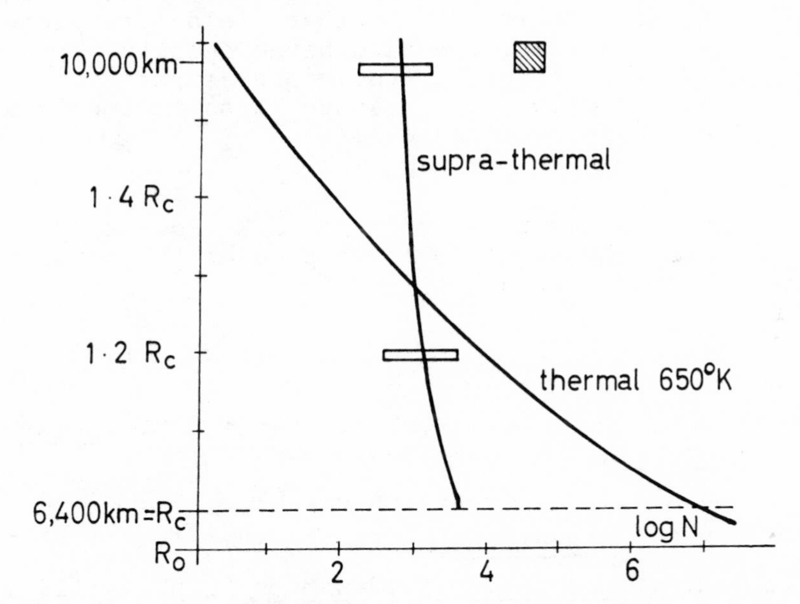

Fig. 2 Densities in the helium exosphere. The thermal atmosphere is normalized to $10^{7.5}\,cm^{-3}$ at 200 km altitude while the supra-thermals of equation (3) are normalized to the incoming energy flux with an uncertainty to a factor 10. The combined density would clearly have a 'two-temperature' appearance. The shaded square represents the estimated density of ref. 4.

After simplifying (2) by taking $\tau = \tau(E)$, power function solutions are found in the steady state for velocities below the escape velocity

$$\psi \sim E^{-\nu+1/2},$$

which lead to the density distribution of the supra-thermal population in the upper exosphere

$$N(r) \sim (1-R_c/r)^{-\nu+1/2} - 1. \qquad (3)$$

Fig. 2 shows this density in comparison with the traditional exospheric distribution. Although the supra-thermals' density at 10,000 km is much larger, it still seems smaller than $10^{4.5}/cm^3$ by a significant factor, 10-100.

Alternative explanations of this density, involving exospheric satellite particles or products of multiple collisions of the high energy particles of ϕ, have not been investigated. Nor can it be ruled out that, because of 'critical velocity' effects as observed directly in the simulation experiment with a helium cloud /2, 8/, the estimate of $10^{4.5}$ atoms/cm^3 is too high. It is clear in any case, that the tail of the velocity distribution in the exospheres of Venus and other field-free planets is far from Maxwellian. The exospheric density variation is quite different from the traditional exponential (Fig. 2) and the atmospheric escape rates seriously different from the Jeans formula /9/.

I acknowledge support from the U.K. Science Research Council through a fellowship under the Royal Society European Programme.

References

1. Biermann L., Brosowski B. and Schmidt H.U.: Solar Physics 1, 254 (1967)
2. Danielsson L.: this conference
3. Wallis M.K.: Planet. Sp. Sci. 16, 1221 (1968)
4. Wallis M.K.: Cosmic Electrodynamics in press (1971)
5. Paul J.W.M.: this conference
6. Wallace L.: J. Geophys. Res. 74, 115 (1969)
7. Chamberlain J.W.: Planet. Sp. Sci. 11, 901 (1963)
8. Danielsson L.: Phys. Fl. 13, 2288 (1970)
9. Hunten D.: Comments Astrophys. Sp. Phys. 111, 1 (1971)

LABORATORY EXPERIMENTS ON THE INTERACTION BETWEEN A PLASMA AND A NEUTRAL GAS

Lars Danielsson

Division of Plasma Physics

Royal Institute of Technology, Stockholm

Introduction

In many plasma experiments it has been observed that there is an upper limit for the energy of the plasma as long as the plasma is not fully ionized everywhere. Attempts to increase the energy beyond this limit will only achieve additional ionization. Further, this ionization seems to be extremely efficient even if the electrons, which must cause most of the ionization, are initially cold.

This phenomenon was discovered accidentally as an unwanted limitation of the velocity of rotation in the early thermonuclear experiments like the Homopolar and the Ixion [1,2,3]. Several years before this a similar phenomenon had been suggested by Alfvén[4] and used in his theory for the origin of the solar system. Alfvén's suggestion can be formulated in the following way: the relative velocity between a plasma and a neutral gas in a magnetic field cannot be increased beyond a critical level - the critical velocity, v_c - which is determined by the relation

$$\frac{1}{2} Mv_c^2 = eV_i$$

M is the mass of the neutral gas atoms or molecules and eV_i is the ionization energy. Any further increase of the energy of the plasma goes entirely into ionization (and losses). It might seem plausible that the plasma-gas ionizing interaction should become appreciable when the relative velocity exceeds v_c. However, in trying to account for the experimental observations insurmountable difficulties seem to arise. In all the cases we are interested in, $\omega_e\tau_e$ is much larger than unity so the drift energy of the electrons is negligible; the electron have just their thermal energy and no drastic changes would be anticipated at the relative velocity v_c. For direct ionization by ion-impact the relative velocity has to be higher, by a factor 1.4 or so; such a discrepancy could probably be experimentally explained in most cases but instead the cross section for this interaction is orders of magnitude too low in most cases.

Thus we are left with an overwhelming experimental evidence from at least a dozen experiments[5] for an abnormally strong ionizing interaction between a plasma and a neutral gas in a magnetic field when the relative velocity exceeds the critical velocity.The application of this phenomenon to cosmic physics is even older than the observations. In Alfvén's theory for the origin of the solar system of 1954 it was introduced to ionize the gas that is falling towards the sun and thereby stopping it at certain distances - determined by the falling velocity in the gravitation field - from the sun. The clouds formed in this way later developed into planets. A similar sequence was applied for the formation of the major satellite system. Other possible applications to cosmic physics include the interaction of the solar wind with comets and certain planetary atmospheres (e.g. Venus') or with the interplanetary background gas.

Major Experimental Investigations

The critical velocity, or as it appears in the experiments, the voltage limitation of an electric discharge across a magnetic field, is a well known observational fact which has been reported from many and different types of experiments. Among them are rotating plasma devices, PIG-discharges, plasma guns, plasma shock tubes and one experiment exhibitting a direct collision between a plasma and a neutral gas. One striking circumstance in these experiments is the wide range of configurations, currents and gas pressures over which the effect is observed: the current and pressure parameters both span five orders of magnitude.

1. The rotating plasma devices can be represented by that of Angerth et al[6]. The experimental arrangement

is shown in Figure 1. The tank is pre-filled gas of 5-200 μHG pressure and a discharge (5-15KV) is generated between the coaxial electrodes, across the magnetic field (1-10kGauss). The Lorentz force will make the plasma rotate in the azimuthal direction. In trying to increase the kinetic energy of the plasma it was discovered that, no matter how much electric energy was applied, the velocity or the burning voltage did not increase beyond a certain value. This limitation was investigated in some detail.

By varying the resistor R(1-200Ω) the current was varied between 30 and 10^4A and the current-voltage characteristic was studied. Results concerning the burning voltage, V_b, in seven light gases are:

1) V_b is almost independent of pressure and current down to minimum pressure and up to a maximum current. At the limit the degree of ionization is approaching 100%.

2) V_b is independent of capacitor bank voltage.

3) V_b is directly proportional to magnetic field and the electrode separation, d:

$$V_b \simeq v_c d B \quad ,$$

additional right-hand-side terms (electrode sheaths, centrifugal effect) being negligible in most cases.

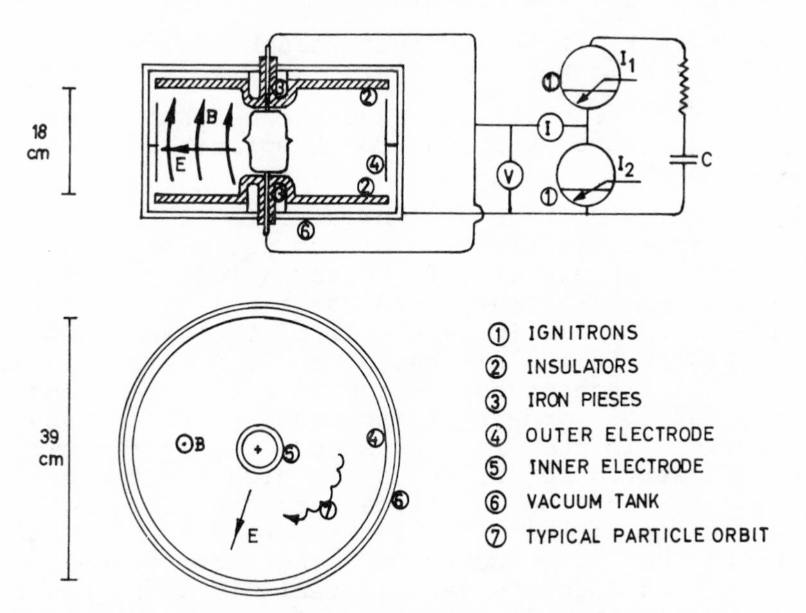

Fig. 1. Block's rotating plasma device

4) The proportionality constant v_c - the critical velocity - depends on, and only on, the gas used and for all seven gases studied it was very close to the prediction.

The necessity of complete ionization everywhere has been clearly demonstrated by Lehnert et al[7]. The plasma was fully ionized except in a small region near the insulators at the magnetic mirrors of a dipole-like field. Thus the velocity was limited at the insulators; corotation gave a higher velocity limit in outer parts of the plasma.

2. The most elaborate (in this connection) experiment with a plasma gun is the one by Eninger[8]. An unusual feature of his coaxial device was a strong (7-13kGauss) azimuthal magnetic bias field from a conductor through the center electrode. The discharge gave rise to an ionizing wave which travelled across the magnetic field along the gun. Also here the system was pre-filled with the gas to be studied.

Pressure, current, magnetic field and applied voltage were varied within wide limits in four different gases. The experimental results can be described in terms of four characteristic velocities, viz.

i) from the burning voltage of the discharge the E/B velocity, u_b, was calculated;

ii) the velocity of the current sheath, u_s, was measured with magnetic probes;

iii) the plasma velocity, u_p, and the neutral gas velocity, u_n, were obtained from Doppler-shift measurements.

The burning voltage was characteristic of the gas used. The radial distribution of the potential was also studied giving an electric field variation $\sim r^{-1}$. Since B also varied as r^{-1} the ratio E/B was independent of r so that this characteristic parameter of the gas was the same everywhere in the discharge. The E/B value was also essentially independent of the plasma current and gas pressure over four orders of magnitude of these parameters. However, in the E/B versus current diagram two levels were observed for H_2 and N_2; a high one for small currents and a low one for large currents. This was assumed to be due to the diatomic structure of these gases. Contrary to the E/B-velocity (u_b) the current sheath velocity (u_s) was found to vary appreciably with pressure, current and magnetic field. It was lower than u_b but higher than the theoretical "snow plow" velocity, u_{sp}, except for the largest currents combined with the lowest pressures. The snow plow model, which presumes that the current sheath acts as an impermeable piston ionizing and accelerating all gas in its way, yields a velocity which is

$$u_{sp} = c(\text{gas})\ (IP/p)^{1/2}$$

For a given geometry the constant c depends upon the gas used, I, B and p are the plasma current, magnetic field (average) and gas pressure. The observation $u_s > u_{sp}$ implies that ions and neutrals slip through the current sheath. Doppler-shift measurements of the plasma (u_p) and neutral gas (u_n) velocities showed that $u_n \ll u_p \simeq u_s$.

The main results of this experiment were discussed by means of a diagram showing u_b and u_s both plotted as a function of u_{sp}. Three different regions of operation could be located: a) for large currents and low pressures it was shown that $u_s \simeq u_b \simeq u_{sp} > v_c$ (v_c = the critical velocity). Here the "snow plow" model was applicable; b) for intermediate currents and pressures, the "plateau region", $u_s \simeq u_b \simeq v_c > u_{sp}$; c) for low currents combined with very high gas pressures, $u_b \simeq v_c > u_s > u_{sp}$. Thus the sheath velocity u_s was connected to the critical velocity only in the plateau region. The point separating region b and c was experimentally found to be situated at a constant value of B/p, at a point where $(\omega_i \tau_{in} \omega_e \tau_{en})^{1/2} = 1$, ($\omega_i$ and ω_e being the gyrofrequencies for ions and electrons τ_{in} and τ_{en} the collision times for ions and electrons with neutral atoms).

3. The experiment with a direct collision between a plasma and a neutral gas[9] represents a completely new

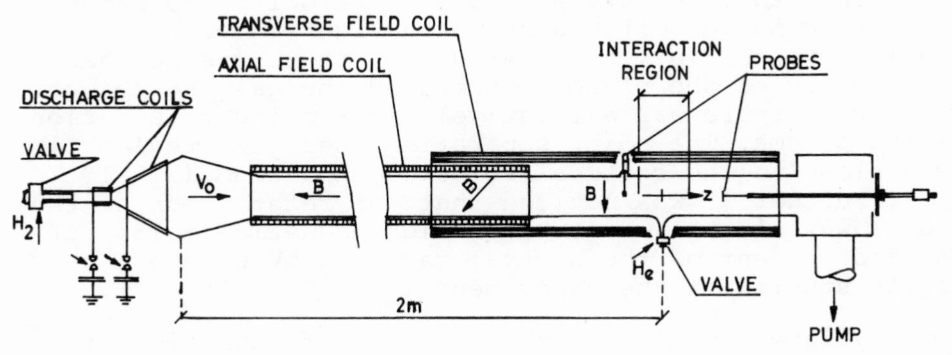

Fig. 2. Experimental setup: plasma-neutral gas impact

approach to the critical velocity phenomenon. Here many of the relevant experimental characteristics are opposite to those of the experiments described in the preceding paragraphs. The plasma is fully ionized and has a supercritical velocity prior to the commencement of the interaction. There is no discharge connected to the interaction and the macroscopic net current is zero. In fact there is no electric connection between the plasma and external power units anywhere in the system.

The experimental arrangement is shown in Figure 2. A hydrogen plasma is generated and accelerated in an electrodeless plasma gun (a conical theta pinch) and flowing into a drift tube along a magnetic field. The direction of the magnetic field changes gradually from axial to transverse. As the plasma flows along the drift tube much of it is lost but a polarization electric field is developed and a plasma of about $3 \times 10^{17} m^{-3}$ proceeds drifting across the magnetic field with a typical velocity of $3 \times 10^5 m/s$. In the region of the transverse magnetic field the plasma penetrates into a small cloud of neutral gas, usually helium, released from an electro-magnetic valve. This helium cloud has an axial depth of 5 cm and a density of $10^{20} m^{-3}$ (3μHg) at the time of the arrival of the plasma. The remainder of the system is under high vacuum (3×10^{-7} mmHg); under these conditions the mean free paths for all important, direct, binary collisions is much more than 5 cm so that the interaction according to common terminology is classified as collisionless.

In the experiment it was observed that the velocity of the plasma was substantially reduced over a typical distance of only 1 cm in the gas cloud. It was also found that this reduction in plasma velocity depends on the impinging velocity as shown in Figure 3. For the smallest velocities, below $4 \times 10^4 m/s$, there was no change in velocity as the plasma penetrated the gas. For higher velocities there was a relatively increasing retardation of the plasma. The point separating these two regions is very close to the critical velocity in helium (3.5×10^4 m/s). Further it was noticed that the retardation of the plasma was slightly faster for high magnetic fields but was independent of the neutral gas density within the limits studied in the experiment (3×10^{19}-$3 \times 10^{20} m^{-3}$).

By investigation of the emission of radiation from the plasma and neutral gas it was found that the electron energy distribution changed drastically at the penetration of the plasma into the gas and that the ioni-

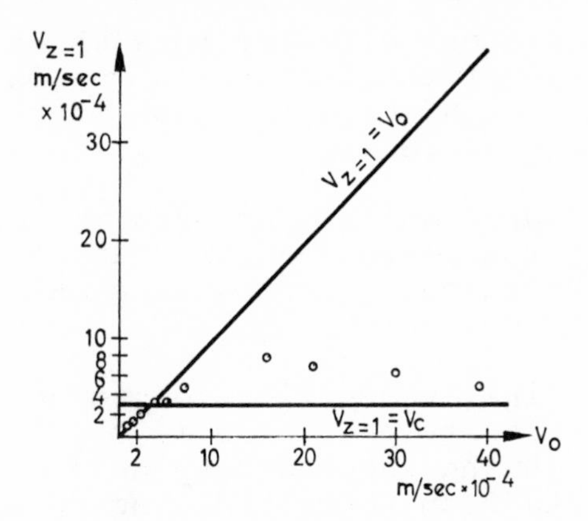

Fig. 3. Plasma velocity 1 cm behind the center of the gas cloud as a function of original velocity

zation of the gas atoms was many orders of magnitude faster than anticipated from the parameters of the free plasma stream. The characteristic electron energy was found to jump from about 5eV to about 85eV at least locally in the gas cloud. This is supposed to be the cause of the ionization and retardation of the plasma but no further clue to the details of the heating mechanism has been found.

So far this experiment has demonstrated that even in a situation where the primary collisions are negligibly few there may be a very strong interaction between a moving plasma and a stationary gas. This interaction is active above $4x10^4$m/s in helium ($v_c=3.5x10^4$m/s) and it leads to

i) local heating of the electrons

ii) ionization of the neutral gas

iii) retardation of the plasma stream.

Conclusions

The theoretical understanding of the critical velocity phenomenon in the individual experiments is poor and the detailed understanding of the phenomenon itself is even worse. No single mechanism has ever been suggested which could explain all experimental observations. In some cases suggested classical interactions possibly could be worked out to account for the observation. In others;notably the direct collision experiment, there doesn't seem to be any well-known process which could work. Here one may infer a collective interaction between heavy ions, protons and electrons which would transfer kinetic energy of the protons to thermal energy of the electrons. Some kind of two stream instability could be involved in this process.

Little is known about what causes the enhanced interaction between the plasma and the gas at the critical velocity. Its existence, however, is a well established experimental fact.

The most imaginative application to cosmical physics of the critical velocity phenomena is the interaction of the solar wind with the gaseous component of cometary atmospheres. It has been suggested long ago[10] that the very fast ionization rate in comets is a critical velocity effect. Maybe this approach can prove fruitful the day we have sufficient knowledge - experimental and preferably theoretical - about the effect. Attempts to simulate this interaction in the laboratory have been made[11, 12].

References

1. U.V. Fahleson, Phys.Fluids 4, 123(1961)
2. A. Bratenahl et al., UCRL-reports 9002, 9106 and 9243 (1959-60)
3. D.A. Baker and J.E. Hammel, Phys.Fluids 4,1549(1961)
4. H. Alfvén, On the Origin of the Solar System, Oxford University Press, Oxford 1954
5. L. Danielsson, Report 70-05,Div. of Plasma Physics Royal Institute of Technology, Stockholm
6. B. Angerth, L. Block, U. Fahleson and K. Soop, Nucl. Fusion Suppl. Pt 1,39 (1962)
7. B. Lehnert, J. Bergström and S. Holmberg, Nucl. Fusion 6, 231 (1966)
8. J. Eninger, Proc. 7 Intern. Conf. Ionized Gases, 1, 520, Beograd 1966.
9. L. Danielsson, Phys. Fluids 13, 2288 (1970)
10. W.F. Hübner, Rev. Mod. Phys. 33, 498 (1961)
11. L. Danielsson and G.H. Kasai, J. Geophys. Res. 73, 259 (1968)
12. H. Kubo, N. Kawashima and T. Itoh, Plasma Physics 13, 131 (1971)

WAVE MOTION IN TYPE I COMET TAILS

Marino Dobrowolny and Nicola D'Angelo

European Space Research Institute

Frascati (Rome)

ABSTRACT

Filamentary structures are observed in type I comet tails, with sometimes a wavelike or helical appearance. Most features are occasionally in rapid motion. This behaviour is analyzed in terms of a Kelvin-Helmholtz instability developing in the tail itself, rather than in the comet head as proposed by Ioffe. Reasonable agreement is obtained with the "parallel" and "perpendicular" wavelengths of the observed features, the periods of wave motion and the growth rates.

1. Comet tails of type I are composed of ionized material. When observed in the visible they appear to consist essentially of CO^+ ions, although ions such as CO_2^+, N_2^+, CH^+, and OH^+ are also present. Type I tails are generally very straight, pointing, to a good approximation, away from the sun, and have typically a filamentary structure, with the filaments tending to orient themselves symmetrically around the tail axis. The filaments'length is of the order of 10^6-10^8 km, their diameter of the order of 2000-4000 km. The presence of a magnetic field, directed along the plasma filaments, appears to be necessary to account for the confinement of the tail plasma for sufficiently long times within regions of such small size. The observed structure, according to the picture proposed by Alfvén (1957), would in fact result in the interplanetary field

lines being swept by the solar wind plasma around the cometary coma and stretched along the filaments.

Type I comet tails present at times a wavelike or helical appearance. Such structural features "are occasionally in rapid motion, so that the aspect of an active tail can change considerably within one hour and look considerably different after the passage of a day" (Biermann and Lüst, 1968).

The wavelike motions of the tails are commonly interpreted as magnetohydrodynamic waves (Marochnik, 1964). A specific mechanism of generation of Alfvén waves has in particular been recently proposed (Ioffe, 1970). This consists of excitation of surface waves through a Kelvin-Helmoltz instability at the interface between the cometary plasma and the solar wind (where a tangential velocity discontinuity is supposed to occur). According to the author such instability cannot develop in the tail but certainly arises in the comet head, where the plasma flow velocity is nearly perpendicular to the ambient magnetic field. The surface waves generated at the head need to transform into Alfvén waves during their motion along magnetic field lines and would finally appear as Alfvén waves in the tail filaments, still propagating along magnetic lines.

We wish to point out that, although a velocity shear is very probably the agent responsible for the observed type I tail motions, a given theory should be able, first, to account for the fact that wave motion is seen occasionally and not at all times, a fact suggesting therefore that the Kelvin-Helmoltz instability conditions might often be only marginally met. Secondly the helical appearance of the motions, which are often clearly indicated by the observations, should also be explained. In both respects (and also with reference to the region of wave generation) the mechanism outlined previously appears open to some doubt. On the other hand it is worth noticing that, for a proper treatment of the Kelvin-Helmholtz instability, one should take into account a) the compressibility of the tail and solar wind plasma, b) the finite ion Larmor radius effects, c) the effect of ion Landau damping on wave motion, d) the presence of finite density gradients across the region of velocity shear, and e) effects of small but finite β. Furthermore, upon comparing typical values for ion Larmor radii (~ 100 km) in the tails with the filament diameters ($(2-4)\times 10^3$ km), it appears that a theory allowing for smooth velocity and density gradients of the plasma is more appropriate than one based on the approximation of discontinuous transitions.

The plasma model we will present and investigate in some detail in the following sections is one in which the flow velocities are parallel to magnetic lines and most of the effects mentioned above are taken into account. Thus, according to the model, excitation of the Kelvin-Helmholtz instability would occur in the tail itself where the parallelism condition ($v_o // B$) is satisfied.

In Sec. 2 we first discuss some treatments of the Kelvin-Helmholtz instabilities which were developed for low β plasmas in connection with observations and measurements in laboratory plasmas (D'Angelo and von Goeler, 1965, 1966). In such cases the perturbations arising from velocity shear are of electrostatic character. We will then partly generalize such treatments to finite β and discuss the modification of the previous instabilities as well as the new possibly unstable branches which are introduced.

Sec. 3 will compare the main results of the theory with cometary tail observations.

2. The model we will refer to is that of an inhomogeneous plasma in a uniform magnetic field B. The plasma flows in the magnetic field direction, which we take to be the z-direction in a Cartesian coordinate system, with velocity v_o. Gradients of both velocity and density are taken to be in the x-direction.

The perturbations over the plasma equilibrium are taken to be of the form

$$\tilde{A}(x)e^{ik_y y+ik_z z-i\omega t}$$

and the problem of linear stability is considered in the W.K.B. approximation (implying wavelengths of the perturbations in the x-direction smaller than the scale lengths of the inhomogeneities).

In the low β limit, and hence for the case of purely electrostatic perturbations, D'Angelo found instability (D'Angelo, 1965) under the condition

$$\frac{1}{\Omega_{ci}} \frac{\partial v_o}{\partial x} > \frac{k_z}{k_y} \left[1 + \frac{T_i}{T_e} + \frac{1}{4} \frac{k_y^2}{k_z^2} \frac{a_i^2}{L_N^2} \right] \qquad (1)$$

where $\Omega_{ci} = \frac{qB_o}{m_i}$ is the ion cyclotron frequency, a_i the ion Larmor radius and L_N the scale length for the equilibrium density variations. Condition (1) was found with a two fluid model, without taking into account finite ion Larmor radius effects, and in the approximation

$$v_i < \frac{\omega}{k_z} < v_e \qquad (2)$$

for the wave phase velocity parallel to magnetic lines (v_i, v_e being the ion and electron thermal speeds respectively). From eq. (1) we see that for very smooth density profiles ($\frac{a_i}{L_N} \rightarrow 0$), even a very small velocity gradient is sufficient to excite the instability. Conversely, a large density gradient tends to stabilize the plasma. Looking at the minimum of the right hand side of (1) (as a function of $\frac{k_z}{k_y}$), it is found that there are always unstable waves provided $\frac{\partial v_o}{\partial x} \gtrsim \frac{1}{L_N} c_s$ where c_s is the ion sound speed. In the case of density and velocity varying on comparable scales this would give a threshold velocity of the order of c_s.

Smith and von Goeler (1968) have extended the instability theory of D'Angelo, properly allowing for ion finite Larmor radius effects as well as for ion Landau damping on the waves. A comparison of their results with those of D'Angelo can be found in Fig. 1 and 2 of their work. The main effect of Landau damping was shown to be a shift of the short wavelength instability onset to smaller values of $\frac{k_z}{k_y}$ and a decrease of the maximum growth rate.

Experiments by D'Angelo and von Goeler (1966) in alkali plasmas are in general agreement with the predictions of these theories.

In relation to the application of the above theories to the wave motions in the comet tails, it appears of interest to investigate the effects of finite β on the instabilities. Although the β for the plasma in the comet tails, computed on the basis of the plasma "thermal" pressure, is smaller than one, we can hardly expect a zero β theory to represent the prevailing physical situation, and some magnetic field perturbation has to be allowed for. We recall that, according to Alfvén's picture (Alfvén, 1957), the directed kinetic energy of the solar wind ions bends the B-field lines into the configuration of a "folding umbrella".

We have started investigating the finite β theory of velocity shear instabilities in the framework of a two fluid theory, as in the case of D'Angelo's electrostatic treatment but including finite ion Larmor radius effects. The ion and electron equations have been again considered in the approximations (2). A great simplification in the problem is introduced by considering the limit of wavelengths perpendicular to the ambient magnetic field much smaller than parallel wavelengths, i.e. $\frac{k_z^2}{k_y^2} << 1$. This approximation, which indeed corresponds to the features of the helical wave motions in type I comet tails ($\frac{k_z}{k_y} \sim 10^{-1}$), allows us to neglect terms of order $\frac{\omega^2}{k_\perp^2 V_A^2}$ in our equations ($V_A = \frac{B}{\sqrt{4\pi n_o m_i}}$ being the Alfvén speed), still keeping terms of order $\frac{\omega^2}{k_z^2 V_A^2}$. By going to the homogeneous plasma limit it is then easily seen that our description reproduces Alfvén waves and slow magnetosonic waves, while the fast magnetosonic branch is out of the picture.

In the simplest case of velocity shear but no density gradient, the dispersion relation takes the form, for any β,

$$\left[1 - \frac{k_z^2 v_i^2}{\Omega_z^2} - \frac{k_z^2 c_{se}^2}{\Omega_z^2}\left(1 - \frac{k_y}{k_z}\frac{v_o'}{\Omega_{ci}} + \beta\right)\right] \cdot \left[1 - \frac{\Omega_z^2}{k_z^2 V_A^2} + \frac{1}{2}\beta_i \frac{k_y}{k_z}\frac{v_o'}{\Omega_{ci}}\right] = 0 \qquad (3)$$

where $\Omega_z = \omega - k_z v_o$ is the frequency Doppler-shifted by the plasma flow velocity; $\beta = 4\pi n_o \frac{T_i + T_e}{B^2}$, $\beta_i = 4\pi n_o \frac{T_i}{B^2}$ and $c_{se}^2 = \frac{T_e}{m_i}$. The first factor in (3) is the slow magnetosonic wave; it becomes unstable owing to velocity shear when

$$\frac{k_y}{k_z}\frac{v_o'}{\Omega_{ci}} > 0 \quad \text{and} \quad \frac{k_y}{k_z}\frac{v_o'}{\Omega_{ci}} \geqslant 1 + \frac{T_i}{T_e} \qquad (4)$$

By comparing with the electrostatic stability

criterion (1) we see that, in the absence of density inhomogeneity, there is no β effect on the stability of this wave. The second factor in (3) is the Alfvén wave. It becomes unstable when

$$\frac{k_y}{k_z}\frac{v'_o}{\Omega_{ci}} < 0 \quad \text{and} \quad \left|\frac{k_y}{k_z}\frac{v'_o}{\Omega_{ci}}\right| > \frac{2}{\beta_i} \cdot \quad (5)$$

Comparing (5) with (4), we see that for a given sign of v'_o and values such that both instabilities are present, the two unstable waves correspond to opposite signs of $\frac{k_z}{k_y}$ (and therefore to opposite sense of rotation for equal direction of parallel propagation or vice versa). We also see that for given wave numbers, the Alfvén wave requires higher velocity shears to be excited for low β's, while the shear requirement tends to become of the same order for $\beta \sim 1$. Finally, upon comparing growth rates for parameter values such that both instabilities are present, one sees that the magnetosonic wave has a larger growth for $\beta<1$, while the two growths tend to become of the same order for $\beta \sim 1$.

Preliminary calculations allowing for density gradients in the plasma show that these have a stabilizing effect on both the magnetosonic instability (4) and the Alfvén instability (5), in the sense that the minimum $\left|\frac{k_z}{k_y}\right|$ value for which the instabilities start goes from zero to a finite value and also, for a fixed value of $\frac{k_z}{k_y}$, the corresponding shear value for instability is increased. It also remains true that, when $\beta<1$, the Alfvén waves are harder to excite than the magnetosonic waves, and, when they are both excited, the Alfvén wave growths remain smaller than the magnetosonic growths. The complete dispersion relation and its numerical solutions will be reported in detail elesewhere.

We therefore see that, in relation to type I comet tails where $\beta \sim 10^{-1}$, the magnetosonic wave instability is the most relevant, and actually the stability features of this wave remain much the same as those derived in the electrostatic theories previously mentioned.

3. On the basis of the results of the previous section we will now compare the observational data on wave motions in the comet tails with the theory of Smith

and von Goeler, which is certainly the most accurate as it includes the effect of ion Landau damping on the waves.

In Table I we list numerical values for some relevant parameters, selecting data from the review papers of Biermann and Lüst (1963), Brandt (1968) and Miller (1969) on the development and kinematics of the type I tail in comet Arend-Roland (1957 III), on 5 May, 1957.

TABLE I

Density of CO^+	$10^2 - 10^3$ cm^{-3}
Larmor radius, a_i, for tail ions ($T \sim 10^4 K$, $B \sim 5\gamma$)	$\sim 10^2$ km
Diameter of tail filaments	$2 \times 10^3 - 4 \times 10^3$ km
Velocity component parallel to tail axis	$\sim$ 100 km/sec
"Parallel" wavelength of perturbations	$\sim 5 \times 10^5$ km
"Perpendicular" wavelength of perturbations	$\sim 3 \times 10^4$ km
Period of wave excitation (order of magnitude)	5×10^3 sec

From these data we infer a value of $\frac{a_i}{L_N} \sim 0.1\text{-}0.2$ and a value of $k_\perp a_i \sim 0.02\text{-}0.04$. For rough estimates we may then use the numerical results in Figs. 1 and 2 of Smith and von Goeler (1968) obtained for $a_i/L_N \approx 0.09$ and $k_\perp a_i \approx 0.08$. For the quantity $A = \frac{1}{\Omega_{ci}} \frac{\partial v_o}{\partial x}$ which represents the strength of the velocity shear, we obtain (assuming a solar wind velocity of a few hundred km/sec) $A \sim 1$.

It seems unlikely that A would exceed unity. However, a value of A as low as 0.1 is conceivable for cometary tails and is probably quite common. It is appropriate to remark here explicitly that, according to Smith and von Goeler (1968), only for strong velocity shear does one obtain unstable wavelengths.

From Figs. 1 and 2 of Smith and von Goeler we see

that:

a) the $k_{\parallel}/k_{\perp}$ corresponding to maximum growth rate is of order 10^{-1}, for values of A between 0.1 and 1. This result compares favourably with the ratio

$\lambda_{\perp}/\lambda_{\parallel} \simeq \frac{3\times10^{4}}{5\times10^{5}} \simeq 0.06$ quoted by Miller (1969) for the observed structures in comet Arend-Roland. It accounts also, of course, for the "helical" structures which are observed;

b) the frequency of the unstable waves turns out to be $\omega \sim 10^{-3}$ sec^{-1}. This corresponds to a wave period $T \sim 5\times10^{3}$ sec which agrees with the order of magnitude of the observed periods quoted by Miller (1969);

c) the (linear) growth rate for the instability is $\tau_{growth} \approx (\sim1 \text{ to } \sim10)\ T$, for $0.1 \lesssim A < 1$. If we take a growth rate $\tau_{growth} \approx 4T$, we obtain $\tau_{growth} \approx 5$ hours. It is worth recalling here that "an active tail can change considerably within an hour and may look considerably different after the passage of a day" (Biermann and Lüst, 1963);

d) the condition of velocity shear and density gradients across the $\underline{B}$ lines required for instability (which can be roughly written $A \gg \frac{a_i}{L_N}$) is certainly not met at all times. It appears then, in agreement with observations, that violent tail motions should not be a feature of comet tails to be expected at all times.

REFERENCES

Alfvén, H., 1957, Tellus 9, 92.
Biermann, L. and Lüst, R., 1963, "The Solar System" Vol. IV, p. 618, Ed. by B.M. Middlehurst and G.P. Kuiper.
Brandt, J.C., 1968, Ann. Rev. Astron. Astrophys. 6, 267.
D'Angelo, N., 1965, Phys. Fluids 8, 1748.
D'Angelo, N. and von Goeler, S., Phys. Fluids 8, 1570, 1965, and Phys. Fluids 9, 309, 1966
Ioffe, Z.M., 1970, Soviet Astronomy 13, 1042.
Marochnik, L.S., 1964, Soviet Phys. Uspekhi 7, 80.
Miller, F., 1969, Astron. Journal 74, 248.
Smith, C.G. and von Goeler, S., 1968, Phys. Fluids 11, 2665.

DIVERS SOLAR ROTATIONS

John M. Wilcox

Institute for Plasma Research

Stanford University, Stanford, California

The subject of solar rotation, or more properly rotations, has become of considerable interest in the last few years. These new developments are primarily associated with spacecraft observations, including observations of the interplanetary medium near the earth, and with improved ground-based solar telescopes and digital data-handling facilities. The physical processes responsible for these divers observed solar rotations and the relationships between them are by no means understood, but may perhaps serve as an interesting challenge to the participants in this conference.

If one looks in a modern textbook on solar physics under rotation one will find the classical results of Newton and Nunn (1951) as shown by the solid curve in Figure 1. They studied the rotation periods of long-lived sunspots which could still be observed when they returned to the visible solar disk one rotation later. In order to acquire enough statistics to reasonably well define a differential rotation curve it was necessary to combine the observations during a complete 11-year sunspot cycle, or at the very least during perhaps half of a cycle. We shall see later that in modern observations this time period necessary to define a differential rotation curve can be reduced from eleven years to one hour, with some interesting consequences resulting. Newton and Nunn found that the differential rotation of the long-lived sunspots was essentially unchanged during several 11-year sunspot cycles.

Sunspots mark the location of small-scale strong magnetic fields having a magnitude of a few kilogauss. By contrast we may inquire about the rotational properties of the large-scale weak

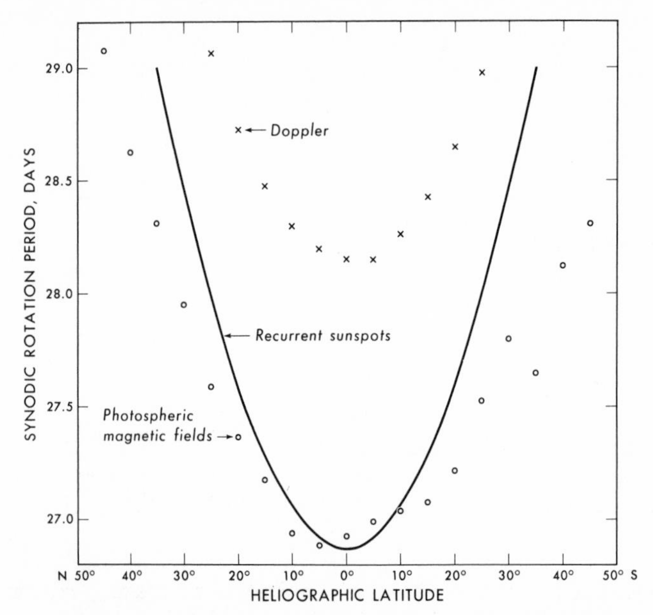

Fig. 1. Solar differential rotation. The solid curve represents the results of Newton and Nunn (1951) for long-lived sunspots. The circles are the average results for large-scale photospheric magnetic fields. The X's are average results obtained by Howard and Harvey (1970) from Doppler observations, with the results from the northern and southern hemisphere observations shown separately (from Wilcox and Howard, 1970).

photospheric magnetic fields whose magnitude is a few gauss. This question was investigated using autocorrelation techniques by Wilcox and Howard (1970) with the results shown by the circles in Figure 1. We see that near the equator the rotation period of the large-scale field is approximately the same as the sunspots, but at higher latitudes the period of the large-scale field becomes slightly less than the period of the sunspots. In the autocorrelation analysis of the periods of the large-scale photospheric field it was necessary to use observations during an interval of approximately six months in order to define a statistically significant differential rotation curve. This is an improvement from the eleven years required for the analysis of the rotation of long-lived sunspots, and we find from one 6-months interval to another a considerable variation in the rotation properties of the field (the results shown in Figure 1 represent an average over several years). During some 6-month intervals the range of latitudes within perhaps 20° of the equator may show an almost rigid rotation.

Since interactions between magnetic fields and plasmas lie at the heart of cosmical plasma physics, we may inquire about the rotational properties of the photospheric plasma. The line-of-sight component of the plasma velocity is observed from the Doppler shifts of Fraunhofer absorption lines. The most recent and comprehensive results have been given by Howard and Harvey (1970) using daily observations of Doppler shifts over the entire solar disk obtained simultaneously with the observations for the daily solar magnetograms at Mount Wilson Observatory. These results are shown as X's in Figure 1. We notice immediately the startling result that the average period of the plasma rotation is approximately $1\frac{1}{4}$ days longer at each latitude than the period of the magnetic fields. If we take these observations literally this means that on the average the photospheric field lines have a rotational velocity about 4 or 5% larger than the photospheric plasma, or that the field lines are plowing through the plasma with a relative velocity of about 100 m/sec. This is surprising since on the large scale we would expect the field to be frozen into the plasma. A beginning toward a physical explanation of this result may be found in the observations of Sheeley (1967) that many of the photospheric field lines do not exist as a relatively uniform large-scale field but instead are clumped into small filaments of cross-section less than 500 kilometers and with field strengths of the order of a kilogauss. Within these filaments the magnetic energy density $B^2/8\pi$ is larger than the plasma energy density.

The time interval required to obtain a complete differential rotation observation from the Doppler shifts is approximately the one hour required to obtain a solar magnetogram. This time is to be compared with the six months required for the autocorrelation analysis of the large-scale magnetic fields, and with the eleven years required for observations of long-lived sunspots. The large improvement comes from the greatly increased number of observations that can be obtained with the solar magnetograph across the entire visible disk during the course of one hour, and is possible only because of modern digital data-handling capabilities. These observations are usually obtained once per day. The variation in the differential rotation which we have already encountered in the above discussion now becomes very large, as is graphically illustrated in a motion picture prepared from the Doppler observations of Howard and Harvey (1970). One frame from this movie is shown in Figure 2. If this figure were the motion picture we would see the lines representing the observations in continual motion from day to day, often deviating from the average curve by 10 or 15%. The observed curve will often be above the average for perhaps half a dozen days, and then below the observed curve for a similar interval. There does not appear to be a precise periodicity associated with these changes.

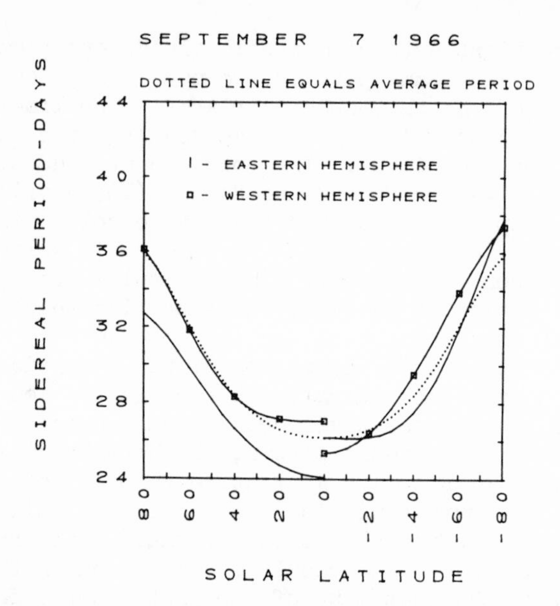

Fig. 2. Sample frame from motion picture representing Doppler shift observations of Howard and Harvey (1970). The results for the eastern half of the visible solar disk and for the western half are shown separately.

Very recent work by Gosling and Bame (1971) suggests that a similar difference between the field and the plasma rotation periods may exist in the interplanetary medium observed by spacecraft near the earth. The variation of the recurrence period of the solar wind plasma during an interval of several years is shown in Figure 3. We note that there is considerable variation in this quantity, just as was the case for the observations of the recurrence period of the photospheric plasma. We should note an important distinction between the two observations. In the case of the line-of-sight Doppler observations of the photospheric plasma we are observing and analyzing an actual rotational component of plasma velocity. In the case of the observations of the solar wind plasma we are doing an autocorrelation of the solar wind velocity, which velocity is predominantly directed in the radial direction away from the sun. Thus the recurrence peak in the autocorrelation is not the result of a rotational component in the solar wind plasma velocity, but rather represents the return of features in the solar wind velocity. Probably the predominant contribution to the recurrence peak comes from the recurring streams of high velocity solar wind plasma. Each of these streams tends to be observed for a few days, corresponding to a longitudinal width of perhaps 50 to 100 degrees.

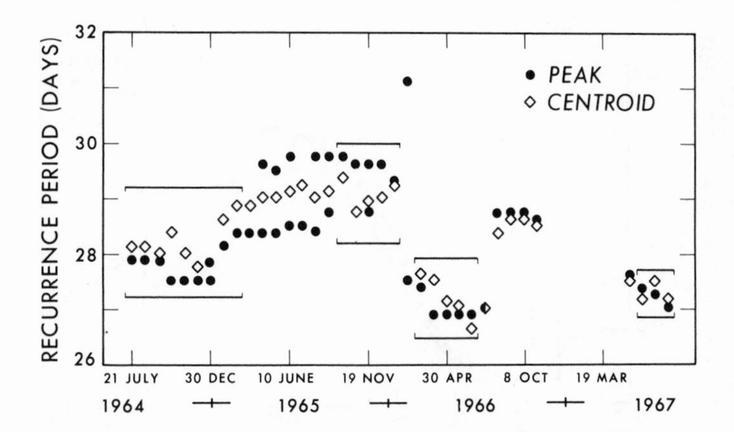

Fig. 3. Estimates of the period of recurrence of stationary solar wind velocity structures. The brackets enclose the most reliable determinations of recurrence period. PEAK and CENTROID refer to the autocorrelation curves from which these periods were estimated (from Gosling and Bame, 1971).

We may contrast the recurrence period of the solar wind velocity shown in Figure 3 with the recurrence period of the interplanetary magnetic field observed by Wilcox and Colburn (1970) and shown in Figure 4, which is also determined with autocorrelation techniques. There is a considerable tendency in the interplanetary medium for the recurrence period of the plasma to be several percent longer than the period of the field, just as was the case in the photosphere as shown in Figure 1.

With direct month-by-month comparisons of the variations in the recurrence periods of the photospheric plasma and of the solar wind plasma it may be possible to investigate the relationships between these quantities and to begin to get an idea of the physical processes involved. Some authors have discussed the large-scale (i.e. several days) variations in the solar wind velocity in terms of channeling effects in the strong magnetic fields of the chromosphere and low corona. To the extent that the recurrence periods of the photospheric and the solar wind plasma may be related as discussed above, and noting that the recurrence period of the magnetic fields tends to be distinctly shorter than the plasma periods, it appears that the most important physical influence on the large-scale solar wind velocity may rotate with the photospheric plasma, not with the fields.

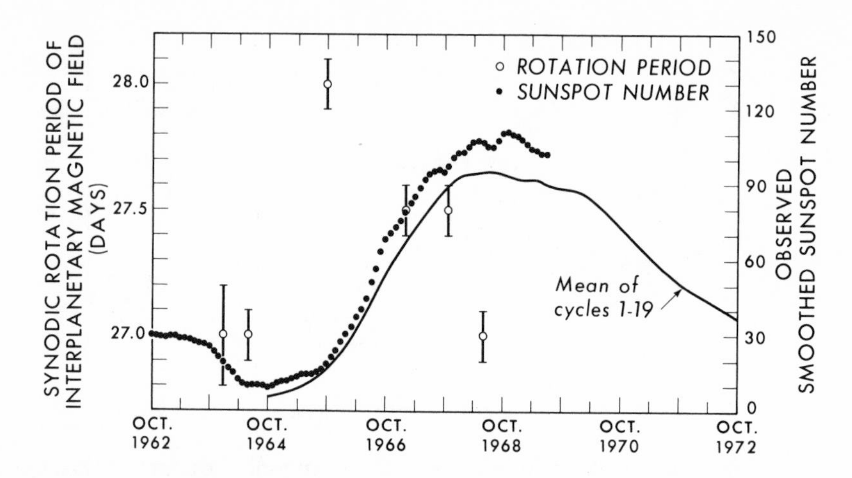

Fig. 4. The synodic rotation period of the interplanetary magnetic field and the observed sunspot numbers during the past several years (from Wilcox and Colburn, 1970).

At a height in the solar atmosphere in between the photosphere and the interplanetary medium near one AU that we have just discussed, namely the chromosphere and the lowest corona, Livingston (1971) has apparently observed a super rotation of the tenuous plasma that surrounds the localized features of strong magnetic field such as the prominences. In these observations the slit of a spectrograph is set perpendicular to the solar limb at the location of a prominence, and the Doppler shift is observed as a function of height above the solar limb. From the limb up to near the top of the prominence the observed wavelength is nearly constant. Presumably within this range of heights the strong magnetic fields associated with the prominence are rooted in the photosphere and cause the dense prominence material to corotate with the photosphere. Above the top there is a considerable change. Wisps of material appear to have Doppler shifts corresponding to an increased rotational velocity 10 or 20% larger than the photospheric value. This may correspond to tenuous plasma super-rotating above the region where the strong magnetic field enforces corotation.

Finally we may discuss the rotation properties of the recently discovered solar sector structure (Wilcox and Howard, 1968), which has been discovered by comparing spacecraft observations of the nearby interplanetary magnetic field with observations of the photospheric magnetic field obtained with the solar magnetograph at Mount Wilson Observatory. Unlike the other solar observations discussed above, the sector structure appears to rotate in a rigidly rotating system with a synodic period near 27 days. A schematic

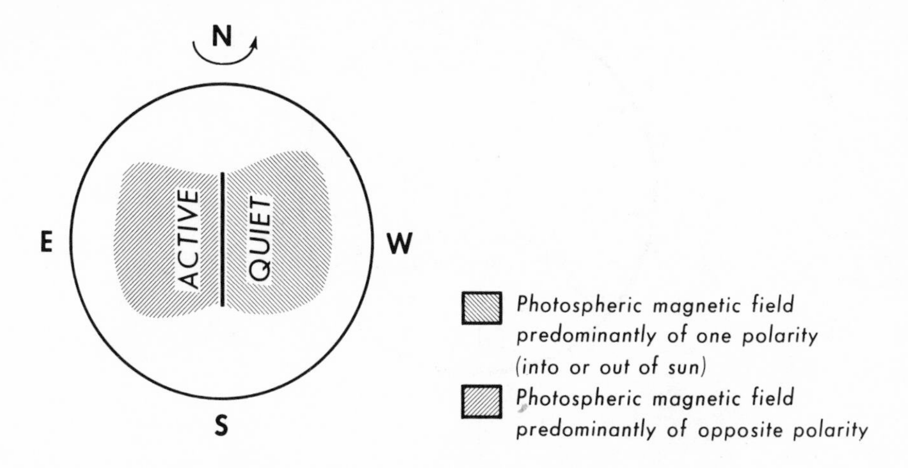

Fig. 5. Schematic of an average solar sector boundary. The boundary is approximately in the north-south direction over a wide range of latitude. The solar region to the west of the boundary is unusually quiet and the region to the east of the boundary is unusually active (from Wilcox, 1971).

of an average solar sector boundary (Wilcox, 1971) is shown in Figure 5. It appears that individual photospheric magnetic features such as bipolar magnetic regions display the shearing effects to be expected from differential rotation. However if one averages the observations over a few solar rotations a pattern similar to that shown in Figure 5 emerges.

The observed sectors may represent variations about a basic "dipole" configuration whose effects were first noticed in observations of polar geomagnetic fields by Olsen (1948). The link between the polar geomagnetic fields and a possible rotating solar magnetic "dipole" comes through a relationship between the polar geomagnetic fields and the polarity of the interplanetary magnetic fields discovered by Svalgaard (1968) and Mansurov (1969) and confirmed by Friis-Christensen _et al_. (1971), and by the link between interplanetary magnetic fields and photospheric magnetic fields demonstrated by Ness and Wilcox (1966). A schematic of the rotating solar magnetic "dipole" (Wilcox and Gonzalez, 1971) is shown in Figure 6.

In summary, we find an interesting variety of rotational properties in the photospheric and solar wind plasma and magnetic fields. In both the photosphere and in the interplanetary medium near the earth there is a tendency for the field patterns to rotate a few percent faster than the plasma patterns. The fields and

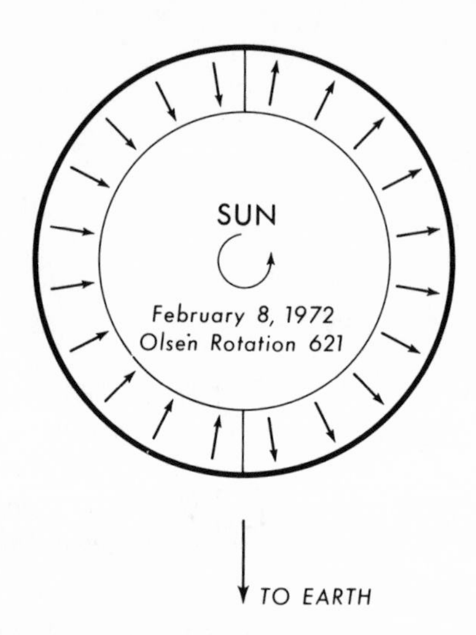

Fig. 6. Schematic of the rotating solar magnetic "dipole" (from Wilcox and Gonzalez, 1971).

plasmas show variability in their rotational properties on time scales of days or months, but averages over a few years tend to become much less variable, as shown by the results for long-lived sunspots, and by the rotating solar magnetic "dipole".

This work was supported in part by the Office of Naval Research under Contract N00014-69-A-0200-1049, by the National Aeronautics and Space Administration under Grant NGL 05-003-230 and by the National Science Foundation under Grant GA-16765.

REFERENCES

Friis-Christensen, E., Lassen, K., Wilcox, J. M., Gonzalez, W., and Colburn, D. S.: 1971, submitted to Nature.
Gosling, J. T. and Bame, S. J.: 1971, submitted to J. Geophys. Res.
Howard, R. and Harvey, J.: 1970, Solar Physics 12, 23.
Livingston, W. C.: 1971, submitted to Solar Physics.
Mansurov, S. M.: 1969, Geomagn. Aeronom. 9, 622.
Ness, N. F. and Wilcox, J. M.: 1966, Astrophys. J. 143, 23.
Newton, H. W. and Nunn, M. L.: 1951, Monthly Notices Roy. Astron. Soc. 111, 413.
Olsen, J.: 1948, Terr. Mag. Atmosph. Elect. 53, 123.
Sheeley, N. R., Jr.: 1967, Solar Physics 1, 171.
Svalgaard, L.: 1968, Danish Meterolog. Inst. Geophys. Papers R-6.
Wilcox, J. M.: 1971, Comments Astrophys. Space Phys., to be published.
Wilcox, J. M. and Colburn, D. S.: 1970, J. Geophys. Res. 75, 6366.
Wilcox, J. M. and Gonzales, W.: 1971, submitted to Science.
Wilcox, J. M. and Howard, R.: 1968, Solar Physics 5, 564.
Wilcox, J.M. and Howard, R.: 1970, Solar Physics 13, 251.

SOFT X-RAY SPECTRAL STUDIES OF SOLAR FLARE PLASMAS

G. A. Doschek, J. F. Meekins, R. W. Kreplin,
T. A. Chubb and H. Friedman

E. O. Hulburt Center for Space Research

Naval Research Laboratory, Washington, D. C.

INTRODUCTION

Many interesting astrophysical problems are concerned with the interaction of partially ionized gases with magnetic fields. The discovery of pulsars, and the subsequent interpretation of these objects as neutron stars surrounded by a relativistic plasma, is an example of such a problem. The origin of cosmic rays, the nuclei of Seyfert galaxies, and the newly discovered X-ray sources such as Sco XR-1 are other obvious problems requiring at least a partial interpretation in terms of plasma physics. It is therefore important that nearby astrophysical plasmas, which can be studied in far greater detail than their more distant counterparts, be investigated as intensively as possible; because at present these plasmas provide the only astrophysical possibility of subjecting the theories of plasma interaction to direct observational tests.

Partially with this viewpoint, the Naval Research Laboratory has been conducting an extensive program investigating the soft X-ray spectrum of solar flares. The flare spectrum has been recorded by Bragg crystal spectrometers flown on the fourth and sixth Orbiting Solar Observatories (OSO). In this paper we will be concerned with results obtained from OSO-6.

INSTRUMENTS

The NRL instruments on OSO-6 consist in part of three

uncollimated Bragg crystal spectrometers, which scan the flare spectrum in the 0.6 A to ~ 14 A region. A LiF crystal spectrometer covers the 0.6 A to 4 A band, an EDDT spectrometer scans from 1.5 A to 8.5 A, and a KAP spectrometer scans from ~ 5 A to ~ 14 A. The detectors are argon filled, bromine quenched Geiger counters with mica windows. The system is shielded from ultraviolet light and low energy particle contamination by aluminum coated Mylar and Parylene filters placed on the entrance aperture. The thin coating of aluminum on the filters also reduces heating of the instruments.

The crystals are rotated on a common assembly by a stepping motor such that a spectral scan is completed in either 2 or 7 minutes. The time is chosen by command from the ground. Each step in angle is 6 arc minutes, and the crystal rocking curves are deformed in a manner insuring that it is not possible to step over a spectral line.

The observed spectrum is a convolution of the true source spectrum at a given time with the instrumental efficiencies. Since for high spectral resolution, 7 minutes are required to complete a spectral scan, the observed spectrum is usually distorted by variations in flux from the flare plasma.

FLARE SPECTRA

The soft X-ray flare spectrum is characterized primarily by resonance lines of elements in hydrogenic and heliumlike ionization stages, and a strong continuum assumed due to free-free and free-bound processes. Figure 1 shows the flare spectrum obtained by the LiF and EDDT crystal spectrometers of an intense flare that occurred on 16 November 1970. The spectra are not corrected for instrumental efficiencies and therefore instrumental edges appear at 3.871 A, 3.436 A, 6.745 A, and 7.951 A due to argon in the detectors, potassium and silicon in the mica windows, and aluminum on the filters, respectively. In addition, the spacecraft was scanning (rastering) the solar disk during this event, and the data are unreliable when the spacecraft returned to start of raster. This effect is marked in the figure. Also note that the wavelength intervals become larger towards longer wavelength. This is because the data have been plotted as a linear function of crystal step (proportional to diffraction angle) and corresponding wavelengths are proportional to the sine of this angle (Bragg's Law). Lines observed in higher orders of diffraction are indicated by the order numbers in parentheses beside the ion symbols, e.g., the three lines at ~ 6.4 A are due to second order diffraction of the blended Ca XIX emission feature at ~ 3.2 A. A detailed listing of all the lines and transitions observed in our data and in data from other groups is given in Doschek (1971). For the purposes of this paper, however, we only discuss some of these emission lines. The Lyman-α lines of hydrogenic ions of abundant solar elements are found at: Mg XII, 8.421 A; Al XIII, 7.173 A; Si XIV, 6.182 A; S XVI, 4.729 A; Ar XVIII, 3.733 A (blended with second order iron); Ca XX, 3.02 A; and Fe XXVI, 1.79 A (usually too weak to be observed). Higher terms of the Lyman series have also been identified and are easily seen for silicon in the EDDT spectrum.

Emission from heliumlike ions is primarily characterized by three lines: the resonance ($1s^2\ ^1S$-$1s2p\ ^1P$), line, the intercombination line ($1s^2\ ^1S$-$1s2p\ ^3P$), and the forbidden ($1s^2\ ^1S$-$1s2s\ ^3S$) line (Gabriel and Jordan, 1969a), in order of increasing wavelength. The $2\ ^1S$ state decays via the two-photon process.

Densities in flare plasmas are sufficiently low to observe the forbidden lines of heliumlike ions although collisional quenching from the $2\ ^3S$ state into the $2\ ^3P$ state is competitive with spontaneous decay of the $2\ ^3S$ state for atomic number less than ~ 16. Gabriel and Jordan (1969a) have utilized this circumstance to determine electron densities in active regions. Examples of these three lines in flare spectra may be seen in Figure 1 for K XVIII, ~ 3.54 A; S XV, ~ 5.04 A; and Ca XIX, seen in second order at ~ 6.4 A.

From the viewpoint of plasma physics, there are two separate aspects of these spectra that can be discussed, i.e., plasma spectroscopy and plasma dynamics. We discuss some of the features of our spectra below which pertain to each of these fields.

Plasma Spectroscopy

The presence of strong resonance lines of hydrogenic ions implies a high electron temperature in the flare plasma or a nonthermal equivalent. Assuming that the lines are formed by collisional excitation and that ionization equilibrium is valid, temperatures may be estimated by comparing the ratios of resonance lines of hydrogenic ions to heliumlike ions of the same element. Temperatures so obtained range from 10×10^6 K to 35×10^6 K. The data

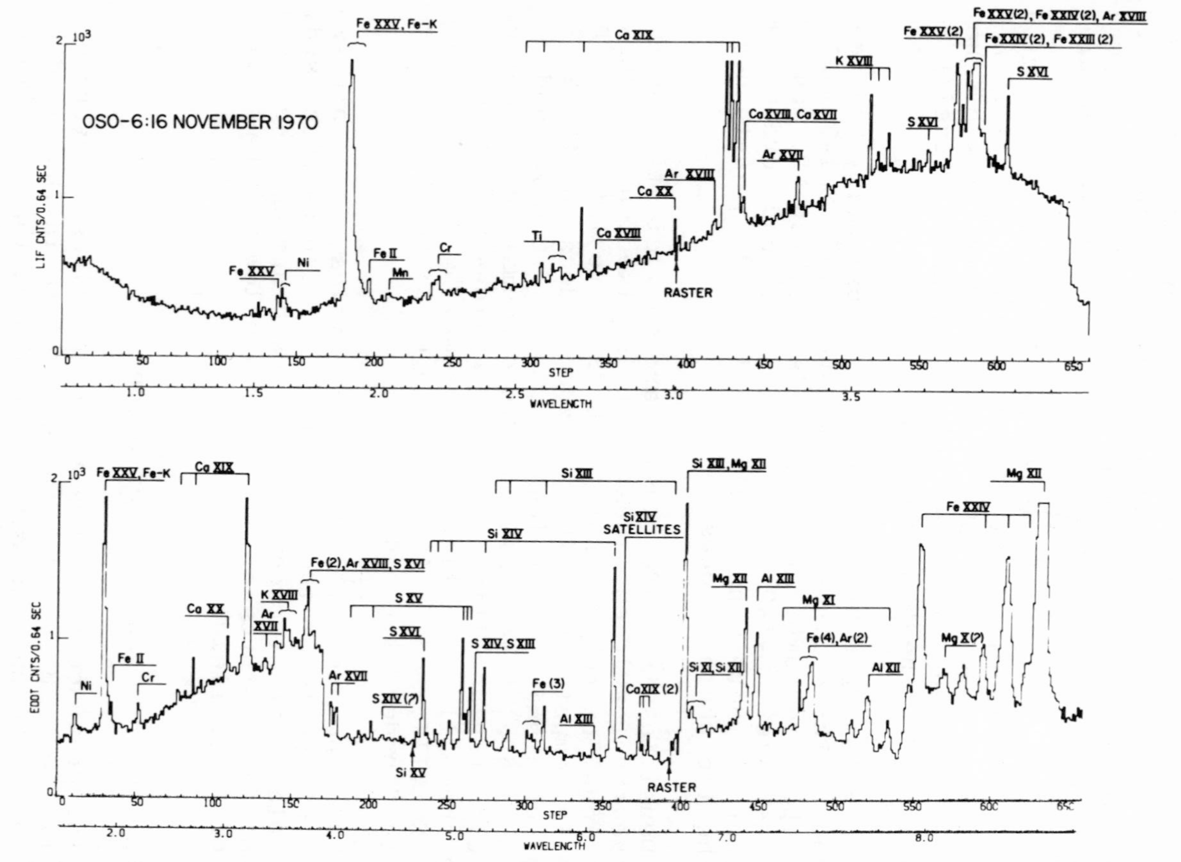

Fig. 1: Solar flare spectra obtained by NRL Bragg crystal spectrometers on OSO-6.

also indicate that the plasma is multithermal, as temperatures obtained from Mg XI and Mg XII lines are considerably less than those obtained from ratios of the S XV lines to S XVI lines and from ratios of the Ca XIX to Ca XX lines.

Besides the resonance lines observed in the spectrum, weaker satellite emission lines of hydrogenlike and heliumlike ions have been observed, e.g., Ca XVIII, Ca XVII, Si XI, and Si XII in Figure 1, and these lines are of interest for determining the basic atomic processes responsible for line emission in the plasma. The strongest satellite line observed is the $1s^2 2s\ ^2S$-$1s2s2p\ ^2P$ transition in the lithiumlike ion. This line falls between the intercombination and forbidden lines of the heliumlike ions and is blended with these lines. Therefore it cannot be observed in the spectra of Figure 1 except in the second order iron feature near 3.70 A in the LiF spectrum. However, the transition has been observed in neon, silicon and sulfur by Walker and Rugge (1971) and in silicon, sulfur, and iron by Neupert and Swartz (1970) and Neupert (1971). The line has been numerically resolved by Doschek et al. (1971a) for calcium and has also been resolved and discussed in the NRL second order iron spectrum (Doschek et al., 1971b). The NRL second order iron spectrum is shown in Figure 2 compared with a recent laboratory spectrum obtained by Lie and Elton (1971). The ^2P satellite is marked. These lines may also be seen in Figure 1 although some of them are saturated.

The strength of the ^2P lithiumlike satellite relative to the resonance line of the corresponding heliumlike ion increases with increasing atomic number; consistent with the hypothesis that the satellite lines are formed by dielectronic recombination of the heliumlike ion (Gabriel, 1971). However, innershell collisional excitation may contribute to the line strength in the case of the heavier elements.

The satellite lines have also been observed in theta-pinch laboratory spectra and in this case it has definitely been established that dielectronic recombination is the dominant mechanism of line formation (Gabriel and Jordan, 1969b). In the NRL iron spectra, the intensity of the ^2P satellite is about 0.6 that of the resonance line of Fe-XXV, which is consistent with dielectronic recombination theory (Gabriel, 1971). From this intensity ratio the temperature of the emitting plasma may be determined (Gabriel, 1971) since the Fe XXV resonance line is formed primarily by collisional excitation, and has a temperature

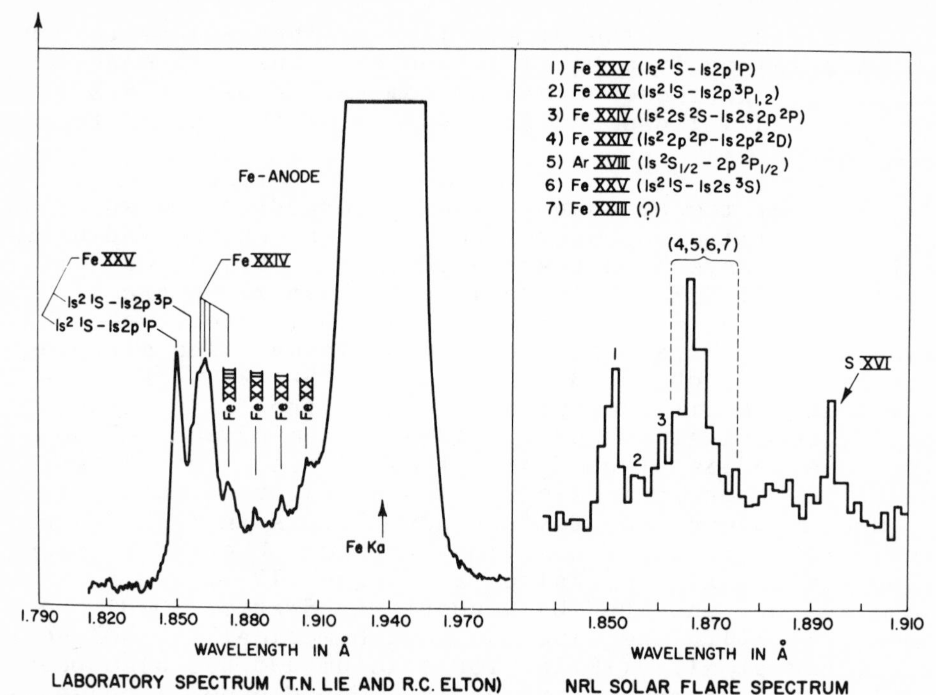

Fig. 2: A comparison of solar flare iron-line emission with a recent laboratory spectrum produced by a vacuum spark.

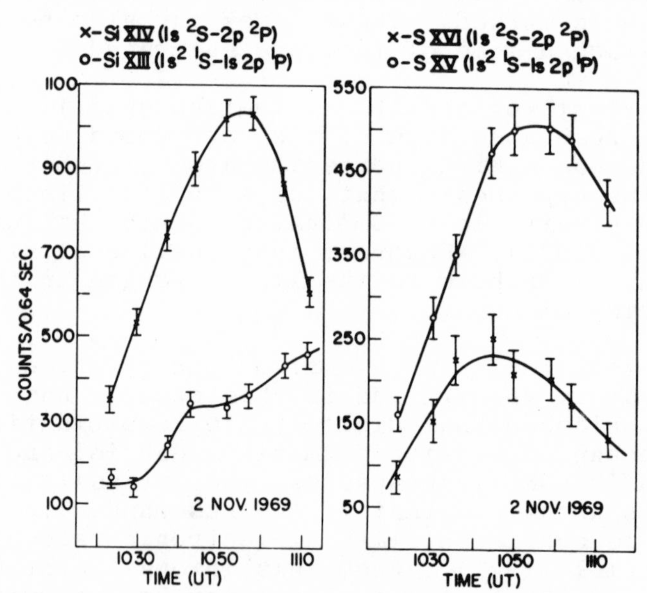

Fig. 3: Time-histories of resonance lines of silicon and sulfur for a large limb flare that occurred on 2 November 1969.

dependence different from dielectronic recombination. Temperatures determined in this manner are independent of ionization equilibrium since both processes begin with the same ion. Gabriel (1971) has determined temperatures from the NRL data and data obtained by Walker and Rugge (1971). His results give temperatures that are in general agreement with those obtained from line ratios of hydrogenic to heliumlike ions and from continuum measurements.

Recently, satellite lines of hydrogenlike ions have also been detected in flare spectra (Walker and Rugge, 1971). An emission feature in NRL data tentatively identified as due to doubly excited heliumlike silicon is shown in Figure 1 at ~ 6.2 A. This feature is not statistical as it appears in a subsequent scan of this event. Unlike the lithiumlike lines, the appearance of transitions in doubly excited heliumlike ions can only be due to dielectronic recombination, since the electron densities in flare plasmas are $\lesssim 10^{13}$ cm^{-3}.

It is desirable to compare the relative solar intensities of satellite lines with laboratory spectra obtained in theta-pinch or vacuum spark devices. However, Figure 2 shows that higher resolution is required both in the laboratory spectra and in the flare spectra before quantitative comparisons are possible. Recently, Schwob and Fraenkel (1971) have obtained iron-line laboratory spectra with greater resolution than Lie and Elton (1971) and Vasiljev et al. (1971) have obtained very high resolution solar iron-line spectra. A comparison of these results promises to lead to a greater understanding of atomic processes both in the laboratory and in solar flares.

Plasma Dynamics in Flares

By observing the time-histories of the resonance lines of hydrogenlike and heliumlike ions, temperature and emission measure variations of the flare plasma can be determined as a flare evolves. Because the NRL instruments are uncollimated, only an average behavior of these quantities can be studied. Nevertheless, the results represent a considerable improvement over broadband observations.

Figure 3 shows the time-histories of the resonance lines of hydrogenlike and heliumlike silicon and sulfur for a large limb flare on 2 November 1969. The data are uncorrected for instrumental effects. The ratio of the flux of Lyman-α S XVI to the flux of the resonance line of S XV decreased steadily from about 1033 UT to the end of the observations, indicating a decrease in temperature of about 5×10^6 K over the observation period. Since the flux in these lines increased substantially over part of this time interval, e.g., from ~ 1020 UT to ~ 1045 UT for S XVI, we can conclude that the emission measure was increasing strongly enough over this time interval to offset the decreasing temperature. This result is in agreement with broadband observations (Horan, 1970).

A consequence of the decreasing temperatures is that the peak flux of S XV occurs later than the peak flux in S XVI, i.e., S XVI recombines to S XV and therefore increases the S XV fractional abundance.

Similar conclusions can be drawn from the silicon data, except that the interpretation is complicated by the apparent jump and leveling off in intensity of the Si XIII line around 1040 UT. The effect appears to be real because higher lines in the Si XIII series show a similar intensity behavior. The intensities of all the Si XIII lines are weak, however, and until similar behavior is observed in other flares, we regard this observation as tentative.

Assuming that the Si XIII variations are real, however, one possible interpretation of the effect is a release of flare energy into surrounding, cooler plasma. The Si XIII lines emit with maximum efficiency $\sim 10 \times 10^6$ K, while the sulfur lines emit strongly $\sim 20 \times 10^6$ K. The near coincidence of the inflection in Si XIII intensity with the maximum emission in the S XVI, Ca XIX, and Fe XXIV lines is consistent with the idea that the high temperature plasma, presumed confined by magnetic fields, heated nearby cooler regions in a time ($\lesssim$ 10 minutes) short compared to the total observation period (~ 50 minutes).

The other lines in Figure 3 would not show a similar behavior as they are formed in the higher temperature regions. Lines of elements lighter than silicon, e.g., magnesium, may not show an intensity increase similar to Si XIII because they are formed in even lower temperature regions ($\sim 6 \times 10^6$ K for Mg XI), which may be too far removed from the magnetic boundary of the hot plasma to be strongly influenced by the heating mechanism. A similar enhancement was noted for the calcium satellite lines at about the same time (~ 1045 UT, Doschek et al., 1971a). Of course, the data under discussion are not sufficiently detailed to permit any conclusions to be drawn concerning the nature of the energy release, e.g., to determine whether or not actual release of particles occurred, or whether other heating mechanisms were involved.

Evidence from photographs and X-ray spectroheliographs from other flares indicates (Vaiana et al., 1968; Beigman et al., 1971) that the soft X-ray region is composed of small, hot loop-like filaments only seconds of arc across interspersed in cooler plasma. Rapid variations in intensity have also been noted (Beigman et al., 1969). The qualitative interpretation of the Si XIII line behavior discussed above seems plausible considering the current picture of the soft X-ray flare plasma as indicated by these observations.

It is obvious that more detailed observations obtained from future instruments, i.e., continued monitoring of lines by collimated spectrometers, will lead to a better understanding of the interaction of high temperature plasmas with magnetic fields, and perhaps shed light on the mechanism responsible for the production of solar flare phenomena, and high energy phenomena in general.

REFERENCES

Beigman, I.L., Grineva, Yu.I., Mandelshtam, S.L., Vainshtein, L.A., and Žitnik, I.A. 1969, Solar Phys., 9, 160.

Beigman, I.L., Vainshtein, L.A., Vasiljev, B.N., Žitnik, I.A., Ivanov, V.D., Korneev, V.V., Krutov, V.V., Mandelshtam, S.L., Tindo, I.P., and Shuryghin, A.I. 1971, preprint.

Doschek, G.A., Meekins, J.F., Kreplin, R.W., Chubb, T.A., and Friedman, H. 1971a, Ap. J., 164, 165.

Doschek, G.A., Meekins, J.F., Kreplin, R.W., Chubb, T.A., and Friedman, H. 1971b, to be published in the Astrophysical Journal.

Doschek, G.A. 1971, The Third Symposium on Ultraviolet and X-Ray Spectroscopy of Astrophysical and Laboratory Plasmas, Utrecht, 24-26 August 1971. To be published in Space Science Reviews.

Gabriel, A.H., and Jordan, C. 1969a, M.N.R.A.S., 145, 241.

Gabriel, A.H., and Jordan, C. 1969b, Nature, 221, 947.

Gabriel, A.H. 1971, The Third Symposium on Ultraviolet and X-Ray Spectroscopy of Astrophysical and Laboratory Plasmas, Utrecht, 24-26 August 1971. To be published in Space Science Reviews.

Horan, D.M. 1970, unpublished Ph.D. Thesis, Catholic University of America, Washington, D. C.
Lie, T.N., and Elton, R.C. 1971, Phys. Rev. A, 3, No. 3, 865.
Neupert, W.M., and Swartz, M. 1970, Ap. J. (Letters), 160, L189.
Neupert, W.M. 1971, Solar Phys., 18, 474.
Schwob, J.L., and Fraenkel, B.S. 1971, The Third Symposium on Ultraviolet and X-Ray Spectroscopy of Astrophysical and Laboratory Plasmas, Utrecht, 24-26 August 1971. To be published in Space Science Reviews.
Vaiana, G.S., Reidy, W.P., Zehnpfennig, T., Van Speybroeck, L., and Giacconi, R. 1968, Science, 161, No. 3841, 564.
Vasiljev, B.N., Grineva, Yu.I., Žitnik, I.A., Karev, V.I., Korneev, V.V., Krutov, V.V., and Mandelshtam, S.L. 1971, XIVth Plenary Meeting of COSPAR, Seattle, Washington, USA, 1971.
Walker, A.B.C., Jr., and Rugge, H.R. 1971, Ap. J., 164, 181.

SIMILARITIES BETWEEN SOLAR FLARES AND LABORATORY HOT PLASMA PHENOMENA

W. H. Bostick, V. Nardi and W. Prior

Physics Department, Stevens Institute of Technology

Hoboken, New Jersey

Summary: Similarities exist among fine structure details (and re-related phenomena) of solar flares and of the current sheath in a plasma coaxial accelerator. The production mechanism of x-ray and high-energy particles can be consistently related to the decay process of a filamentary magnetic structure in both solar and laboratory phenomena. Build-up and explosive decay of this structure are described by a self-consistent theory for fields and phase-space densities.

Elements of Similarity

By considering similarities between solar flares and laboratory phenomena, major attention is paid here to the occurrence of a similar chain of events, with the same pattern for both solar and laboratory scales. No effort is made here to extend the interpretation of a specific event from one scale to the other simply on the basis of an invariance principle[1] for typical (adimensional) quantities [this standard procedure would lead also in our case to electric and magnetic fields, $\underline{E}$, $\underline{B}$, which seem too small for the solar scale or alternatively too large for laboratory phenomena; e.g., if $(E,B)_{solar} = \eta^{-1} (E,B)_{lab}$, where $\eta = (\text{length, time})_{solar}/(\text{length, time})_{lab}$]. Solar flare events of particular interest are: [1] Sudden onset of radiation emission-flash phase-on a time interval, say, t_o. [2] Large amounts of energy emitted within the relatively short length of time, say t_f, of the flare.[2] In the Sweet mechanism[3] for flares, the small value of t_f leads to the well known difficulty of explaining by theory the high-diffusion rate of oppositely directed magnetic fields for the required annihilation. [3] Failure to observe any rapid variation in large-scale magnetic field,[4] although there is some indication of appreciable decreases with the time scale of a day or two. [4] Ejecta (puffs, surges, sprays), disruption of prominences and in most cases inward motions as shown by Balmer-line profile asymmetry[2] (red wing brighter than blue wing in 90 percent of flares). [5] Diversity and complexity of flare shapes with much fine structure including dark and luminous filaments parallel to the isogauss line[4] in regions where the radial field component (parallel to the line of sight) $B_{||} \cong 0$. [6] Severny's results (well established) which indicate that large flares have the tendency to occur between regions of strong and opposite magnetic polarity, at points where $B_{||} = 0$ at the time of the flare.[5] It seems reasonable to consider the possibility that the luminous filaments - parallel to, and in the proximity of, the null line of the observed $B_{||}$ - are the product of a dynamic interaction of the inward flowing particles with the large-scale B parralel to the sun surface in this region. [7] Anisotropy and nonthermal energy spectrum of particle and x-ray radiation.[6] Corresponding elements can be located - and similar phenomena occur - on the current sheath with a fine structure produced by plasma coaxial accelerators (CA) which are normally used in the so-called plasma-focus experiments. The CA used in our experiments has been described elsewhere.[7] The current sheath (CS) has a typical filamentary structure in particle and electric current densities and in the magnetic field which shows a strong component of B in the direction of the filament axis on a filament region.[8] The luminous filaments are embedded in the less luminous front (~ 0.5 mm thick) of the current sheath (~ 3 mm thick) which carries most of the current ($\geq 90\%$) through the plasma.[9] A simple method for observing this filamentary structure is by image converter (IC) photographs in the visible spectrum (5 nsec exposure with neutral filters) under specific conditions of operation (maximum potential U=11-13kV on the electrodes with radii $r_e = 5$ cm, $r_i = 1.7$ cm; peak current $I_{max} \cong 0.4 - 0.5$ MA, 8 torr D_2 or H_2). When these optimum conditions are not satisfied (e.g. for a sufficiently low gas pressure or a sufficiently high I) then IC photographs may show a uniform CS but schlieren photographs and shadowgraphs (by ruby laser light) can still show in most cases a filamentary structure for particle density-gradient variations.[10] The filaments form when the CS is moving between the electrodes with a velocity $u_o \sim 5 \cdot 10^6$ cm/sec and become brighter, with sharply defined boundaries (cylinders with ~ 0.5 mm radii), during CS radial collapse- with u_o increasing up to a factor 2-on the axial region at the electrode ends when $I \sim I_{max}$ (see Fig. 1). Since no biasing field is used, the CS separates the region of vanishing field (ahead of CS) from the region of high azimuthal field B_θ due to the electrode current (behind, or down-stream, CS). These radial filaments (orthogonal to B_θ) have steady configuration and steady azimuthal position during the CS motion. Slow variation of the CS filamentary structure (merging of neighboring filaments, branching, growth or fading of a single filament) can be observed before CS collapses only on the relatively long time scale τ_f ($\sim 10^{-7} - 10^{-6}$ sec) of the CS motion between and off the electrodes. When CS radial collapse reaches the stage of maximum compression near the face of the centre electrode (anode) a new kind of event occurs with onset of an explosive decay of the filamentary structure. The filament decay can take place in localized regions (e.g. filament segments of a few mm's length) and can be observed by a localized increase of luminosity on a time scale, say τ_o ($\sim 10^{-9}$ sec) followed within a time of $\sim$ 20-30 nsec or larger by the onset of hard x-ray ($\gtrsim 10 - 10^2$keV) and neutron (if deuterium is used) emission[11] which can last up to $1\text{-}5 \cdot 10^2$ nsec. The decay can occur also in extended regions (of a few cm^2) when a shock disrupting CS can trigger the decay of many filaments with a pattern resembling the sympathetic flare phenomenon.[7] Purpose of this study is to analyze some of the details of the hard-radiation emission during the filament decay and to relate this emission with specific elements of the magnetic structure as it is described by a self-consistent theory for fields and particle phase-space density. As a conclusion, it seems a consistent approach to extend to solar flare phenomena the mechanisms of build-up (by supersonic plasma flow against a strong orthogonal B-field) and decay (by a high-rate annihilation process) of the magnetic structure in the laboratory experiment.

Laboratory Observations

We have used a variety of diagnostic methods for observations on plasma filaments during build-up and stationary regime of the CS: [A] The electron density in a filament ($\sim 10^{18}$ cm^{-3} by H_β line Stark broadening) is about three times the initial particle density.[7] [B] The magnetic field component (say B_z) along the filament axis on a filament region (by magnetic probes) is $\sim 10^3 - 10^4$ G, i.e. of the same order as the maximum value $B_o \sim 1.5\text{-}2 \cdot 10^4$G of B_θ before the CS axial collapse.[8] With the same method it is verified that the luminous front CS coincides - within the experimental uncertainty $\sim$ 1 mm - with the highest current-density region. B_z and the large mass flow related to u_o in the plane orthogonal to a filament axis indicate that filaments should not be considered the same phenomenon as the streamers due to electron avalanches in spark discharges or as the anode spots in glow discharges.[9] [C] By simultaneously recorded IC photographs and magnetic probe signals, CS appears as the foremost luminous face of the ionizing shock wave with B_θ increasing from 0 to B_o in a region $\sim$ 3 mm thick (shock thickness).[9] From CS non-planarity and CS typical curvatures and from de Hoffmann-Teller shock conditions it follows that a strong velocity-field vorticity exists in the direction of the filament axis on a filament region.[13] [D] By observing at different times the interaction of CS with solid obstacles conveniently inserted between the electrodes, we have no indication that the CS-driven shock has a precursor. [E] A crucial point is to verify that the structural details of CS as recorded by IC photographs have corres-

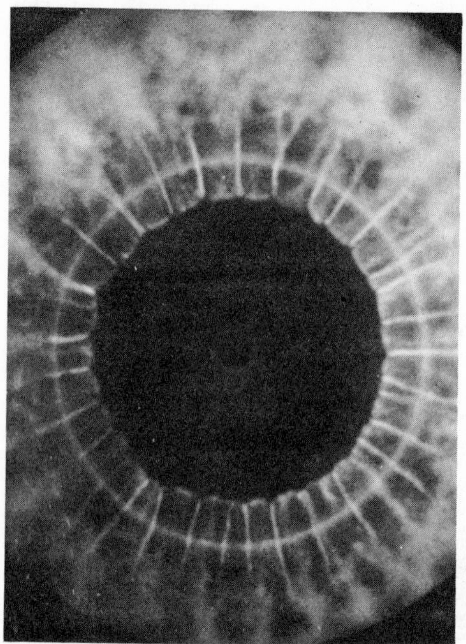

Fig.1 IC photo (5 nsec exp.). Circular edge of solid anode is visible, front view. Conditions: 8 Torr D_2, U=11 kV

Fig.2 Same conditions as Fig.1, at a time closer to CS collapse

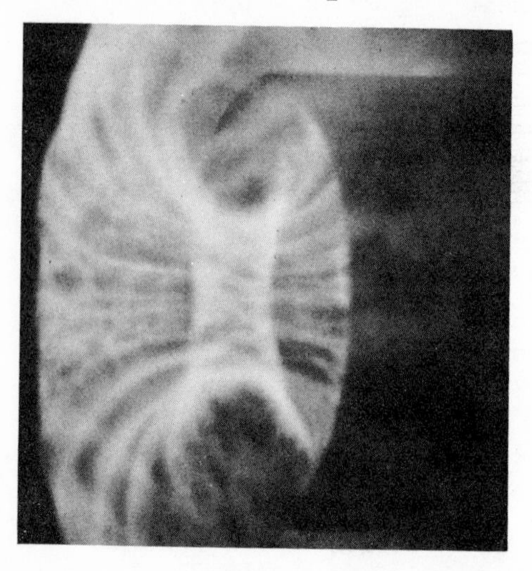

Fig.3 Same conditions as Fig.1, oblique view, hollow anode

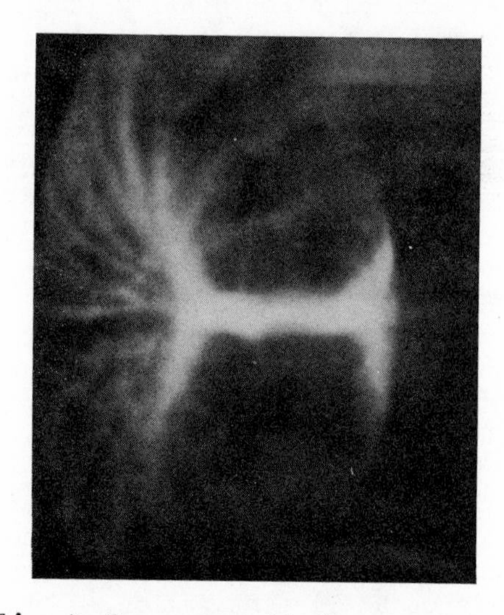

Fig.4 Same conditions as Fig.1, $\sim 60 \pm 20$ nsec after max. compression, time $\tilde{t}_o$

ponding fine-structure details in particle density distribution. This correspondence has been found by a systematic comparison of IC photographs, schlieren photos and shadowgraphs, simultaneously taken.[10] The conclusion is that all three recording methods give a mutually consistent rendering of the plasma density distribution in the CS. [F] The electrons are the current carriers on CS in the direction of the filament axis.[14] Both hollow and solid center electrodes have been used to be sure that filaments, even during CS axial collapse, are not affected by irregularities on the anode-face surface. The decay stage of the filamentary structure has been studied mainly by IC photographs, x-ray pinhole (time integrated) photographs and different observations of x-ray and neutron production rate - in deuterium - with plastic scintillators of different thickness (small thickness for x-rays) to discriminate between x-ray and neutron emission from the plasma. The CS collapses in the electrode axial region and forms a plasma column ($\sim$ 4 mm diameter) starting - at a specific recorded time $\tilde{t}_o$ - from the center of the anode circular face and reaching its maximum length (2-3 cm) in a time of 60-100 nsec[14]. The filaments are still visible in the off-axis part of CS which branches off at the end of the column (or at both ends of the plasma column if a hollow anode is used, see fig. 4) but not-by IC photos-on the column. Resolution of the filamentary structure on the column is still possible by IC photos if an external magnetic field is inserted on the axis in order to reduce the plasma axial compression.[7] The localized region of decay of the filaments on the column can be observed as a bright spot which moves from the anode to the end of the column with a velocity of $5 \cdot 10^7$ cm/sec[7,14] (see Fig. 4). The neutron production (after time-of-flight correction) and hard x-ray (> 10 keV up to and above 10^2 keV) emission start with a delay of about 30 nsec or larger from $\tilde{t}_o$, i.e. from the appearance of the bright spot. Our measurements on collimated x-rays indicate that the region of maximum x-ray production moves on the column with the same velocity of the bright spot.[7] This velocity also agrees, within the experimental uncertainty, with the displacement velocity of the region of maximum neutron production along the column as obtained by collimated neutron measurements.[15] We conclude that a clear correlation exists in space and time between filament decay and hard radiation emission in the plasma column. A second peak for neutron and hard x-ray production may occur for high values of U about 10^2 nsec or more from the first peak. This second peak can be related with the decay of many filaments on the off-axis part of CS, when an outward moving shock hits and disrupts this part of the CS at about the same time of the second peak (see Fig. 5,6). The decay process for a single filament or for filament pairs can be followed at this later stage on CS. Electron bursts ejected tangentially to CS during this shock-induced decay of the off-axis part of CS (and produced by a sequence of other disturbances) are clearly observed on Fig. 7 via the x-ray emission from the anode-inside-wall bombardment. Filament regions of increased luminosity correspond to decaying regions (Fig.5,6).

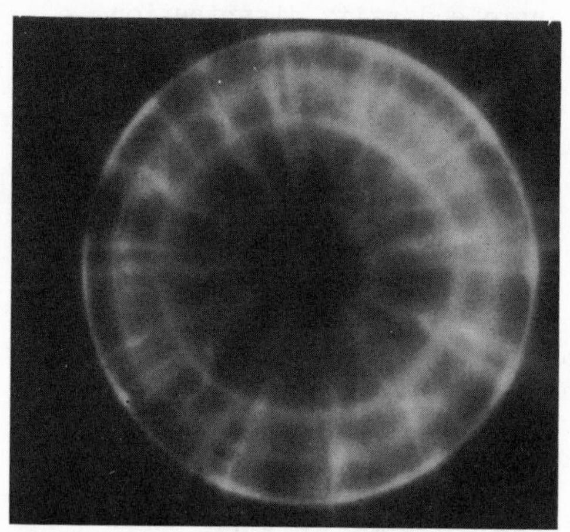

Fig.5 IC photo, front view, ~250 nsec after maximum axial compression (U=11 kV)

Fig.6 Same conditions as Fig.5 (but U=14 kV) near time of maximum compression, oblique view, hollow anode

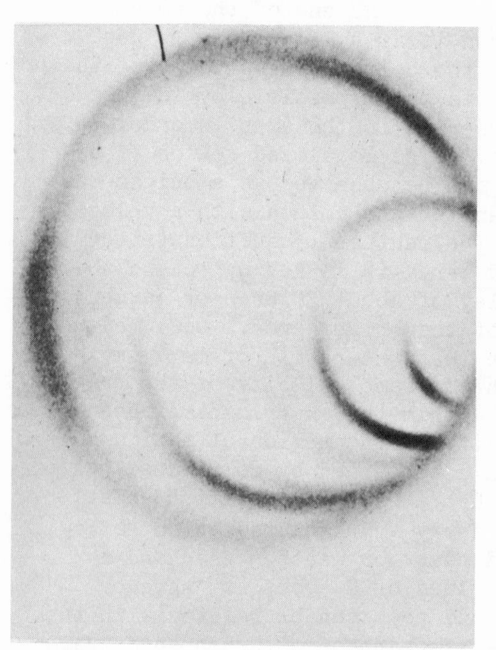

Fig.7 X-ray pinhole camera photo (time integrated) taken at 15° with respect to the electrode axis; hollow anode. Electron bursts tangential to CS last $\leq 10^{-8}$ sec (from CS speed and width of x-ray arcs)

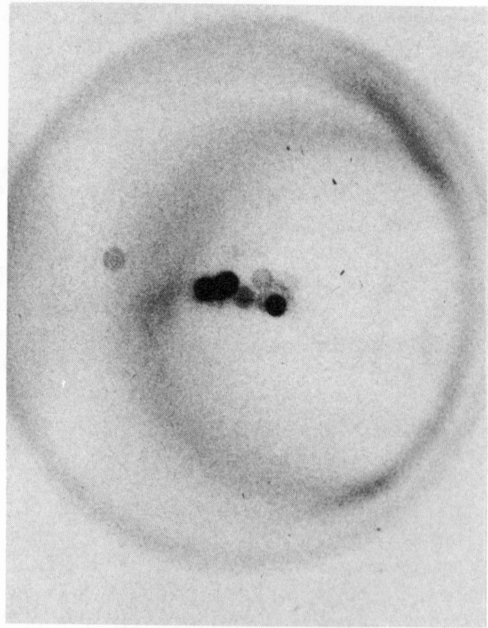

Fig.8 X-ray pinhole camera photo, same conditions as Fig.7 (8 Torr D_2; U=14 kV) but with 1% Ar to enforce x-ray emission, localized regions of emission are visible, single discharge

The explosive decay of the magnetic structure in the plasma column can be observed also as a multiplicity of localized regions of x-ray emission (dots of linear dimensions $\lesssim$ 0.5 mm) by x-ray pinhole photographs (see Fig. 8). The detailed filamentary structure of the off-axis part of CS (with spokes similar to the IC photo filaments) has been observed by hard-x-ray (> 10-10^2 keV) pinhole photographs.[15] Hard-x-ray intensifiers were used and the soft x-rays (produced mainly on the anode surface or from anode vapor near the anode) were screened by absorbing material (10 mm thick plastic)[16]. This fine structure of the x-ray source indicates that a simple pinch effect (with a quasi-adiabatic plasma compression) in the axial region cannot give a satisfactory account of the production of high energy ($\gg |eU|$) electrons and ions. (For other experiments with a capacitor-bank energy higher than in our 4 kJ experiment, the discrete localized x-ray sources on the axial region may become a continuous pattern. This could lead ultimately to a misconception of the hard-radiation production mechanism.) During the decay of the magnetic structure the energy is transferred from field to particles under conditions far from thermal equilibrium. This is well indicated by space anisotropy and by the energy spectrum for neutrons[17] and hard x-rays[16,18]. (The photon-distribution $N(\varepsilon)$ per unit energy (ε) range is $N(\varepsilon) \propto \varepsilon^{-n}$, $n = 4 \pm 1$ for $150 \leq \varepsilon \leq 300$ keV, by photo-electric or Compton electron tracks on Ilford emulsions[18].) The discrepancies among experimental data and the proposed models which neglect the filamentary structure of the CS are recognized and well illustrated in the literature.[17]

Magnetic Field Structure and Decay

A self consistent theory for $\underline{E}$, $\underline{B} = \nabla \times \underline{A}$ fields and for phase-space density of ions, f_+, and of electrons, f_-, has been developed for the region of the ionizing shock (driven by CS) with some generalization of previous results[7,12] (all filaments are considered parallel and variations along the filament axis are still neglected, i.e. $\partial/\partial z=0$, x-axis orthogonal to CS, y, z on CS; stationary conditions are considered in the CS frame of reference, i.e. $\partial/\partial t=0$ in the "fast" time scale τ_o). Since $\underline{E} + \underline{u}_- \times \underline{B} = \nabla \psi(x,y)$ ($\underline{u}_-$ is the electron mean velocity, ψ is a scalar function), the electrons are tied to the magnetic field lines; the ion orbits are nonadiabatic on CS (ion Larmor radius $\sim$ filament radius). The expressions for three dimensional flow and fields have been derived in terms of arbitrary complex functions $g(\eta)$, where $\eta\bar{\eta} = x^2+y^2$, which can be used to match the periodicity of the filamentary structure on CS, as well as all physical conditions at the boundaries, in particular $B_\theta \equiv B_y \to$ const$=B_o$ for $x \to +\infty$ and $B_y \to B_1$ for $x \to -\infty$ i.e. ahead of CS; $B_x(x \to \pm\infty)=0$. By considering charge neutrality or the case of a net charge density proportional to the mass density, we have[12]

$$A_z=(1/\alpha_+ + 1/\alpha_-)[e(c_- - c_+)]^{-1}\ln[|\partial g/\partial\eta|^2 (1+|g|^2)^{-2}]$$

where the constants $\alpha_+, \alpha_- > 0$ and c_+, c_- are characteristic of a com-

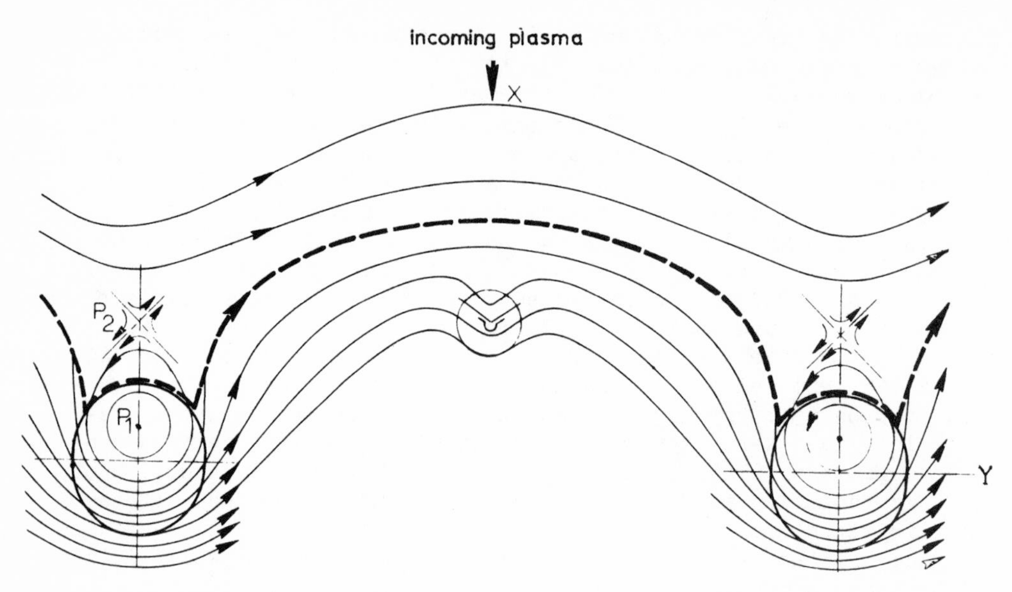

Fig.9 (B_x, B_y)-field lines for a,b,c satisfying a convenient set of conditions [e.g. the maximum possible distance between $P_2(x_2,o)$ and $P_1(x_1,o)$ is given by $|\exp(mx_2) - \exp(mx_1)| \leq \sigma = [|a_1|+|a_2|+..|a_{16}|]/|a_o|$ where the a_i's are the coefficients of the algebraic equation $\Sigma\lambda^i a = o (i=o,1,..16)$, $\lambda = \exp(mx_2)$, obtained by putting $B_x = B_y(x_2^i, o) = o$]. The dashed line represents the optical profile of CS as it is shown by IC photographs (e.g. Fig.1). P_1 may or may not coincide with the current-filament axis on Y (depending on the choice of the free parameters in the source terms of $f_{\pm}$ equations[12]). The experimental evidence (magnetic probe data and IC photo, simultaneously taken) indicates simply that optical profile and current regions (regions of steepest B-variations) are coincident. No attempt to preserve proportionality between line density and field strength was made in the drawing. Further complications - essentially other singular points - may appear and disappear between two large filaments for different values of a,b,c on the "slow" time scale τ_f of CS particle flow.

ponent of $f_{\pm} = f_{v\pm} + f_{s\pm}$ (i.e. $f_{v\pm}$ which describe particles not affected by collisions; $1/\bar{k}\alpha_{\pm}$ give the temperatures and c_+, c_- the mean velocities along the filament axis of ions and electrons of these components, k=Boltzmann constant, -e=electron charge; the signs of $c_{\pm}$ are related to electrode polarities)[12]. The choice $g = 1 + a^2 + be^{mn}$ ($\bar{a}$,b,m=real constants) has been discussed elsewhere and is convenient for the case $B_o > 0$, $B_1 < 0$ (e.g. for a CA with a biasing field)[12]. The case $B_o>0$, $B_1=0$ for our experiments requires at least one more term in g, i.e. $g=a + be^{m\eta}+ce^{n\eta}$. By taking m>n and m>o (the roles of m,n can be interchanged) then $\lim B_y(=-\partial A_z/\partial x)$, $x\to-\infty = \ell\,(n+3m)$ and $\lim B_y, x\to+\infty = \ell\,(m+3n)\equiv B_o$, where $\ell=(\alpha_+ + \alpha_-)/\alpha_+\alpha_- e(c_+ - c_-)$ which give $n = -3m$ and $B_o = -\ell 8m$. The period T of the magnetic structure (essentially the distance between two filament axes) is then $T= 2\pi/m=\pi\, 16(\alpha_+ + \alpha_-)/\alpha_+\alpha_-(c_- - c_+)eB_o$. The determination of the flow velocities in the x,y plane and consequently the calculation of the self-consistent component B_z can be treated to some extent as independent from the other components (this is possible because of the free parameters which control the source terms in the $f_\pm$ equations[12]). The B-field lines in the x,y plane are plotted in Fig. 9. P_1 is not a neutral point because of the strong B_z component inside a filament. P_2 is a neutral point (X-type) which can be associated with the filament annihilation process in the following way. The instability at two neighboring neutral points P_2 in the periodic structure can bring together two filament segments within a time interval $\lesssim$ 5 nsec (i.e. on the τ_o time scale). The instability in P_2 can be studied by a method quite similar to that followed by J.W. Dungey[19] for this kind of neutral points. Under conditions of stationary regime for the CS motion the incoming flow of a large number of neutral particles sweeps the charged particles from the P_2-point region in the direction of the filament. By removing charged particles the local value of the current remains low enough to prevent the destruction of the X-type neutral point (P_2 can be considered a stagnation point only for charged particle flow in the x,y plane). The slowing down of the CS (at a maximum of axial compression in the plasma column) or CS disruptions due to plasma ejected from the axial region (at a late stage of the discharge) change the incoming-flow regime and consequently are a cause of the instability at the P_2 points which can trigger the filament annihilation [the B_z component may have opposite directions in two neighboring filaments, depending on ion-vorticity orientation[7]].

Work supported in part by U.S.A.F. Office of Scientific Research.

References

1. Alfven, H.: Cosmical Electrodynamics, Ch. 3, Oxford, 1950.
2. Smith, H. J., Smith, E.v.P.: Solar Flares, McMillan Co. N.Y. 1963.
3. Sweet, P.A.: 1958 N. Cim. Sup. 8-S.X, 188. Parker, E.N.: Ap. J. Suppl. 8, No. 77, 177, 1963.
4. Howard, R., Harvey, J.W.: Ap. J. 138, 1328, 1968.
5. Severny, A.B. : Proc. AAS-NASA Symp. on Solar Flares, Oct. 28, 1963, NASA SP-50, p. 95.
6. Fichtel, C.E., McDonald, F.B.: Ann. Rev. Astron. Astroph. 5, 531, 1967.
7. W. H. Bostick, V. Nardi, L. Grunberger, W. Prior; Proc. IAU Symp. N. 43, Paris, 1970, Howard edit., Reidel Pub. Co., p. 443.
8. Bostick, W. H., Prior, W., Grunberger, L., Emmert, G.: Phys. Fluids 9, 2078, 1966.

9. W. H. Bostick, V. Nardi, L. Grunberger, W. Prior; Proc. X-th Int. Confer. Phenomena in Ionized Gases, Oxford 1971, Parsons & Co. Publ., Oxford, p. 237.
10. W. H. Bostick, V. Nardi, W. Prior; Proc. IUTAM Symposium on the Dynamics of Ionized Gases, Tokyo, 1971.
11. W. H. Bostick, L. Grunberger, W. Prior, V. Nardi; Proc. 4th European Conf. Controlled Fusion and Plasma Physics, C.N.E.N. edit., Roma 1970, p. 108.
12. V. Nardi; Phys. Rev. L., 25, 718, 1970 .
13. W. H. Bostick, L. Grunberger, V. Nardi, W. Prior; Proc. IX-th Int. Confer. Phenomena in Ionized Gases, Bucharest 1969, Musa et al. Edits. Acad. Rep. Soc. Romania, Bucharest, 1969, p. 66.
14. Vorticity and Neutrons in the Plasma Focus, W. H. Bostick, L. Grunberger, V. Nardi, W. Prior; Proc. 5th Int. Symposium on Thermophysical Properties, Amer. Soc. Mechanical Engineers edit., New York 1970, p. 495.
15. Bernstein M. J., Meskan D. A. and van Paassen H. L. L.: Phys. Fluids 12, 2193, 1969.
16. Lee, J. H., Conrads, H., Williams, M. D., Shomo, L. P., Hermansdorfer, H. , and Kim, K.: Bull. A.P.S. 13, 1543, 1968.
17. Patou C., Simmonet A., and Watteau J.P.: Phys. Letters 29A, 1, 1969.
18. Lee J. H., Loebbaka D.S. , Roos C.E. : Plasma Physics 13, 347, 1971. See also Meskan D.A., Van Passen H.L.L. and Comisar G.F.: Aerospace Rep. Tech. Rep.-0158 (3220-50)-1, 1967.
19. Dungey J. W.: Cosmic Electrodynamics, Ch. 6, Cambridge Un. Pr., 1958.

GYROMAGNETIC RADIATION FROM BUNCHED ELECTRONS

André Mangeney

Observatoire de Paris, Meudon

STATISTICAL BUNCHING OF ELECTRONS

The theory of coherent emission by bunches of electrons spiralling in a magnetic field has attracted the attention of many authors since the discovery of pulsars (1, 2, 3). Some doubt has been expressed about the possibility that such a mechanism could work under cosmic conditions (3). In the present paper it is proposed, on the contrary, that such a radiation mechanism could be invoked for the interpretation of a particular kind of solar radio burst, the so-called "stationary" type IV burst (4).

A convenient way of describing the bunching of electrons is to introduce the quantity

$$N(\underset{\sim}{R},\theta,p_{\parallel},p_{\perp}) = \sum_{j}\delta(\underset{\sim}{R}-\underset{\sim}{R}_j(t))\delta(\theta-\theta_j(t))\delta(p_{\parallel}-p_{\parallel j}(t)) \times \delta(p_{\perp}-p_{\perp j}(t))$$

which is the phase-space density of electrons having $\underset{\sim}{R}$ as center of gyration, $2\pi\theta$ as phase, $p_{\parallel}$ and $p_{\perp}$ as parallel and perpendicular momentum. N is a random function, whose average value for a homogeneous and gyrotropic plasma is

$$\langle N\rangle = df_o(p_{\parallel},p_{\perp})$$

where

d	is the number density of the electrons,
$f_o(p_{\parallel},p_{\perp})$	is the distribution function of the electrons.

The autocorrelation function of the random part of N, say δN, can be written:

$$
\begin{aligned}
&<\delta N(\underset{\sim}{R}_1,\theta_1,p_{\parallel 1},p_{\perp 1})\delta N(\underset{\sim}{R}_2,\theta_2,p_{\parallel 2},p_{\perp 2})> \\
&= d\delta(\underset{\sim}{R}_1-\underset{\sim}{R}_2)\delta(\theta_1-\theta_2)\delta(p_{\parallel 1}-p_{\parallel 2})\delta(p_{\perp 1}-p_{\perp 2})f_o(p_{\parallel 1},p_{\perp 1}) \\
&\quad + d^2G(\underset{\sim}{R}_1-\underset{\sim}{R}_2,\theta_1,\theta_2,p_{\parallel 1},p_{\perp 1},p_{\perp 2},p_{\parallel 2})
\end{aligned}
$$

This equation defines the correlation function of two electrons, $G(\underset{\sim}{R}_1-\underset{\sim}{R}_2,\theta_1,\theta_2,\underset{\sim}{p}_1,\underset{\sim}{p}_2)$. There will be a statistical bunching when for sufficiently small $|R_1-R_2|$, G is essentially a function of $\theta_1-\theta_2$. This means that, if $G_{s_1,s_2}(K)$ denotes the Fourier transform of G with respect to $\underset{\sim}{R}_1-\underset{\sim}{R}_2$, θ_1 and θ_2, there is a domain in $\underset{\sim}{K}$ where $G_{s_1,s_2}(\underset{\sim}{K})\sim 0$ when $s_1+s_2\neq 0$. As may be expected, the equilibrium correlation function does not have this property.

Suppose now that at time t=0, the Fourier transform of the correlation function has a given value $G_{s_1,s_2}(\underset{\sim}{K}, t=0)$. Then if one takes into account only the free motion of the particles in the static magnetic field $\underset{\sim}{B}_o$, the value of G_{s_1,s_2} at time t is

$$G_{s_1,s_2}(\underset{\sim}{K},t)=\exp\left[-i(k_{\parallel}(v_{1\parallel}-v_{2\parallel})+s_1\Omega_1+s_2\Omega_2)t\right]G_{s_1,s_2}(\underset{\sim}{K},t=0).$$

Clearly, the time scale of evolution of G_{s_1,s_2} is

$$T = \left[k_{\parallel}\Delta v_{\parallel} + \sigma\Omega_e + s_1\Delta\Omega_e\right]^{-1}$$

where

$\Delta v_{\parallel}$ is a measure of the spread in parallel velocities,
$\sigma = s_1 + s_2$,
$\Delta\Omega_e$ is a measure of the spread in gyrofrequencies,
$\Delta\Omega_e = \Omega_e(\Delta E/E)$; E = energy of the particles.

If the particles are injected in the form of a quasi-monochromatic beam (i.e. $\Delta E/E<<1$, $k_{\parallel}\Delta v_{\parallel}<<\Omega_e$), T can be quite large when $s_1+s_2=0$ ($T>>\Omega_e^{-1}$). This is the case considered by Caroff and Scargle (1), who assume that such a beam is periodically injected perpendicularly to the magnetic field.

Another case will be considered here, where G is quasi-stationary, as a result of a balance between a phase diffusion mechanism and an instability which tends to bunch the electrons. Let us assume that the distribution function of the electrons is sufficiently anisotropic to drive the electron cyclotron waves unstable (these waves are longitudinal waves propagating almost perpendicularly to the magnetic field, see for example (5)). It is easy to show that such waves induce a bunching of the particles. If one takes into account, in a phenomenological way, a mechanism of diffusion of the phase of the particles (for example due to the reflection at mirror points), Δ being the corresponding diffusion coefficient, a simple calculation of the quasilinear type leads to:

$$G_{s,-s}(\underset{\sim}{K},p_1,p_2) \sim \frac{8\pi e^2}{k^2} \frac{s^2\Omega_e^2}{(s\Omega_e-\omega_s(K))^2+4s^4\Delta^2} \frac{W_s(K)}{\Theta_\perp^2}$$

$$\times J_s\left(\frac{k_\perp v_{\perp 1}}{\Omega_e}\right) J_s\left(\frac{k_\perp v_{\perp 2}}{\Omega_e}\right) f_o(p_1) f_o(p_2)$$

where

Ω_e = gyrofrequency of the electrons,

$\Theta_\perp$ = perpendicular temperature of the electrons,

(the distribution function has been assumed to be of the bi-Maxwellian type),

$W_s(K)$ is the energy density of the unstable longitudinal mode oscillating with a frequency $\omega_s(K) \sim s\Omega_e$.

EMISSIVITY OF BUNCHED ELECTRONS

To calculate the emissivity a convenient starting point is the work by Heyvaerts (6) on the statistical mechanics of charged particles interacting with electromagnetic waves in the presence of a static magnetic field. A straightforward application of his techniques leads to the following expression for the emissivity:

$$a_\lambda = \frac{e^2 d\nu_\lambda^2}{2\pi c^3} \sum_s \iiiint dp_{\parallel 1} dp_{\perp 1} dp_{\parallel 2} dp_{\perp 2}\, \delta(k_{\lambda\parallel} v_{\parallel 1} + s\Omega_e - \nu_\lambda)$$

$$\times \left[\underset{\sim}{e}_\lambda \cdot \underset{\sim}{\Gamma}_s^+(K_\lambda, 1) \right] \left[\underset{\sim}{e}_\lambda \cdot \underset{\sim}{\Gamma}_s^-(K_\lambda, 2) \right] \Big\{ f_o(p_1)\, \delta(p_{\parallel 1} - p_{\parallel 2})\, \delta(p_{\perp 1} - p_{\perp 2})$$

$$+ 8\pi^3 G_{s,-s}(\underset{\sim}{K}, p_1, p_2) \Big\}$$

where

ν_λ is the frequency, $\underset{\sim}{e}_\lambda$ the polarisation vector and $\underset{\sim}{K}_\lambda$ the wave vector of the mode considered.

$\underset{\sim}{\Gamma}_s^\pm$ is a vector whose components in a system whose z-axis is directed along the magnetic field are

$$\Gamma_{s,x}^\pm = \frac{\Omega_e}{k_{\lambda\perp}} sJ_s \cos 2\pi\xi_\lambda + i\varepsilon v_{j\perp} J_s' \sin 2\pi\xi_\lambda$$

$$\Gamma_{s,y}^\pm = \frac{\Omega_e}{k_{\lambda\perp}} sJ_s \sin 2\pi\xi_\lambda - i\varepsilon v_{j\perp} J_s' \cos 2\pi\xi_\lambda$$

$$\Gamma_{s,z}^\pm = v_{j\parallel} J_s$$

(J_s is the Bessel function of order s, its argument is $k_\perp v_{j\perp}/\Omega_e$, and $2\pi\xi_\lambda$ is the polar angle of $\underset{\sim}{K}_{\lambda\perp}$).

Let us note that the above expression for the emissivity neglects the effect of the plasma on the emitted waves; it is valid only for $\nu_\lambda \gg \omega_{pe}$. The inclusion of the effect of the plasma does not lead to difficulties.

One may recognize the ordinary expression for the emissivity in the part containing only f_o, whereas the part containing $G_{s,-s}$ represents the contribution of the bunches of electrons. This last part vanishes identically for an equilibrium correlation function.

APPLICATION TO TYPE IV RADIO BURSTS

We shall now apply these results to the interpretation of the "stationary" type IV m burst, whose description may be found in (4). The wavelengths lie in the meter-decameter range; at one altitude the frequency band emitted by the source is relatively narrow; the sources are often double, each component being almost fully circularly polarised.

There are good reasons to think that the source of these bursts is located in magnetic loops which are the extension in the corona of the loop prominences system described for example by Bruzek (7). One may assume that the electrons in these loops have an anisotropic distribution function due either to their injection in the loop or to the existence of a loss cone. This leads to the excitation of unstable electron cyclotron waves.

Assuming now that

a) the magnetic field in the source is so strong that Ω_e is close to ω_{pe}; for example taking

$$n_e \sim 5\times10^8 cm^{-3} \text{ (i.e. } \omega_{pe} \sim 10^9 sec^{-1})$$

which corresponds to an altitude of some 10^5km in the corona, we choose somewhat arbitrarily the magnetic field to be of the order of 30G, giving a gyrofrequency $\Omega_e \sim 6\times10^8 sec^{-1}$;

b) the linear dimensions of the source are of the order of 10,000 km;

c) the energy density of the electron cyclotron waves is one part in a thousand of the energy density of the electrons ($T_e \sim 5\times10^6$ °K);

d) the source is optically thin,

one obtains an emitted energy flux of about 10^{-16}ergs/sec/cm^2/Hz which is of the order of what is typically observed. This model can explain other observed properties of these bursts such as the limited bandwidth which is due to the fact that only a limited number of harmonics are unstable when $\Omega_e \sim \omega_{pe}$; a detailed account of this model will be given in another paper. Let us note, to conclude, that this model does not require any maser mechanism to explain the observed brightness temperature as is the case in most of the theoretical work devoted to this kind of bursts (see for example (8) where one finds a detailed bibliography on the subject).

REFERENCES

1. Caroff L.J., Scargle J.D., 1970, Nature 225, 168.
2. Eastlund B.J., in "The Crab Nebula", Symp. I.A.U. No. 46, ed. by Davies R.D. and Smith F.G., 1971, p.443.
3. Ginzburg V.L., preprint No. 93, Lebedev Physical Institute, 1970.
4. Kundu M.R., 1965, Solar Radio Astronomy, John Wiley and Sons, Inc., New York.
5. Crawford F.W., Tataronis J.A., 1970, J. Plasma Physics 4, 231.
6. Heyvaerts J., 1969, Ann. Astrophys. 30, 925.
7. Bruzek A., 1964, Astrophys. J. 140, 817.
8. Fung P.C.W., 1969, Can. J. Phys. 47, 179.

OBSERVATIONS OF CORONAL MAGNETIC FIELD STRENGTHS AND FLUX TUBES AND THEIR STABILITY

Hans Rosenberg

Astronomical Institute, Zonnenburg , Utrecht

At the Utrecht Observatory a 60-channel solar radio-spectrograph is operating. In the relevant period the frequency bandwidth was 160 - 320 MHz. The instrument is particularly well suited for the study of short-time scale fluctuations in weak or strong continua, due to a differentiating technique (De Groot and Van Nieuwkoop, 1968; Van Nieuwkoop, 1971). In addition to some unclassified phenomena, all well known types of solar radio bursts have been observed, and in particular a wealth of detailed structure in these bursts. We concern ourselves at present with the details in type IV continuum emissions. Type IV storms are generally believed to be due to synchrotron radiation of mildly relativistic electrons. Two details are prominent:

a) <u>Harmonic patterns</u>:
   Parallel ridges in frequency, that drift in time but stay parallel. Their frequency separation ranges from $\sim$ 15 MHz in a particular case to a few MHz observed in some more recent events. Sometimes several of these patterns are present forming a more or less Moire-like fringe pattern.
   (fig. 1 bottom)
b) <u>Pulsating structure</u>:
   A wide-band fluctuation in time: in general quasi-periodic with characteristic time scales in the order of 0.1 - 0.3 s. In a few cases periodic, with periods around 1^s.
   (fig. 1, top)

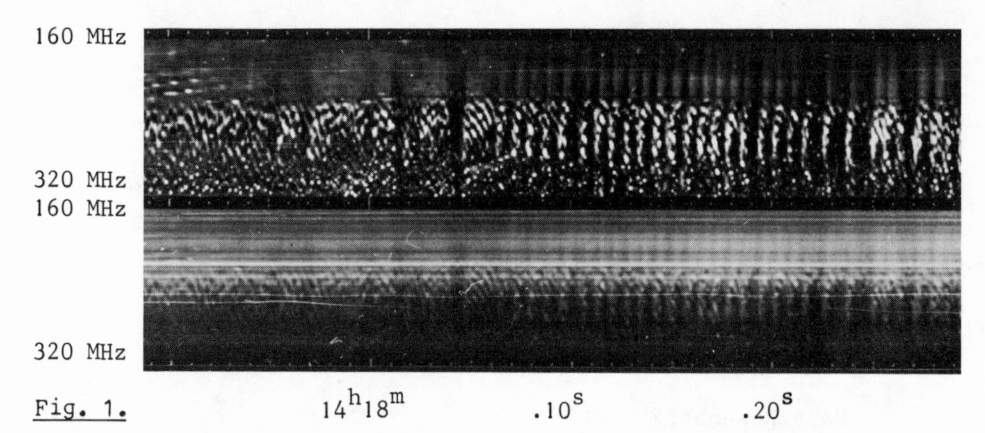

Fig. 1.

Radiospectrographic record for March 2nd 1970. The pulsating structure can be seen in the upper portion around $14^h18^m10^s$ U.T. and some evident examples of the harmonic structure at $14^h18^m05^s$, $14^h18^m25^s$ U.T.

Proposed explanation:

Harmonic patterns.

Since the ridges occur at frequencies with equal frequency intervals (~ 15 Mz), this suggests that this interval is some characteristic frequency of the plasma.

It is very unlikely that the local plasmafrequency plays this role, since in that case the source would have to be very far out in the solar corona (i.e. several to many solar radii), which in general is not observed for type IV storms. Secondly, no emission or absorption mechanism is known which operates at such high harmonics of the plasma frequency ($n \sim 10 - 20$, in some cases even 40) and only there.

The local gyrofrequency is a much more likely candidate. However transverse emission or absorption at such high harmonics with small bandwidths remains a problem, even though an enhanced absorption can be expected for those harmonics which are close to the plasmafrequency: $nf_{gyro} \sim f_{plasma}$ and $n \lesssim 10$ (Zheleznyakov and Zlotnik, 1971). One argument is that we did observe ridges where $n \gg 10$ under the hypothesis of harmonic transverse emission or absorption. A second argument follows also from the observations: if f is the frequency of the ridge then $f = n \times f_{gyro}(t)$; due to magnetic field variations f_{gyro} will change, thus causing the pattern to drift. In that case $\frac{df}{dt} = n \times \frac{df_{gyro}}{dt}$, hence the slopes of the ridges should yield the same sets of n, as the frequency of the ridges themselves. Even though the determination of the slopes is rather inaccurate, we have some evidence that the thus determined sets of n are distinctly smaller than those determined from the ridge frequency.

We therefore propose a third mechanism based on observations in laboratory plasmas (Landauer, 1962; Bekefi et al., 1962) of longitudinal ($k \parallel E$; $k \perp B_o$) electrostatic cyclotron waves at many successive harmonics of the gyrofrequency. The transverse waves can then result from a non linear coupling between the upper hybrid wave (also slow, electrostatic, and perpendicular to B_o) and these cyclotron waves, resulting in emission at approximately $f_{plasma} + nf_{gyro}$; in a manner similar to the emission at $2f_{plasma}$ due to the coupling between two plasma waves. There is both theoretical and experimental evidence (Canobbio and Croci, 1963; Dreicer, 1964) that the upper hybrid frequency does play a role in the conversion of these electrostatic waves into transverse ones. At present we are studying this coupling mechanism in more detail.

From the observations and the hypothesis $f = f_{upper\ hybrid} + nf_{gyro}$, we can deduce in a similar manner as indicated above that the $f_{upper\ hybrid}$ equals approximately 160 MHz, and n ranges from 2 to 10, and from the observed period in frequency one finds in the case of March 2nd 1970 a magnetic field strength of about 5G, and for several other events field strengths in the order of 1G.

Pulsating structure.

In a previous paper (Rosenberg, 1970) we tried to explain this on the basis of radial fluctuations in the magnetic field strength of a flux tube propagating perpendicular to the magnetic field direction. Due to these field strength modulations, the synchrotron emission is modulated. One then can derive a relation between the radius of the flux tube r, the Alfvén speed v_A and the period p :

$\frac{2r}{pv_A} \sim 1.8$. From the record shown in fig. 1 the pulsating structure seems to be correlated with the fluctuations in time in the harmonic patterns, thus indicating a common origin. Using the above derived values for the plasma frequency and the magnetic field strength and a characteristic time of the pulsating structure of $0^s.8$, we find for the radius of the flux tube approximately 200 - 300 km. For further details we refer to (Rosenberg, 1971).

References

Bekefi, G. et al., Phys. Rev. Letters 9, 6 (1962)
Canobbio, E., and R. Croci, Phys. Fluids 9, 549 (1966)
De Groot, T., and J. van Nieuwkoop, Solar Phys. 4, 332 (1968)
Dreicer, H., Bull. Am. Phys. Soc. 9, 512 (1964)
Landauer, G., Plasma Phys. (J. Nucl. Energy C) 4, 395 (1962)
Rosenberg, H., Astron. and Astrophys. 9, 159 (1970)
Rosenberg, H., Solar Phys. to be published 1971.
Van Nieuwkoop, J., thesis, Utrecht (1971)
Zheleznyakov, V.V., and E.Ya. Zlotnik, Solar Phys. to be published 1971.

THE DYNAMICAL BEHAVIOR OF THE INTERSTELLAR GAS, FIELD, AND COSMIC RAYS

E. N. Parker

Department of Physics

University of Chicago

Cosmical gas dynamics is a remarkable display of plasma instabilities (see for instance, the review edited by Wentzel and Tidman, 1969; and Lerche, 1969; Kadomtsev and Tsytovich, 1970). The present review is limited to a brief summary of the large scale instability associated with the galactic magnetic field and cosmic rays, and to a few remarks on the origin of the magnetic field. The field and cosmic rays are described in detail elsewhere (Parker, 1968, 1969a, 1970b). For our purposes here it is sufficient to note that the gaseous disk of the galaxy is composed of a highly conducting gas in the form of a turbulent disk with a thickness of some 300 pc and a radius in excess of 10^4 pc. The disk rotates with the brighter stars (which were formed recently from the gas in the disk) with a period of 2.5×10^8 years at the position of the Sun. The rotation is more rapid toward the center of the galaxy and less rapid farther out. The nonuniform rotation continually stretches the magnetic fields embedded in the gas into the azimuthal direction. The gas is turbulent and quite inhomogeneous. Densities range typically from 0.1 to 10 hydrogen atoms/cm^3 over scales of 10-200 pc (Heiles, 1967; Weaver, 1970) and temperatures from 10^4 °K to 10^2 °K, respectively. The mean density is estimated to be 2/cm^3 (Schmidt, 1963). The rms turbulent velocity u in any one direction is of the order of 6 km/sec and varies over the same 10-200 pc scales as the density. The gas is heated by UV, X-rays and cosmic rays, as well as turbulent dissipation. The heating and cooling mechanisms are of such a nature that the temperature and pressure are unstable (Savedoff and Spitzer, 1950; Parker, 1953a, b; Field, 1970). Increasing the density leads to cooling, sometimes to such a degree that the pressure declines with increasing density.

The galactic magnetic field B lies mainly in the azimuthal direction with a mean strength believed to be of the order of $3\text{-}5 \times 10^{-6}$ gauss. But local fluctuations in the field are large, $\Delta B \cong B$ (Berge and Seielstad, 1967; Appenzeller, 1968; Jokipii and Lerche, 1969) and the field reverses sign in 10^3 pc in several directions from the Sun. The general configuration of the field, apart from its tendency to be stretched along the azimuthal direction, is not known. Turbulent diffusion leads to a free decay time of the field of some 10^8 years. Calculations (Parker, 1971a) show that the field is evidently maintained by the combined shear of the nonuniform rotation and the cyclonic turbulence of the gas.

For our purposes the cosmic rays may be considered simply as a relativistic gas (see review, Parker, 1968). The energy density U is approximately 1.5×10^{-12} ergs/cm^3, with about half the energy in particles above 10^{10} ev and half below.* The pressure P of the cosmic rays is very nearly isotropic, with $P_{\parallel} \cong P_{\perp}$, presumably because of some vigorous relativistic instabilities (Lerche, 1969). Because the gas is relativistic it follows that $P \cong \frac{1}{3} U$. A typical cosmic ray particle of 10^{10} ev has a cyclotron radius of the order of 10^{13} cm in the galactic field. So the particles are constrained very closely to motion along the lines of force. Streaming along the lines of force is limited only by the excitation of hydromagnetic waves when the streaming velocity sufficiently exceeds the Alfven speed in the ionized component of the gas, about 10^2 km/sec (Wentzel, 1968, 1969; Kulsrud and Pearce, 1969).

It is known from the degree of fragmentation of the heavy nuclei among the cosmic rays that the individual particles have spent some 10^6 - 10^7 years in the disk of the galaxy, where

*The major uncertainty in U is in the low energy cosmic ray particles, below about 2×10^8 ev, which are excluded from the solar system by the solar wind, and hence, are not observed directly.

they collide with the nuclei of the interstellar gas. One problem is to understand how the cosmic rays can escape from the galactic field in so short a time. Bohm diffusion is entirely inadequate.

Now the dynamical balance of forces within the galactic disk is between the expansive forces of the cosmic ray pressure P, the magnetic pressure $B^2/8\pi$, and the gas pressure, p (including the turbulent pressure ρu^2) against the inward gravitational force ρg of the gas. The galactic disk is a giant air mattress pumped up by $p + B^2/8\pi + P$ and compressed by ρg.

The gravitational acceleration g is perpendicular to the plane of the disk, inward from both sides, and vanishing on the central plane z = 0. Equilibrium requires that

$$(d/dz)(p + B^2/8\pi + P) = -\rho g$$

Thus if p, $B^2/8\pi$, and P all vary with z in about the same proportion, the scale height Λ of the gas in the disk is

$$\Lambda \cong (p + B^2/8\pi + P)/\rho\langle g\rangle$$

where $\langle g\rangle \cong 1.6 \times 10^{-9}$ cm/sec is the mean gravitational acceleration over one scale height ($0, \Lambda$). With $= 3 \times 10^{-24}$ gm/cm, we calculate an equilibrium scale height of 100 pc, in general agreement with the values 100-200 pc inferred from direct observation of the gas.

The equilibrium is, however, unstable, basically for three reasons. First of all the interstellar gas is thermally unstable, for the reasons already mentioned. The gas tends to be either cold and dense, or hot and tenuous. Second, any dense portion of the gas tends to sink, relative to the surrounding tenuous gas, carrying the lines of force with it. Hence the neighboring gas finds itself resting on sloping lines of force and consequently slides downward into the region of depressed field, thereby adding to the weight and causing further sagging.

It is the same phenomenon as occurs when two heavy stones are placed on a hammock. They tend to roll together. The field between the regions of dense gas is unloaded and expands upward. The third effect, then, is the inflation of the regions of expanded, weakened field by the freely streaming cosmic rays.

The characteristic scale of the instability is $10\text{-}10^3$ pc, the characteristic time is 1-3 x 10^7 years. Presumably the enormous inhomogeneity of the interstellar gas is the result of this general instability of the gas, field, and cosmic ray system.

We suggest that cosmic rays escape from the galaxy by inflating the raised expanded portions of the field, literally pushing their way out of the galaxy by blowing bubbles in the field. I know of no way to compute how far out the bubbles extend while still remaining attached nor to what extent the bubbles form a "halo" around the galaxy.

It is worth noting that the cosmic ray input can be estimated from the cosmic ray energy density U and estimated dwell time $\tau = 10^6 - 10^7$ years in the disk. The input is 0.5 - 5 x 10^{-26} ergs/cm sec. Altogether it appears that cosmic rays are a major source of turbulence in the interstellar gas, not only because of their impulsive injection from supernovae and pulsars, etc., but because of their continual inflation of the expanded portions of the field, at 10-100 km/sec.

The reader interested in the formal theory of the magnetic field and cosmic ray instability of the galactic disk is referred to the review of Parker (1969a and references therein) which discusses the general problem with emphasis on the dynamical effects of the cosmic rays and field, and to the review by Field (1970 and references therein) which gives a general discussion but with more attention to the thermal instability.

A recent analysis by Appenzeller (1971) of polarization data in the α - Persei cluster shows the magnetic field configuration in the local dense gas cloud. The lines of force sag toward the central plane of the galaxy, presumably under the weight of the gas cloud, in the manner one would expect from theory (see Figure 1 for a reproduction of Appenzeller's results).

While we are on the subject of turbulence and the galactic magnetic field, it should be noted that the turbulent diffusion coefficient for eddies of scale l = 10 pc, and velocities u = 6 km/sec in any one direction, is $\eta = 0.3ul = 10^{26}$ cm^2/sec. Hence the turbulent diffusion time of the galactic field over the scale Λ of the disk is of order of $\Lambda^2/\eta = 10^8$ years. Maintenance of the galactic field, therefore, must involve generation at a

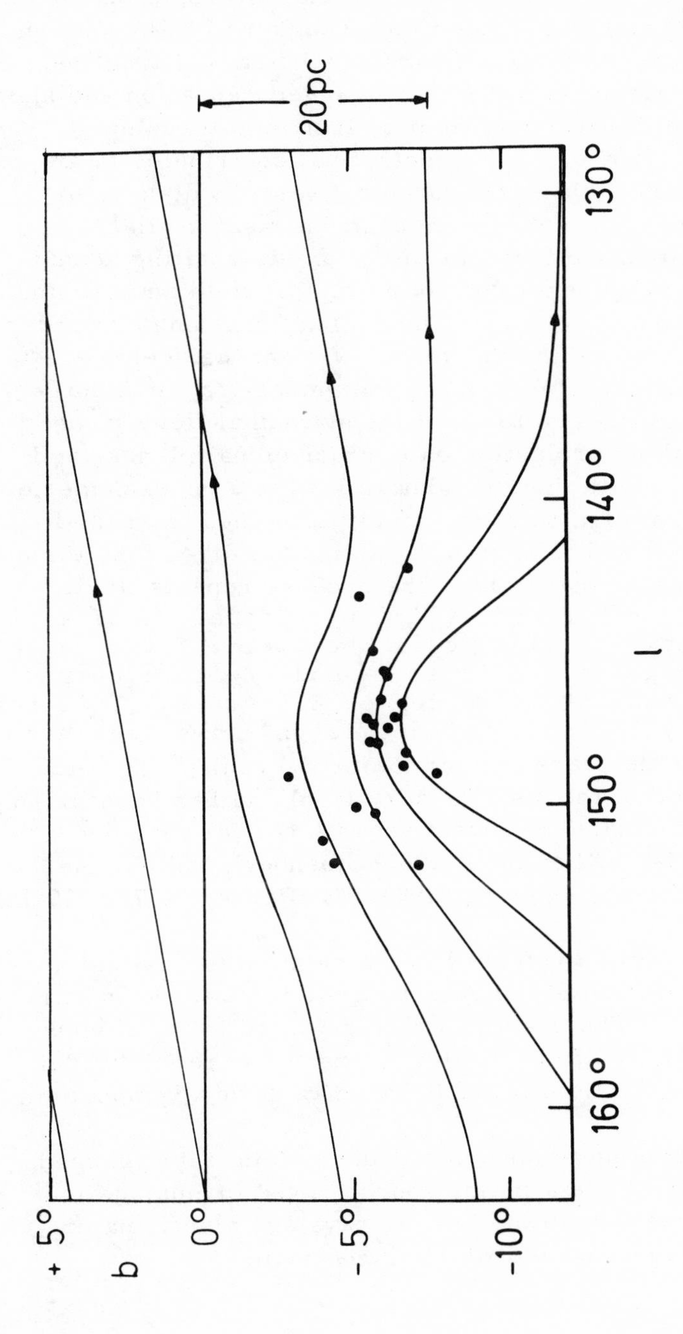

Fig. 1 A plot of the magnetic lines of force in the α – Persei cluster, worked out and published by Appenzeller (1971) from polarization observations. The cluster lies below the central plane of the galaxy, so that gravity at the location of the cluster is directed upward in the figure.

comparable rate. We have pointed out that isotropic turbulence appears to produce large-scale fields (Parker, 1969b). Recently Lerche has carried through a formal mathematical treatment of the problem, starting with the hydromagnetic equation and showing the generation of large-scale field by isotropic turbulence (Lerche, 1971). Hence, such effects must contribute to the generation of the galactic field in the turbulent gaseous disk. But there appears to be a more efficient mechanism available in the galaxy based on the fact that the galaxy is rotating nonuniformly, giving rapid generation of azimuthal field and causing the turbulence to be cyclonic. The cyclonic motions are ordered (by the Coriolis force) and are more efficient than completely random isotropic turbulence. The individual cyclonic eddies twist loops in the lines of force of the azimuthal field. The loops have a nonvanishing projection on the meridional planes, and upon coalescence with their neighbors lead to a general meridional field. The nonuniform rotation rapidly stretches the radial component of the meridional field into the azimuthal direction, forming, then, a strong azimuthal field. The process repeats itself, giving rise to rapid and efficient generation of magnetic field (Parker, 1971a, c). The field of the galaxy appears to be generated in a low mode, with a characteristic growth time of the order of 10^8 years. This combination of cyclonic turbulence is the same mechanism as was proposed some years ago for the origin of the magnetic field of Earth and of the sun (Parker, 1955) and has been much investigated and studied in recent years (see, for instance, Braginskii, 1964a, b; Leighton, 1969; Steenbeck and Krause, 1969; Gilman, 1969; Deinzer and Stix, 1971; Parker, 1970c, 1971b).

The reader is referred to the detailed formal calculations of the generation of the galactic field to be found in Parker (1970c, 1971a, b). The reader is also referred to the reviews by Parker (1970a) and Roberts (1971), and to the paper by Bel, Fitremann, Frisch, Leorat and Schatzman in this Symposium.

It is amusing to note that if the cosmic rays trapped in the galactic field are a major source of the turbulence in the interstellar gas, then the cosmic rays are also a major source of the magnetic field which traps them.

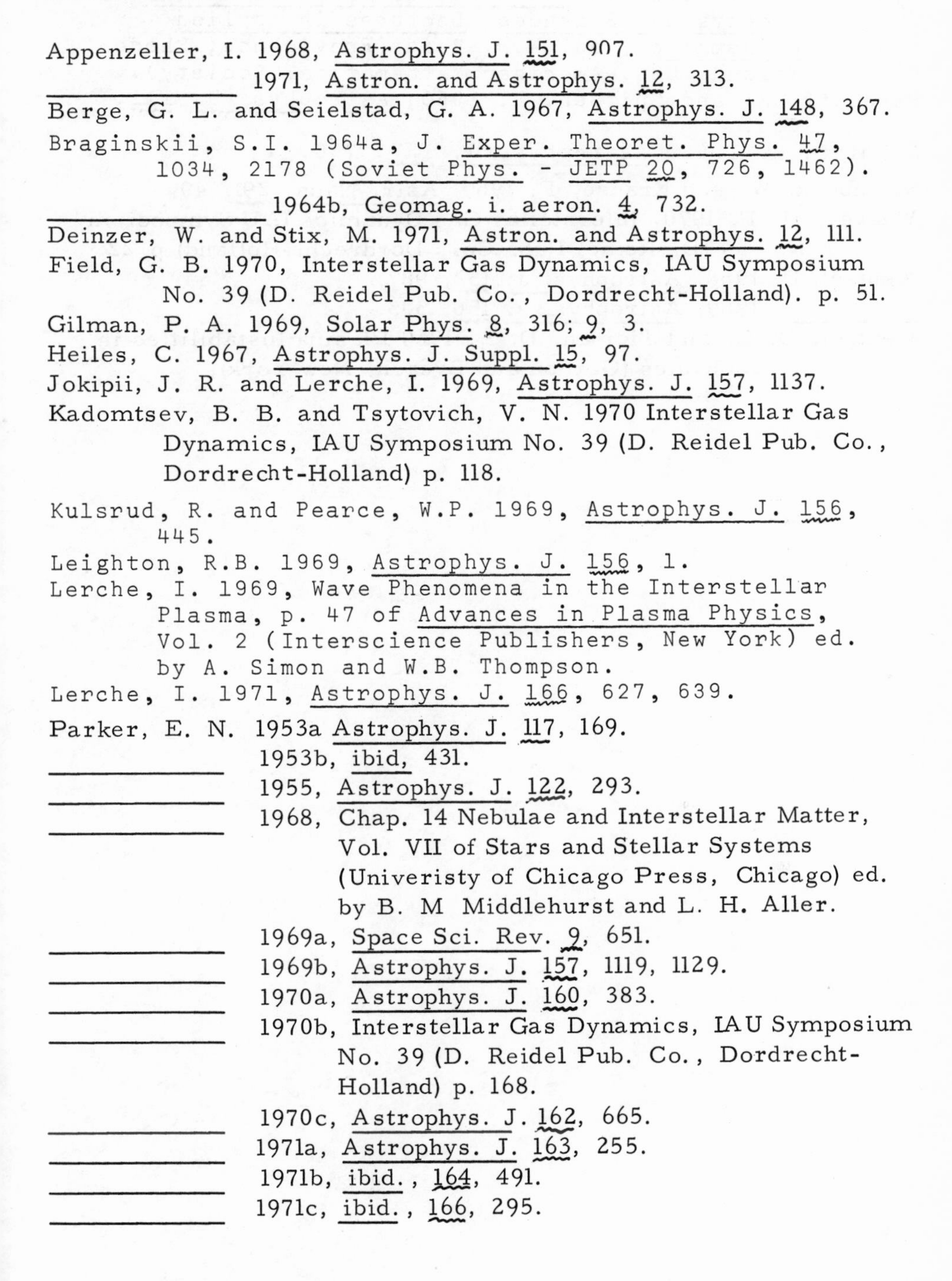

REFERENCES

Appenzeller, I. 1968, Astrophys. J. 151, 907.

——— 1971, Astron. and Astrophys. 12, 313.

Berge, G. L. and Seielstad, G. A. 1967, Astrophys. J. 148, 367.

Braginskii, S.I. 1964a, J. Exper. Theoret. Phys. 47, 1034, 2178 (Soviet Phys. - JETP 20, 726, 1462).

——— 1964b, Geomag. i. aeron. 4, 732.

Deinzer, W. and Stix, M. 1971, Astron. and Astrophys. 12, 111.

Field, G. B. 1970, Interstellar Gas Dynamics, IAU Symposium No. 39 (D. Reidel Pub. Co., Dordrecht-Holland). p. 51.

Gilman, P. A. 1969, Solar Phys. 8, 316; 9, 3.

Heiles, C. 1967, Astrophys. J. Suppl. 15, 97.

Jokipii, J. R. and Lerche, I. 1969, Astrophys. J. 157, 1137.

Kadomtsev, B. B. and Tsytovich, V. N. 1970 Interstellar Gas Dynamics, IAU Symposium No. 39 (D. Reidel Pub. Co., Dordrecht-Holland) p. 118.

Kulsrud, R. and Pearce, W.P. 1969, Astrophys. J. 156, 445.

Leighton, R.B. 1969, Astrophys. J. 156, 1.

Lerche, I. 1969, Wave Phenomena in the Interstellar Plasma, p. 47 of Advances in Plasma Physics, Vol. 2 (Interscience Publishers, New York) ed. by A. Simon and W.B. Thompson.

Lerche, I. 1971, Astrophys. J. 166, 627, 639.

Parker, E. N. 1953a Astrophys. J. 117, 169.

——— 1953b, ibid, 431.

——— 1955, Astrophys. J. 122, 293.

——— 1968, Chap. 14 Nebulae and Interstellar Matter, Vol. VII of Stars and Stellar Systems (Univeristy of Chicago Press, Chicago) ed. by B. M Middlehurst and L. H. Aller.

——— 1969a, Space Sci. Rev. 9, 651.

——— 1969b, Astrophys. J. 157, 1119, 1129.

——— 1970a, Astrophys. J. 160, 383.

——— 1970b, Interstellar Gas Dynamics, IAU Symposium No. 39 (D. Reidel Pub. Co., Dordrecht-Holland) p. 168.

——— 1970c, Astrophys. J. 162, 665.

——— 1971a, Astrophys. J. 163, 255.

——— 1971b, ibid., 164, 491.

——— 1971c, ibid., 166, 295.

Roberts, P.H. 1971, in Mathematical Problems in the Geophysical Sciences, Lectures in Applied Mathematics, ed. W.H. Reid (Providence, Rhode Island, USA, American Mathematical Society).

Savedoff, M. and Spitzer, L. 1950, Astrophys. J. **111**, 593.

Schmidt, M. 1963, Astrophys. J. **137**, 758.

Steenbeck, W. and Krause, F. 1969, Astr. Nach. **291**, 49.

Weaver, H. F. 1970, Interstellar Gas Dynamics IAU Symposium No. 39 (D. Reidel Pub. Co., Dordrecht-Holland) p. 22.

Wentzel, D. 1968, Astrophys. J. **152**, 987.

_________ 1969, Astrophys. J. **156**, 303.

Wentzel, D. G. and Tidman, D. A. 1969 Plasma Instabilities in Astrophysics (Gordon and Breach, New York).

STELLAR MAGNETOHYDRODYNAMICS

L. Mestel

Dept of Mathematics, The University, Manchester

The outstanding features of the strongly magnetic stars are (e.g. Ledoux and Renson 1966):

(1) Their apparent confinement mainly to stars of type A and earlier.

(2) The presence of spectral anomalies: the magnetic A stars seem to be almost co-extensive with the peculiar A_p stars, with their abnormally high surface abundances of the rare earths, silicon, chromium, manganese, strontium, yttrium and zirconium.

(3) The high field-strengths, inferred from the Zeeman effect: typical polar fields are 10^3-10^4 gauss, with the strongest known being 35,000 gauss.

(4) The variability of the fields, spectra and luminosities. The fields are often found to reverse in sign. Typical periods are 5-9 days, but periods both shorter and very much longer are found, a few stars having periods of several years.

(5) The low rotations of the A_p stars as compared with normal A stars. This is inferred from statistical analysis of spectral line-widths.

We shall adopt the *oblique rotator* model for these stars (Stibbs 1950 ; Deutsch 1958, 1970; Preston 1967), with the comparatively long magnetic period an additional indicator of the abnormally low rotation rate. The field is assumed tentatively to be a slowly-decaying relic (Cowling 1945), either from the star-formation epoch, or possibly from an earlier phase in which dynamo action built up a quasi-permanent large-scale field. Much of the justification of the model must come from detailed comparison with observation of plausible surface flux distributions. Recent phenomenological work of this type (e.g. Böhm-Vitense 1967;

Landstreet 1970; Hockey 1971) has done much to encourage confidence in this essentially geometric model. The other main contender - the oscillating magnetic star - seems incapable of yielding either the correct periods or fields that reverse (Cowling 1965).

We shall be concerned with some of the theoretical problems raised by this identification; in particular, the origin of the slow rotations and of the large angles of obliquity χ between the rotation and magnetic axes, required to explain field reversal; and also why the magnetic star phenomenon appears confined to early-type stars. But first we note the contrast with the nearest star to show striking hydromagnetic phenomena - our sun. This has strong local fields, especially in sunspots, but only a very weak general field of 1-2 gauss, which appears to reverse in sign along with the 11-year sunspot cycle (Babcock 1959). A similar magnetic cycle in other solar-type stars has been inferred from the long-term periodic behaviour of the calcium H and K lines (Wilson 1971). Further, there is the steady decline in calcium activity with age (Wilson and Woolley 1970) and the associated correlation of slow rotation with age (Kraft 1967). We are led to postulate that *stars with strong outer convection zones* generate external fields by an *oscillatory dynamo process* (Babcock 1961; Leighton 1969), with the strength of the field - as measured by the chromospheric calcium activity - dependent on the stellar rotation (Cowling 1965), and hence declining as the star is magnetically braked (cf. below). But - certainly in the solar case - the large-scale dynamo-built field is not strong enough to yield an integrated Zeeman effect observable at interstellar distances; and equally it is very implausible (Cowling 1965) to explain the strong magnetic variables in solar-cycle terms, with the 22-year period reduced typically to 9 days, and the 1-gauss polar field increased to 10^3 gauss. Thus we have noted that the magnetic A stars are slow rotators, and since they have hotter surfaces than the sun, the sub-photospheric convection will be weaker (and in any case will probably be effectively suppressed by the magnetic field - see below): whereas a stepped-up solar-type dynamo process would probably require both much more rapid rotation and more violent convection. It seems far more promising to think of the fields of the strongly magnetic stars as primeval, and explain the variations (even those with periods of the order of years) as a simple consequence of obliquity rather than as a dynamo effect.

This sharp division between on the one hand those early-type stars which are observed to be strongly magnetic, and on the other the solar-type stars with dynamo-built fields, raises in its turn some pertinent questions. First of all, what is it that determines when an early-type star retains a strong magnetic flux; and what is the maximum that the star can retain? The answer is part of the whole involved and only partially understood problem of star formation from magnetic interstellar gas clouds. The first conclusion (Mestel 1965) is that if the flux-freezing constraint is *not* relaxed, then although proto-stellar masses of stellar order can form by preferential flow of gas down the field-lines, they are likely to have far too much flux for consistency with observation: on reaching the main sequence they should show surface fields of order 10^5-10^6 gauss. This embarrassing result can be avoided if the ion-electron density in dusty clouds of neutral hydrogen falls to abnormally low values during the collapse and break-up of the cloud (Mestel and Spitzer 1956; Spitzer 1968). According to the modified hydromagnetics appropriate to a lightly ionized gas, the magnetic field is frozen into the ionized component, but the field-plus-ions-and-electrons will drift relative to the neutral bulk at a rate dependent on the magnetic forces and on the strength of the frictional coupling between ions and neutral particles. If the ion density is low enough the neutral gas can collapse under its self-gravitation without dragging the field with it. The exact density epoch at which this decoupling occurs depends quite critically on various astronomical factors, such as the intensity of starlight, sub-cosmic rays and X-rays, all of which tend to maintain ionization, and on the proportion of dust-grains, which allow exponential decline of the ion-electron density. Flux-freezing within a proto-star is re-established at higher densities, when the cloud opacity has forced up the internal temperature and so restored a more normal ion density.

However one wonders whether conditions will *always* be favourable for flux-loss by this process; for example, early in the galactic lifetime, when there was probably far less dust. It would be valuable to find an alternative process of flux destruction which does not depend on adventitious circumstances. More work could be done on *hydromagnetic instabilities*, in the hope that if the star contains more flux than a new critical value - much below the maximum possible within a newly forming proto-star - then it will become unstable according to the energy principle (Bernstein *et al* 1958), in spite of the stabilizing effect of the star's gravitational field. The non-linear development of the unstable modes could lead to locally large magnetic field gradients, so that Ohmic resistivity becomes important and destroys excess flux. It could even turn out that only topologically complex fields - with toroidal flux loops linking the poloidal loops, as in some projected thermonuclear devices - will be stable; if so, this could link up very satisfactorily with current ideas on the oblique rotator (cf. below). Work so far (Wright 1970) - on an analogue of the sausage instability of a toroidal discharge (Tayler 1957) - has in fact yielded only rather modest results: a local criterion on the structure of the field near the ring of neutral points, rather than one depending on the overall ratio of magnetic to gravitational energy. However, the study is still in its infancy.

It is believed (Hayashi 1961) that during their contraction

to the main sequence most stars pass through a phase in which they are wholly turbulent. The estimated field within an A_p star has an energy density that is not small compared with the turbulent energy density, so we may expect the field to survive the attempts of the turbulent eddies to tangle it up. And even if the primeval field within a star is so weak that it offers no resistance to tangling, it is by no means clear that the inevitable increase in Ohmic dissipation with reduction of the scale of the field will lead to *destruction* of the field. It is certainly arguable that the flux will merely be concentrated at the stellar surface (Spitzer 1957), to leak back into the bulk of the star as the turbulence decays. If such a weak field is preserved in the radiative core of the sun, it would probably wreck a recently proposed solar model with the core rotating much faster than the convective envelope (Dicke 1970, 1971).

However, the tentative picture outlined above raises an immediate astronomical problem: since it is surely unlikely that *no* late-type stars succeed in attaining their main-sequence structure - with a strong outer convection zone - while retaining a strong primeval flux, why then do we not see late-type stars with strong primeval fields (as opposed to solar-cycle, dynamo-built fields)? This leads in turn to the whole question of the relation between the total flux within a star and the emergent flux seen at the surface. A large-scale laminar circulation superposed on the turbulence will tend to concentrate flux in the high density regions deep in the convective zone. Biermann (1958) has suggested that such an inexorable circulation necessarily flows in a rotating convection zone, driven by the non-conservative centrifugal field set up by the anisotropic turbulent viscosity. And even if there were no rotation, one could perhaps appeal to the giant turbulent cells postulated by Simon and Weiss (1968) to concentrate the flux deep down.

The normal early-type stars have only very weak sub-photospheric convection zones, and a magnetic field of 10^3 gauss or more has enough energy to stifle the turbulent motions. (Since most of the energy transport is by radiation, suppression of the turbulence will make little difference to the temperature structure of the star). In fact, the persistence in the A_p stars of local abundance anomalies is a very strong hint, since the anomalies could hardly survive turbulent mixing. According to our picture the magnetic field is basic to the A_p phenomenon, in that by stabilizing the atmosphere it allows abundance anomalies to develope and persist instead of their being redistributed both vertically and horizontally. Several processes are currently being discussed as the origin of the anomalies, e.g. a combination of gravitational settling and selective radiation pressure (Michaud 1970); or selective accretion from the interstellar medium (Havnes and Conti 1971); or nuclear reactions in an advanced evolutionary phase (Fowler, Burbidge, Burbidge and Hoyle 1965): all would seem to require a stabilized surface.

However, even with suppression of the convection by the field, there remains the possibility of large-scale laminar circulation in the radiative envelope. In a uniformly rotating star (the rotation being kept nearly uniform in the mean by the magnetic stresses), we have the circulation (Eddington 1929; Sweet 1950), driven by the breakdown in local radiative equilibrium due to the centrifugal forces. This is much slower than the Biermann circulation (1958) in a convective zone, but will also be "inexorable" if the centrifugal forces dominate over the magnetic forces. One can picture the circulation steadily distorting a given primeval field until Ohmic diffusion disconnects the internal field from the external, which may either decay or be blown away from the star, leaving the star *apparently* non-magnetic. The circulation speeds increase like Ω^2, so that we may expect this effect to be more pronounced in rapid rotators. Thus tentatively we picture a two-way interaction between Ω and B: a strong field, with lines emerging from the star is likely to reduce the star's rotation (cf. below), but also a star which has maintained a high enough angular momentum may trap its field within it via the circulation it generates. In order that a star should permanently retain its surface flux and associated spectral peculiarities, we should strictly demand that the distortion to the thermal field by the centrifugal forces is off-set by the corresponding distortions by the magnetic forces, so that radiative equilibrium holds strictly, without any circulation: i.e.

$$\nabla . \underset{\sim}{F} \equiv (\nabla . \underset{\sim}{F})_{\Omega} + (\nabla . \underset{\sim}{F})_{\underline{B}} = 0$$

in an obvious notation. This problem has been studied by Davies (1968) and Wright (1969) for the special case with the magnetic field dipolar and symmetric about the rotation axis (the "non-oblique rotator"). The most striking result is that with a prescribed total flux $\underline{F_t}$ for the field, an increase in Ω reduces the flux $\underline{F_s}$ which crosses the photosphere and is therefore observable. At a finite value of Ω, $\underline{F_s}$ *vanishes*; and with Ω still higher, there are no circulation-free solutions. Alternatively, with Ω prescribed, there is a minimum value $\underline{F_c}$ for $\underline{F_t}$ in order that $\underline{F_s}$ should be non-zero. If these models are at least qualitatively applicable to the observed magnetic stars, then they imply that the flux $\underline{F_t}$ is usually no more than ten-percent above $\underline{F_c}$. The surface flux actually seen would depend very sensitively on the closeness of the total flux $\underline{F_t}$ to $\underline{F_c}$: a slight degree of Ohmic destruction of $\underline{F_t}$ would lead to a disproportionately large decrease in $\underline{F_s}$. The observations of the A_p stars do in fact show a bewildering lack of correlation between $\underline{F_s}$, Ω and spectral type, pointing against dynamo maintenance of the fields - which one would expect to yield a strong (and universal) correlation between Ω and $\underline{F_s}$ - and towards a fossil theory.

It is not clear whether a (non-axisymmetric) extension of these models is in fact applicable to the obliquely rotating magnetic stars. The condition $\nabla . \underset{\sim}{F} = 0$ almost certainly holds to a high approximation in the outer regions, but deeper down it may be more correct to think in terms of the slow Eddington-Sweet circulation steadily distorting the field and so indirectly affecting the observable flux. However, Wright's models do suggest strongly that (i) $F_s \ll F_t$ - the surface flux is the "tip of the iceberg"; and (ii) rapid rotators tend to be *visibly* non-magnetic because of internal trapping of flux.

The generally low angular velocities of the magnetic stars are presumably due to magnetic torques. The magnetic braking process that has been studied in most detail recently (e.g. Mestel 1966, 1968a; Weber and Davis 1967) is by coupling with a wind emitted from the star. Satellite probes and observations of cometary tails show that the sun is steadily losing angular momentum by this process.(Extrapolation back to the geophysically-estimated epoch of the formation of the solar system shows that the sun began its life as a rapid rotator, in agreement with observations of rotation in young solar-type stars; this supplies an important boundary condition on future models of the origin of the solar system). For solar-type stars the essential features are the formation of a hot expanding corona by the damping of waves emitted from the sub-photospheric convection zone, and the dynamo generation of a surface magnetic field. The net torque on the star can be computed by assuming that the expanding gas co-rotates with the star out to the Alfvénic surface, where the wind speed equals the local Alfvén speed. Near the star almost all the angular momentum is carried outwards by the twisted magnetic field; at the Alfvénic surface comparable amounts are carried by the magnetic and wind fields; while ultimately all the angular momentum is carried by the streaming gas.

It is tempting to argue that the A_p stars have suffered excess braking because of their having primeval fields which are much stronger than any likely dynamo-built field. The absence of a strong outer convection zone in A stars on the main sequence is a difficulty (though there are workers who argue that all stars are surrounded by hot coronas); however, stars of this mass do have Hayashi and post-Hayashi phases during which they should emit violent winds. As long as the primeval field has lines that emerge from the star, then we may certainly expect excess braking (though it should be noted (Mestel 1966, 1968a) that a strong field tends to trap near the star gas that would otherwise expand, so that the increased angular momentum loss is unlikely to be as great as a naive application of the theory suggests). But we have earlier argued that strong primeval fields are not observed in late-type stars because circulation in the convection zone has concentrated the flux deep down; for consistency, we would need to show that in spite of this, enough flux-lines emerge from the star for sufficiently long for excess braking to occur.

Even if the star does not have a hot corona there will be some outflow of gas driven by the magnetically-controlled centrifugal forces (Mestel 1968a). However, this is an efficient process for angular momentum removal only as long as the centrifugal forces are comparable with gravity near the star, so that the gas density is comparatively high at the point where it attains the sound speed. Thus we may expect a "centrifugal wind" to limit the increase in the rotation of a contracting proto-star, but not to yield *slowly* rotating stars.

Perhaps the most promising line of attack is to forget about stellar winds in this context, and to study direct transfer of angular momentum as the rotating magnetic star ploughs its way through the interstellar gas. Order-of-magnitude estimates suggest that adequate braking can be achieved; and one model that has been studied in detail (Kulsrud 1971) yields promising numbers.

As our last topic we note two alternative processes which may spontaneously yield a large angle χ between the magnetic and rotation axes, as is apparently required for the majority of the magnetic variables if they are to be interpreted as oblique rotators (Preston 1967). The first is an extension of the magnetic braking process: we inquire whether the mechanism responsible for the low rotation can simultaneously yield a large angle χ. The only case discussed in detail so far is via coupling with a stellar wind. It can be shown (Mestel 1968b) that when $\chi \neq 0$, so that the system is essentially non-axisymmetric, the tensions along the rotationally-distorted field-lines have integrated components of torque not only about the angular momentum axis (the braking component) but also about the two perpendicular axes, so that the instantaneous axis of rotation precesses through the star, and the magnetic axis rotates in space. Detailed analysis for one special case (Mestel and Selley 1970) confirms a qualitative result that emerges from picturing the distorted field-lines: the rotation axis seeks out the region on the star's surface where the field is strongest. Thus if the surface flux is more concentrated to the magnetic poles, the angle $\chi \to 0$ and an initial obliquity tends to disappear; while if the field is stronger at the magnetic equator $\chi \to \pi/2$, and any initial obliquity is systematically increased.

If this were the dominant cause of changes in obliquity, then we would need to study further the internal stellar hydrodynamics, with a view to explaining the necessary equatorial flux concentration on the stellar surface. However, the precessional torque component is always smaller than the braking component, and it is not clear that a sizeable change in χ can be achieved without an embarrassingly large loss of angular momentum being required. Nor is it clear what criterion on the surface magnetic flux would result if the magnetic braking process is by direct coupling with the interstellar medium.

A more promising mechanism ignores coupling with the external world, and considers the star as a body with an invariant angular momentum vector $\mathbf{h}$, but with a density distribution that is not symmetric about $\mathbf{h}$. We write

$$\rho = \rho_0 + \rho_\Omega + \rho_{\underline{B}} ,$$

where ρ_0 is the zero-order, spherically-symmetric density field, and ρ_Ω and ρ_B are respectively the first-order perturbations due to centrifugal force, symmetric about the direction $\mathbf{h}$, and the magnetic force, symmetric about the magnetic axis. If ρ_Ω could be ignored, then the motion of the star could be described simply by the Eulerian nutation: an observer, rotating with the angular velocity component in the direction of $\mathbf{h}$, would see the star rotate slowly about the magnetic axis at a rate $\omega \simeq (F^2/\overline{GM^2})\Omega$. But because of the term ρ_Ω, the nutation alone would lead to a density-pressure field which did not satisfy hydrostatic equilibrium: subsidiary motions $\boldsymbol{\xi}$ with the same frequency ω therefore arise so as to maintain equilibrium (Mestel and Takhar 1971). The radial component is

$$\xi_r = \left[\rho_\Omega(\underline{r}, \theta, \lambda - \omega\underline{t}) - \rho_\Omega(\underline{r}, \theta, \lambda)\right]/(\underline{d}\rho_0/\underline{dr}),$$

where $(\underline{r}, \theta, \lambda)$ are spherical polar coordinates referred to the magnetic axis: the horizontal components are restricted by the condition $\nabla.\boldsymbol{\xi} = 0$. If these motions were strictly adiabatic they would persist indefinitely, and would be important only in so far as they caused mixing of matter between convective and non-convective regions. But dissipative processes acting on these motions steadily drain energy from the only available source - the kinetic energy of rotation (Spitzer 1958). With $\underline{h}$ conserved, this is of the form $\underline{h}^2/2\underline{I}$, where $\underline{I}$ is moment of inertia about the instantaneous axis of rotation. Hence the angle will change until the star rotates about the maximum principal axis; if the star is oblate about the magnetic axis, $\chi \to 0$; if prolate, $\chi \to \pi/2$.

Preliminary estimates suggest that this mechanism is likely to be much more powerful than the wind-coupling process. It should be noted that the criteria for the generation respectively of high or low obliquity χ now depend on the flux distribution throughout the star, not just on the surface distribution. The most obvious way of building a star permanently prolate about the magnetic axis is to have a strong toroidal flux linking the poloidal loops and maintained by currents that flow parallel to the poloidal loops. We have already suggested tentatively that fields of such topology may be necessary for stability. It should also be noted that there is a small group of A_p stars which appear to require a *small* angle χ (Preston 1971, private communication); this would result if the toroidal component is not strong enough to off-set the oblateness caused by the poloidal component (which must be present if the magnetic flux is to be observable). If the necessity for linked poloidal and toroidal flux could be established by two quite distinct types of argument - to ensure stability, and to yield the required distribution of angles χ - one would begin to feel that the theory had moved away from the tentative towards the definitive end of the credibility spectrum; especially as the interaction of even a moderately strong poloidal field with the Hayashi turbulence in a rotating proto-star is likely to complicate the field topology and yield the required linked flux.

In conclusion, we remark that some of the above arguments may be relevant to the pulsar problem. Goldreich and Julian (1969) have shown that even though the thermal scale-height is negligibly small, a neutron star with a magnetic field symmetric about its rotation axis is necessarily surrounded by an electrically-supported corona, and steadily loses electrons and ions via an electrically-driven wind. Their arguments apply equally to an obliquely-rotating neutron star (Cohen and Toton 1971; Mestel 1971). The significant point is that - at least within the light-cylinder - the electric field and the associated net charge density are still described approximately by the hydromagnetic condition $c\mathbf{E} + \mathbf{v}\times\mathbf{B} \simeq 0$ rather than by the vacuum condition $\nabla.\mathbf{E} = 0$. Thus deductions from the results of classical radiation theory should be treated with some caution; for example, one may get an incorrect sign for the precessional torque exerted on the star as it loses angular momentum (Mestel 1971).

REFERENCES

Babcock, H.W., 1959. *Astrophys. J.*, 130, 364.

Babcock, H.W., 1961. *Astrophys. J.*, 133, 572.

Bernstein, I.B., Frieman, E.A., Kruskal, M.D. and Kulsrud, R.M., 1958. *Proc. R. Soc. A.*, 244, 17.

Biermann, L., 1958. *Electromagnetic Processes in Cosmical Physics*, ed. B. Lehnert, 248. Cambridge University Press, Cambridge.

Böhm-Vitense, E., 1967. *Modern Astrophysics*, ed. M. Hack, 97. Gauthier-Villars, Paris.

Cohen, J.M. and Toton, E.T., 1971. *Astrophys. Lett.*, 7, 213.

Cowling, T.G., 1945. *Mon. Not. R. astr. Soc.*, 105, 166.

Cowling, T.G., 1965. *Stellar and Solar Magnetic Fields*, ed. R. Lüst, 405. North-Holland, Amsterdam.

Davies, G.F., 1968. *Austr. J. Phys.*, 21, 294.

Deutsch, A.J., 1958. *Electromagnetic Processes in Cosmical Physics*, ed. B. Lehnert, 209. Cambridge University Press, Cambridge.

Deutsch, A.J., 1970. *Astrophys. J.*, 159, 985.

Dicke, R.H., 1970. *Astrophys. J.*, 159, 1.

Dicke, R.H., 1971. *A Rev. Astr. Astrophys.*, 8, 297.

Eddington, A.S., 1929. *Mon. Not. R. astr. Soc.*, 90, 54.

Fowler, W.A., Burbidge, E.M, Burbidge, G.R. and Hoyle, F., 1965. *Astrophys. J.*, 142, 423.

Goldreich, P. and Julian, W.H., 1969. *Astrophys. J.*, 157, 869.

Havnes, O. and Conti, P.S., 1971 (preprint).

Hayashi, C., 1961. *Publ. Astr. Soc. Japan*, 13, 450.

Hockey, M.S., 1971. *Mon. Not. R. astr. Soc.*, 152, 97.

Kraft, R.P., 1967. *Astrophys. J.*, 150, 551.

Kulsrud, R.M., 1971. *Astrophys. J.*, 163, 567.

Landstreet, J.D., 1970. *Astrophys. J.*, 159, 1001.

Ledoux, P. and Renson, P., 1966. *A.Rev. Astr. Astrophys.*, 4, 293.

Leighton, R.B., 1969. *Astrophys. J.*, 156, 1.

Mestel, L. 1965. *Q. Jl. R. astr. Soc.*, 6, 265.

Mestel, L., 1966. Liège Symposium on *Gravitational Instability*.

Mestel, L., 1968a. *Mon. Not. R. astr. Soc.*, 138, 359.

Mestel, L. 1968b. *Mon. Not. R. astr. Soc.*, 140, 177.

Mestel, L., 1971. *Nature (Physical Science)*, 233, 149.

Mestel, L. and Selley, C.S., 1970. *Mon. Not. R. astr. Soc.*, 149, 197.

Mestel, L. and Spitzer Jr., L., 1956. *Mon. Not. R. astr. Soc.*, 116, 583.

Mestel, L. and Takhar, H.S., 1971. *Mon. Not. R. astr. Soc.*, (in press)

Michaud, G., 1970. *Astrophys. J.*, 160, 641.

Preston, G.W., 1967. *Astrophys. J.*, 150, 547.

Simon, G.W. and Weiss, N.O., 1968. *Z. Astrophys.*, 69, 435.

Spitzer Jr., L., 1957. *Astrophys. J.*, 125, 525.

Spitzer Jr., L., 1958. *Electromagnetic Processes in Cosmical Physics*, ed. B. Lehnert, 169. Cambridge University Press, Cambridge.

Spitzer Jr., L., 1968. *Diffuse Matter in Space*, 238. Interscience, New York.

Stibbs, D.W.N., 1950. *Mon. Not. R. astr. Soc.*, 110, 395.

Sweet, P.A., 1950. *Mon. Not. R. astr. Soc.*, 110, 548.

Tayler, R.J., 1957. *Proc. Phys. Soc. B.*, 70, 31.

Weber, E.J. and Davis Jr., L., 1967. *Astrophys. J.*, 148, 217.

Wilson, O.C., 1971. Report to *Solar Wind Conference*, Asilomar.

Wilson, O.C. and Woolley, R., 1970. *Mon. Not. R. astr. Soc.*, 148, 463.

Wright, G.A.E., 1969. *Mon. Not. R. astr. Soc.*, 146, 197.

Wright, G.A.E., 1970. Ph.D. dissertation, Manchester University.

PLASMA TURBULENT HEATING AND THERMAL X-RAY SOURCES

B. Coppi and A. Treves*

Massachusetts Institute of Technology

Cambridge, Mass.

ABSTRACT

A class of thermal X-ray sources is associated with rotating collapsed stars surrounded by an extended plasma shell. Scorpio-X1 is considered a characteristic example of this type of stars and the heating mechanism of its plasma-sphere is discussed in detail.

INTRODUCTION

We refer to compact objects, typically neutron stars, whose mass is $M \simeq 1M_{\odot} \simeq 2\times10^{33}$gr, and radius $a \simeq 10^{6}$ cm. The magnetic field is assumed to be dipolar on the star surface, with an intensity $B_o \simeq 10^{12}$ G. We assume, for simplicity, that the dipole axis coincides with the axis of rotation, and we consider a region defined by closed magnetic field lines that are contained within the speed of light cylinder, of radius $R_{SL}=c/\omega_o$, where ω_o is the star's angular velocity. In this region, the magnetic field is dipolar, there are no d.c. electric fields and we assume that no emission occurs. In a second region the lines of force are open, an electric field $E_{\parallel}$ can be maintained along them and emission processes can be sustained. Here the magnetic field configuration is determined by the current due to the electric field $E_{\parallel}$. This region is within co-latitude $\Theta_c=(a\omega_o/c)^{1/2}$ (Sturrock

* Permanent address: Istituto di Fisica dell'Università, Milano

1971), and we can argue that the width of the active region on the star is of order

$$\ell_o \sim \Theta_c a = (a^3 \omega_o / c)^{1/2} .$$

The potential difference that is seen in an inertial frame between a pole and a point at approximately co-latitude Θ_c is (Deutsch 1955; Goldreich and Julian 1969)

$$(\Delta\Phi)_M = \frac{1}{4} \frac{\omega_o}{c} a^2 B_o (\cos 2\Theta_c - 1),$$

and, for the parameters indicated above,

$$(\Delta\Phi)_M \sim 1.7 \times 10^{11} \omega_o^{\ 2} \text{ V} .$$

We consider a current line that in the vicinity of the star overlaps with the magnetic field lines of the active regions. If this line closes itself in a medium such as the interstellar plasma which is at rest in the inertial frame, the electric field E will result from a total potential drop of the order of $(\Delta\Phi)_M$.

An important threshold for the electric field $E_{\parallel}$ is the classical runaway field E_R (Gurevich, 1961) defined as $E_R = \nu_{ei} m_e v_{th}/e$, where ν_{ei} is the electron-ion collision frequency and v_{th} the electron thermal velocity. This gives the threshold above which the plasma, under the electric field influence and in the absence of collective effects, would lose its cohesiveness and the electrons (so-called runaway electrons) would tend to undergo almost free acceleration. In reality when $E_{\parallel} > E_R$, plasma collective modes are excited and their effects are equivalent to that of a strongly enhanced collision frequency. Thus the electron distribution function is kept from running away and a strong anomalous resistivity sets in. This involves considerable ohmic heating of the plasma, which is commonly called turbulent.

We propose that an important class of thermal X-ray sources is characterised by a regime of this kind, and consider, in particular, Sco-X1 as the best-known example of this type of star.

THE CASE OF SCORPIO-X1

The observed spectrum of Sco-X1 between 1 and 40 KeV corresponds to bremsstrahlung emission of a plasma with electron temperature

$$T_e = \alpha_1 \times 5 \times 10^7 \ {}^{o}\text{K}$$

(Gorenstein et al. 1968, Meekins at al. 1969), where α_1 accounts for experimental uncertainty. The received intensity is

$$I = \alpha_2 \times 3 \times 10^{-7} \text{ erg/cm}^2\text{sec.}$$

The star distance d is still uncertain and we shall consider

$$d = \alpha_3 \times 6 \times 10^{20} \text{ cm}$$

a value deduced by proper motion observations (Sofia et al. 1969).

The flux F_ν measured at infrared frequencies ($\nu \sim 10^{14}$ Hz) was interpreted by Neugebauer et al. (1969) and by Kitamura et al. (1971) as being due to black body emission of the plasma at the same temperature as deduced from the X-ray spectrum. In particular, according to Neugebauer et al. (1969),

$$F_\nu = \alpha_4 \times 10^{-53}\nu^2 \text{ erg/Hzcm}^2\text{sec.}$$

By equating this flux to the one emitted by a spherical black body with radius R and at a distance d, we obtain

$$R = (F_\nu/2\pi kT)^{1/2} cd = 2.7\times10^8 [\alpha_4{}^{1/2}\alpha_3/\alpha_1{}^{1/2}] \text{cm.} \quad (1)$$

The bremsstrahlung emissivity of a plasma is given by (Ginzburg 1967) $\varepsilon = 1.6\times10^{-27}n_e^2T^{1/2}$ erg/cm^{3}sec, where n_e is the electron density, and the corresponding observed value can be taken as

$$\varepsilon = 3Id^2/R^3 .$$

Then we have a total emitted power

$$P_{obs} = 1.3 \times 10^{36}\alpha_2\alpha_3{}^2 \text{ erg/sec}$$

and a density

$$n_e = 3.8 \times 10^{16} [\alpha_2\alpha_1/\alpha_4{}^{3/2}\alpha_3]^{1/2} \text{ cm}^{-3} . \quad (2)$$

The thermal power generated by a plasma current density $\underset{\sim}{J}$ under the influence of $\underset{\sim}{E}$ is

$$P_{tot} = \int_{volume} \underset{\sim}{E} \cdot \underset{\sim}{J} \, dV. \quad (3)$$

We consider a region close to the star, and assume that the field B_J produced by the current is smaller than the star's dipolar magnetic field B_P. In all the region within the light speed cylinder, since the force $\underset{\sim}{J} \times \underset{\sim}{B}$ is compensated by gravity, pressure gradients and centrifugal effects, for reasonable values of n and T_e we shall have $\underset{\sim}{J} \,//\, \underset{\sim}{B}$. Thus the total magnetic field is approximately force-free so that $\underset{\sim}{J} \approx \sigma\underset{\sim}{B}$ and, from the charge conservation equation, $B \cdot \nabla\sigma = 0$. In particular, in the close vicinity of the star $J \propto r^{-3}$ and, if $J_o = J(r{=}a)$, the total flowing current will be about $\mathcal{J} = J_o \pi \ell_o^2$ and the rate of energy production

$$P_{tot} \approx \mathcal{J} \int E\, ds = \mathcal{J}\, \overline{\Delta\Phi} \; . \tag{4}$$

Here s is a coordinate following the magnetic field lines, and ℓ_o is assumed to be of the order of 10^5 cm, consistently with the estimate of ω_o that will be given later.

An axially symmetric magnetic force-free field asymptotically approaches a dipolar field for $4\pi\sigma r/c < 1$ (Woltjer, 1958). We assume that this limit is obtained for $r \approx a$ so that $\sigma \approx c/(4\pi a \alpha_5)$ with $\alpha_5 > 1$. We have then

$$J_{oP} = \frac{c}{4\pi a \alpha_5} B_{oP} \sim 8 \times 10^5 \frac{1}{\alpha_5} \text{ A/cm}^2 . \tag{5}$$

This current density corresponds to a toroidal field B_{oT} which can be evaluated as

$$B_{oT} = \frac{2\pi}{c} J_{oP} \ell_o = \frac{5}{\alpha_5} \times 10^{10} \text{ gauss.}$$

Then the total current is about

$$\mathcal{J} \sim \frac{2.5}{\alpha_5} \times 10^{16} \text{A}$$

and from the observed value of the power and eq.(4) the average potential drop is

$$\overline{\Delta\Phi} \sim \alpha_5 \times \alpha_2 \times \alpha_3^2 \times 5 \times 10^{12} \text{V}.$$

The corresponding electric field will be

$$\overline{E} \approx \frac{\overline{\Delta\Phi}}{L} = \alpha_5 \times \alpha_2 \times \alpha_3^2 \times 2 \times 10^4 \left|\frac{R}{L}\right| \text{ Volt/cm} \tag{6}$$

where $L\overline{E} = \int E\, ds$ so that L gives approximately the dimension of the region where most of the plasma heating occurs.

TURBULENT HEATING

For the assumed electron density and B field the Langmuir frequency ω_{pe} is much smaller than the electron gyrofrequency $\Omega_e (\omega_{pe} \approx 10^{-6} \Omega_e)$, and the collisional runaway field is $E_R \sim 40$ Volt/cm so that $E_\parallel >> E_R$. In this connection we recall that in laboratory experiments where electric fields $E_\parallel >> E_R$ have been applied to plasma with $\omega_{pe} \sim 10\Omega_e$ no electron runaway process has been observed for $E_\parallel / E_R \sim 10^3$ (Hamberger and Friedman 1968,

Hamberger and Jancarik 1970). On the other hand, another set of experiments for plasmas with $\omega_{pe} < \Omega_e$, electric fields $E_R \lesssim E_{\parallel} \lesssim 10E_R$ and a relatively small level of fluctuations, has shown that a turbulent resistivity increases proportionally to the applied electric field while the velocity $u_{e\parallel} = J_{\parallel}/ne$ increases with the temperature and remains about equal to $1/3 - 1/5\ v_{the}$ (Coppi and Mazzucato 1971), where $v_{the} = (2kT_e/m_e)^{1/2}$ is the electron thermal velocity. A more recent series of experiments in which $\omega_{pe} \lesssim \Omega_e$ and $10 \leqslant E_{\parallel}/E_R \leqslant 80$ has also indicated that the resistivity increases proportionally to the applied electric field but the value of $u_{e\parallel}$ tends to remain limited to about $0.08\ v_{the}$ (Burchenko et al. 1971). Another important parameter of turbulent heating of plasmas with electron temperature considerably larger than the ion temperature is the ion-sound wave velocity $v_s = (ZkT_e/Am_p)^{1/2}$ where Z and A are respectively the atomic number and weight and m_p is the proton mass. With the assumed value of T_e and if we take $Z \approx A/2$, we have $v_s \approx 4.5 \times 10^7$ cm/sec.

In the limit of small amplitude fluctuations it is easy to show that ion-sound electrostatic waves can be excited and lead to momentum transfer between electrons and ions, providing a non-collisional resistivity, when $u_{e\parallel}$ is slightly larger than v_s and corresponds to fields $E_{\parallel} < E_R$. In our case $E_{\parallel} >> E_R$ and large amplitude electrostatic fluctuations are to be expected. More precisely, if $\tilde{\phi}$ indicates the fluctuating potential we consider $\tilde{\phi} \approx kT_e/e$ and expect that the trapping of electrons by the excited waves has a dominant role in determining the resistivity. We shall assume that, as the temperature rises, the maximum current density J_o will be settled at a value of the order of the ion-sound velocity following the trend indicated by the experiments of Burchenko et al. 1971. In this case the condition

$$\frac{J_{\parallel}}{ne} \lesssim \left|\frac{Ze\tilde{\Phi}}{Am_p}\right|^{1/2} \tag{7}$$

is satisfied everywhere, and the situation is similar to that of the theoretical turbulent heating model analyzed by Drummond et al. (1971).

We notice that for a given electric field $E_{\parallel}$ the electron energy balance equation will determine the value of T_e and therefore of $J_{\parallel}$ at which equilibrium between turbulent ohmic heating and losses is reached. In our case the prevalent loss process is by bremsstrahlung and the balance equation is represented by (6). An uppermost limit for the applied electric field can be introduced, taking it as of the order of the maximum fluctuating electric field. For electrostatic plasma modes the largest wave number in the direction of the magnetic field is about $1/\lambda_D$ where λ_D is the Debye length. So the largest fluctuating field will be of the order of $E_M \approx 2\pi\tilde{\Phi}/\lambda_D \approx 2\pi kT_e/(e\lambda_D)$. We assume that the largest value of the current density which occurs in the vicinity of the star corresponds to the ion sound velocity, that is to the condition for marginal stability of current driven ion sound waves. Therefore $J_o \approx n_o e v_s$, and we obtain

$$n_o \approx \frac{1}{\alpha_5} 1.1 \times 10^{17} cm^{-3}$$

a value to be compared with the one given by (2). Then $\lambda_D \approx (\alpha_5 \times \alpha_1)^{1/2} \times 2.6 \times 10^{-4}$cm and we verify that the applied electric field $E_{\parallel}$ as resulting from Eq.(6) is below $E_M \approx \alpha_5^{-1/2}\, 10^8$ Volt/cm for a reasonable choice of α_5 and $L \gtrsim a$.

The non-thermal tail of the Sco-X1 spectrum observed for energies above 40 KeV (Agrawal et al. 1970) can be associated with a very small tail of highly energetic electrons ("super-runaway") riding over the strong electric field fluctuations mentioned above.

The radiation emitting volume, that was taken to have a radius R, corresponds to that part of the magnetosphere which is heated by the high plasma thermal conductivity along the magnetic field lines if $a \lesssim L < R$. Since the thermal photons are radiated isotropically, even if the magnetic axis is not aligned with the rotation axis, no beaming of radiation in the high energy band should be present and no pulsation should be expected.

THE LIFE-TIME AND FLARE PROBLEMS

To find an order of magnitude of the angular velocity of the star, we can assume that the emitting region has a radius of the same order as the speed of light cylinder radius, that is $R = \alpha_6\, R_{SL}$, so that most of the emission occurs around this cylinder. Then from Eq.(1) we have $\omega_o = \alpha_6 c/R \sim \alpha_6 \times 100$ rad sec^{-1} corresponding to a period of $\alpha_6^{-1} \times 60$ msec. From the given expression for $(\Delta\Phi)_M$ we can argue that the current lines will close in a medium whose angular velocity differs by about 1/100 from the angular velocity of the star. The decoupling of the plasma motion from the magnetic field lines is allowed by the combined effects of the large turbulent resistivity and of the enhanced electron inertia at the light speed cylinder, and the slippage in angular velocity results from momentum conservation between plasma and emitted photons.

The corresponding rotational energy of the star is $E_{rot} = 2/5\ M\omega_o^2 a^2 \simeq \alpha_6^2 \times 10^{49}$ erg. Thus a rough estimate of the life-expectancy of the source is given by

$$\tau_E = \frac{E_{rot}}{P_{obs}} \sim 3\times10^4 \text{ years.}$$

In the absence of any indication of the variation of ω_o with time, it is difficult to make an estimate of the star's age. We consider, however, that the star's rotational energy E_{rot} at the moment of its creation could be as high as 10^{52} erg, as suggested for instance by Gold (1969). Then it is possible, on the basis of a plasma model, to have an energy emission rate and therefore a time evolution for $\omega_o(t)$ which would lead to an age $\tau > 10^5$ yrs and would be consistent with the absence of any observable supernova remnant. As indicated above, the energy output will involve a loss of angular momentum by the star to photons and relativistic particles, of the order $\Omega = P_{obs}/\omega_o \sim 1.2 \times 10^{34}$ dyne cm. Neglecting the stellar wind contribution, the loss of angular momentum can be attributed to a Poynting-Robertson torque, as described by Davidson at al. (1971).

Sco-X1 is also known to exhibit bursts of radio, optical and X-ray emission (flares) (see, for example, Evans et al. 1970). For this we notice that the plasma in the regions close to the light speed cylinder can be subject to instabilities which alter the magnetic configuration (e.g. Coppi and Friedland 1971). When the associated perturbations change drastically the medium in which the current lines close, the total overall X-ray emission is affected. Otherwise these instabilities will mainly induce a change of local emission as in the case of solar flares.

The model proposed here for Sco-X1 appears to be consistent with the observations in the radio band by Hjellming and Wade (1971) and by Braes and Miley (1971) which seem to indicate significant similarities with the emission of pulsars such as, for instance, PSR 0329 +54. In particular, the non-periodic variations observed for pulsars and those observed for Sco-X1 appear to be alike. In addition, the point-like radio source associated with Sco-X1 is surrounded by two other sources at 1!3 and 2' symmetrically located. If these two side sources are physically related to the central one, the system exhibits an axial symmetry which can be attributed to a magnetic configuration with a strong dipolar component.

We point out that the rotation frequency proposed for the neutron star in Sco-X1, is close to the one of PSR 0833 (Vela Pulsar). On the other hand no optical counterpart of this radio-pulsar has been detected. We consider this fact as an indication that the rotation frequency alone is not sufficient to determine the emission characteristics of a star. A difference in the plasma density profile and the electron distribution function, as well as a variety of magnetic configurations, are in fact to be considered in assessing the characteristic emission of a collapsed star. For instance, if in the case of PSR 0833 the plasma which surrounds the neutron star extends only for 10 stellar radii ($R \sim 10^7$ cm) and, besides the non-thermal pulsed emission, there is a thermal emission at a temperature $T \sim 10^6$ °K, the object would have visual magnitude $m_v \sim 25^m$ and hence it would be hardly visible.

ACKNOWLEDGEMENTS

It is a pleasure to thank R.H. Cohen, W.E. Drummond, M. Oda, F. Pacini, B.B. Rossi, E.E. Salpeter and P.A. Sturrock for their timely comments.

One of us (A.T.) has been supported by a joint ESRO-NASA fellowship, and this work was sponsored in part by the U.S. Atomic Energy Commission (Contract AT-30-1 3980).

REFERENCES

Agrawal, P.C., Biswas, S., Gokhale, G.S., Iyengar, V.S., Kunte, P.K., Manchanda, R.K., Sreekantan, B.V.,1971, Astroph. Space Science 10, 500.

Braes, L.L.E., and Miley, G.K., 1971, Astr.and Ap.14,160.

Burchenko, P.Y., Volkov, E.D., Rudakov, V.A., Sizonenko, V.L., and Stepanov, K.N., Paper CN-28/H-9, Int. Conf. on Plasma Physics and Controlled Thermonuclear Research (Madison, Wisc., 1971), I.A.E.A., Vienna.

Coppi, B. and Friedland, A. 1971, Ap. J. 169, 379.

Coppi, B. and Mazzucato, E. 1971, Phys. Fluids 64, 134.

Davidson, K., Pacini, F., and Salpeter, E.E., 1971, Ap. J. 168, 45.

Drummond, W.E., Thompson, J.R., Sloan, M.L., and Wong, H.V., Paper CN-28/E-12 presented at the International Conference on Plasma Physics and Controlled Thermonuclear Research (Madison, Wisc., 1971). Proceedings publisher I.A.E.A. (Vienna).

Deutsch, A.J. 1955, Ann. Astrophys. 18, 1.

Evans, W.D., Belian, R.D., Conner, J.P., Strong, I.B. Hiltner, W.A., Kunkel, W.E. 1970, Ap. J. (Letters) 162, L115.

Ginzburg, V.L. 1967, in High Energy Astrophysics, ed. C. DeWitt, E.Schatzman, and P. Veron (Gordon and Breach)

Gold, T., 1969, Nature 221, 25.

Goldreich, P. and Julian, W.H. 1969, Ap. J. 157, 869.

Gorenstein, P., Gursky, H. and Garmire, G. 1968, Ap.J. 153, 885.

Gurevich, A.V. 1961, Soviet Phys. - J.E.T.P. 12, 904.

Hamberger, S.M. and Friedman, M. 1968, Phys. Rev. Letters 21, 674.

Hamberger, S.M. and Jancarik, J.1970, Phys.Rev. Letters 25, 999.

Hjellming, R.M. and Wade, C.M. 1971, Ap. J.(Letters) 164, L1.

Kitamura, T., Matsuoka, M., Miyamoto, S., Nakagawa, M., Oda, M., Ogawara, Y., Takagishi, K., Rao, U.R., Chitnis, E.V., Jayanthi, U.B., Prakasa-Rao, A.S., and Bhandari, S.M., 1971, to be published in Astroph. Space Science.

Meekins, J.G., Henry, R.C., Fritz, G., Friedman, H., and Byram, E.T. 1969, Ap. J. 157, 197.

Neugebauer, G., Oke, J.B., Becklin, E., and Garmire, G., 1969, Ap. J. 155, 1.

Sofia, S., Eichorn, H., and Gatewood, G., 1969, Astron. J. 74, 20.

Sturrock, P.A., 1971, Ap. J. 164, 329.

Woltjer, L., 1958, Bull. Ast. Inst. Netherlands 14, 39.

RADIO AND OPTICAL OBSERVATIONS OF PULSARS

F.D. Drake

National Astronomy and Ionosphere Center

Cornell University, Ithaca, New York

I shall assume that you are familiar with the most basic characteristics of pulsars; there is strictly periodical radiation of discrete pulses of energy, the pulses varying in shape and intensity from one pulse to the next and from one radio frequency to another. There are now some sixty pulsars known with periods ranging from 33 msec to 3.75 sec (Fig.1), all very similar in their characteristics. With 20 of these the rate of change of the period is known and in every case the period lengthens with time. In one case, that of the famous Crab Nebula pulsar, the second derivative is known. This second derivative leads to a law for the deceleration which is very close to that expected if the primary braking torque on the pulsar is magnetic dipole radiation[1,2]. The deviation from a strict magnetic dipole

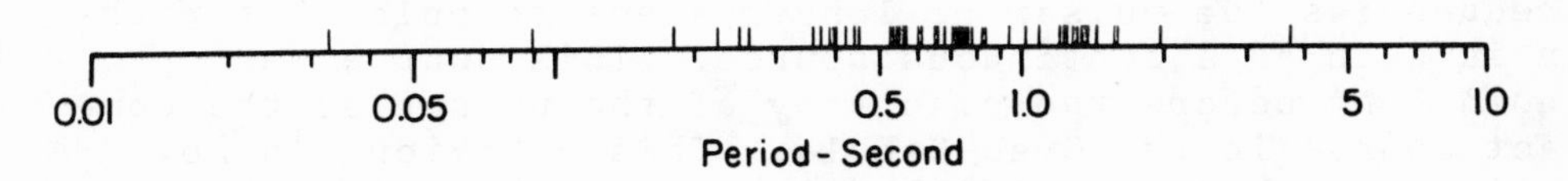

Fig. 1 Distribution of pulsar periods

braking law is in the sense to be expected if the field geometry is not strictly dipolar but the field is stretched radially by the outflowing plasma.

A compelling case now exists that pulsars are rotating neutron stars[3]. Careful calculations of the equation of state for neutron matter have given a rather precise value for the moment of inertia of a neutron star. This value, taken with the observed slowing down rate of the Crab Nebula pulsar, leads to a release of energy which matches the total electromagnetic radiation from the Crab Nebula, about 10^{38} ergs sec $^{-1}$. This agreement could hardly be fortuitous and seems simultaneously to show that the pulsar is a rotating neutron star and that all the rotational energy is converted into relativistic particle energy, which in turn is released as electromagnetic radiation from the Supernova remnant. The magnetic field at the poles of the neutron star which may be inferred from this picture is around 10^{12} gauss, consistent with the field expected if the normal stellar flux is conserved as a star contracts to neutron star dimensions.

One of the most striking characteristics of pulsars is that the emission occurs during a time interval with very well defined boundaries. Within a given pulse there may be highly variable radiation, with no one pulse being even similar to another pulse from the same source. However, the mean of a large number of pulses is a pulse shape which is well defined and characteristic for each source and distinct from the pulse of all other pulsars.

The pulse shape of the Crab Nebula pulsar is of particular interest since it contains three components, including a strong precursor, a counterpart of which has not been found in any other source. The precursor has the characteristic of being 100% or nearly linearly polarized, with no change in the position angle as a function of time. Fig. 2 shows the Crab Nebula pulsar pulse shape as a function of the radiofrequency. At the lower frequencies the pulse components become very broad and finally merge together causing the pulse shape to become nearly sinusoidal. As a result, at the lowest frequencies the pulsar no longer seems to pulse but rather is seen as a continuous source. This source was known long before the discovery of the pulsar as the compact source in the Crab Nebula. This behavior is now believed to be a result of multipath propagation through the inhomogeneous interstellar medium.

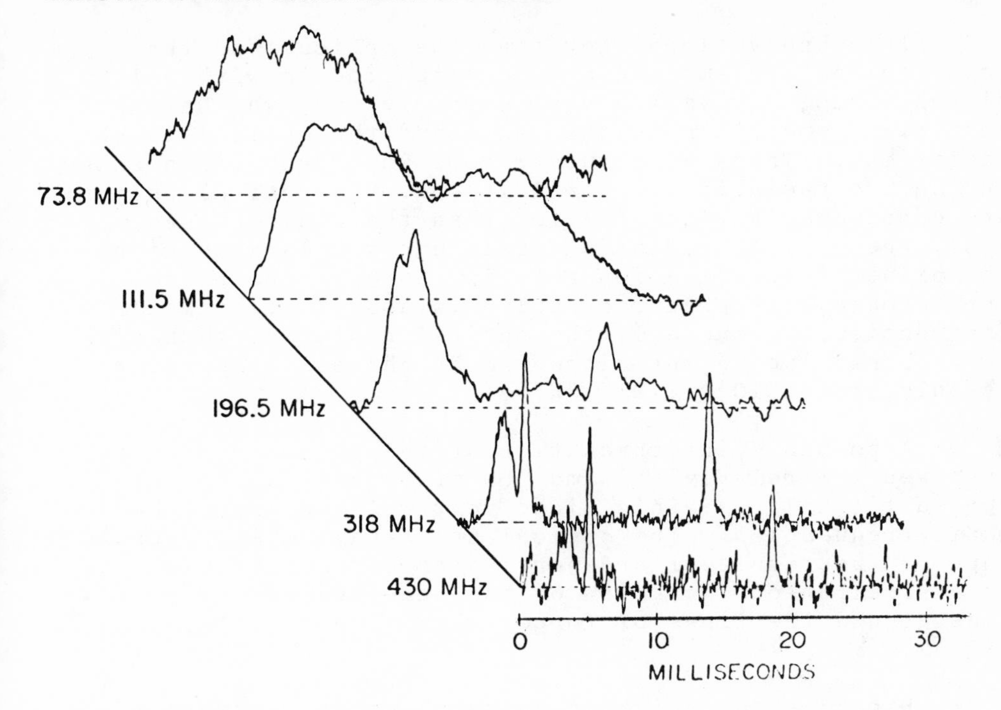

Fig. 2 Pulse shapes of NP 0532 at different frequencies

When individual pulses of the Crab pulsar are examined in detail, it is found that each pulse consists of a few very brief but intense components lasting only hundreds of microseconds. This implies that the size of the radiating region is only a few tens of kilometers at most and therefore the brightness temperature is huge, occasionally reaching 10^{31} °K. An important consequence of this fact is that the radiating entity in the pulsar must have an energy of at least $kT_b \sim 10^{27}$ eV. This could not be a single electron or proton because such a particle would never radiate primarily at radiofrequencies (even if it existed, which is very unlikely). There is a clear implication that the radiation process is of a coherent nature, possibly due to coherently moving bunches of particles. The typical number of particles in a bunch should be in the range 10^{10}- 10^{15}, perhaps even more[4,5].

The optical observations are limited to the Crab pulsar, the only one known to radiate at other than radio frequencies.

The observations show that the optical and the radio pulses are emitted at the same time to within 100 microseconds, suggesting very strongly that the same particles generate both the radio and the optical pulse radiations. There are however some important differences in the two radiations. For instance, the optical pulses are considerably wider in time than the radio pulses. Also, the optical radiation is linearly polarized with the plane of polarization rotating through the pulse. Furthermore, the precursor pulse is absent at optical frequencies. Above all, the optical radiation mechanism is probably incoherent since the brightness temperature is only around $10^{11} - 10^{12}$ °K.

A possible interpretation of the pulsar radiation involves the relativistic motion of bunches along the curved field lines: this might very well generate the radio frequencies. The optical and X-ray emission could then be the result of ordinary synchrotron radiation due to a simultaneous gyration of the same individual particles around the lines of force. For more details, see Ref. 4.

Fig. 3 shows the mean pulse shape of a number of pulsars at many radio frequencies, as marked in the upper right hand of each drawing, the frequencies being in MHz. A typical doubling in the pulse shape is seen. The spectrum of various components in a pulse shape is different from one to another. There is no general rule governing this phenomenon but one sees that the pulse shapes broaden at the lower frequencies. Some of this is explicable in terms of multipath propagation but some must be intrinsic to the source. The origin of the specific pulse shape is still essentially unexplained. However T. Gold[5] has noted that we can see only those bunches of particles whose motion is directed almost precisely towards us. It is possible that the complex motion of particles in the pulsar magnetosphere might lead to a situation where only in certain portions of this magnetosphere are the trajectories aimed in our direction and therefore visible to us.

Most pulsars are well known to exhibit marked changes in intensity over times of the order of only a few pulses. These variations are partly intrinsic and partly due to the interstellar scintillation.

A phenomenon which is extremely provocative and almost certainly related to plasma physics is that of

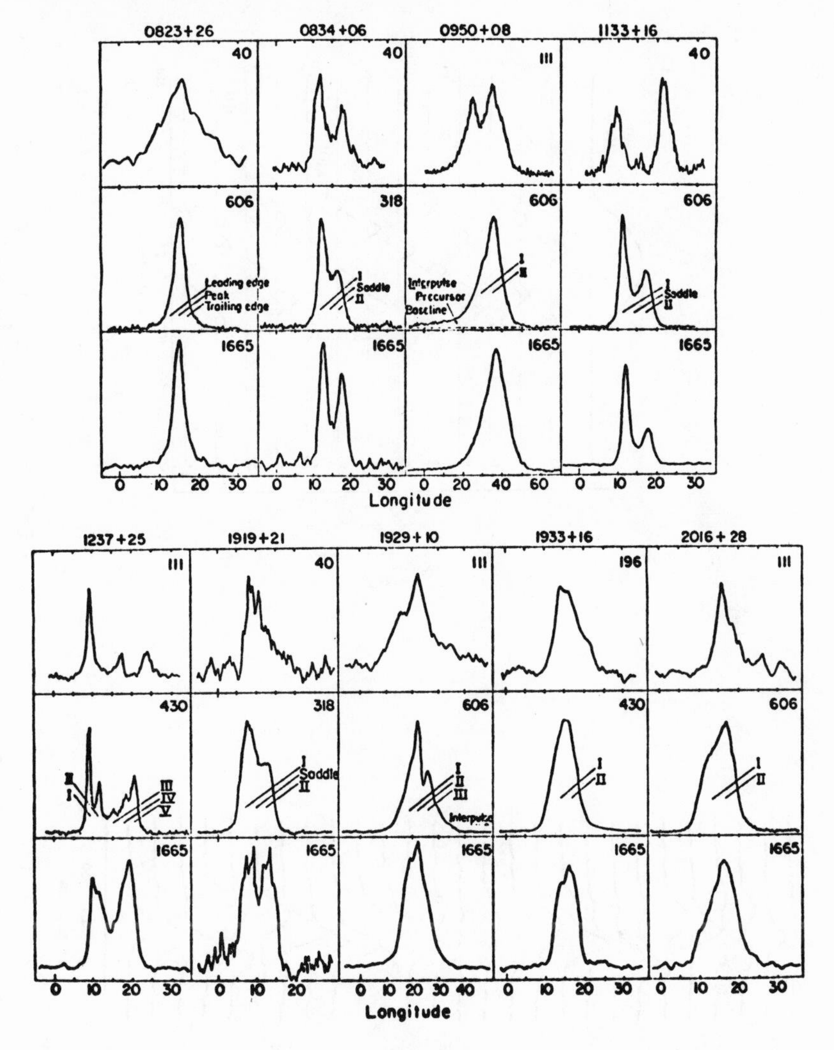

Fig. 3 Shapes of pulsar pulses at different frequencies

the marching subpulses and effects related to it. Fig.4 shows the basic manifestation of this phenomenon. Here in the right column of pulses we see a sequence of consecutive pulses with time increasing from bottom to top It is obvious from the figure that there is a pulse component which is repeating from one pulse to the next but marching forward slightly in time with each successive pulse. The terminology that goes with this phenomenon is shown in Fig. 5. P_1 is the primary period of the pulsar. P_2 is the mean time between subpulse components within an individual pulse. P_3 is the mean time interval between sequences of marching subpulses. In addition to

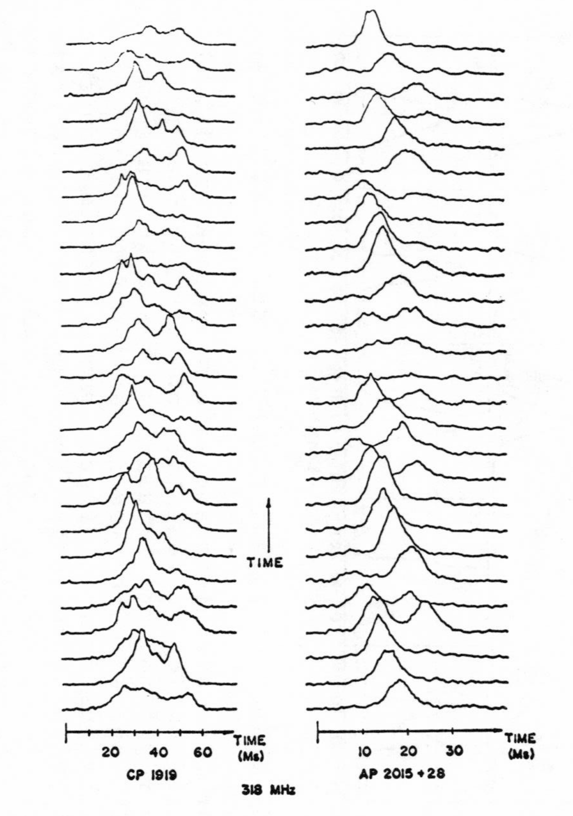

Fig. 4 The marching subpulses phenomenon

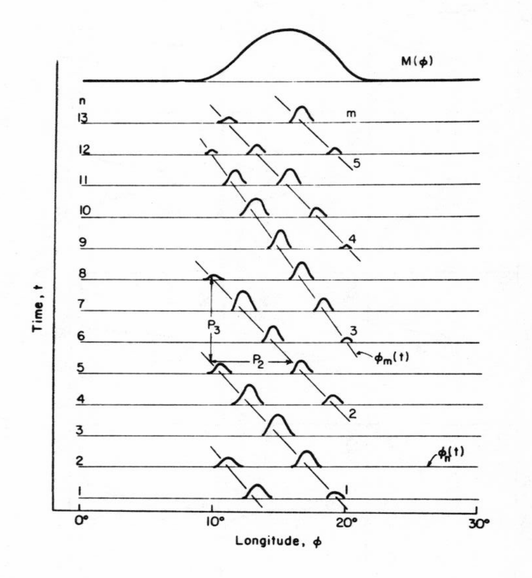

Fig. 5 Definition of different periodicities

these periods a fuller analysis of the intensities of pulses in the pulse train shows the existence of a much longer period in the pulse intensities which has come to be known as P_4. There is clear evidence that in the sequence of marching subpulses we are seeing the same radiative structure over and over each time the pulsar rotates. However, the interpretation of this phenomenon has not been established. One suggestion is that we are seeing a drift of the particles in the inhomogeneous magnetic field, in the same manner as the drift of the Van Allen belt particles around the earth.

Undoubtedly, our understanding of pulsars has made great progress in the little more than three years since their discovery. However, we are still very far from a complete theory for all aspects of this phenomenon.

REFERENCES

1) Pacini, F., Nature 219, 145 (1968)
2) Ostriker, J. and Gunn, J., Ap. J. 157, 1395 (1969)
3) Gold, T., Nature 218, 731 (1968)
4) Pacini, F. and Rees, M., Nature 226, 622 (1970)
5) Gold, T. (private communication)
6) Drake, F. and Craft, H.D., Nature 220, 231 (1968)

PROPAGATION OF RELATIVISTIC ELECTROMAGNETIC WAVES IN A PLASMA

F. W. Perkins and C. E. Max

Plasma Physics Laboratory, Princeton University

Princeton, N. J.

Two new results concerning the propagation of electromagnetic waves with a strength parameter $\nu = eE/m\omega c$ sufficiently large ($\nu \gg 1$) to cause relativistic electron velocities are reported. The first is an analytic solution of the nonlinear equations for linearly polarized waves in a uniform medium. Secondly, propagation in a nonuniform medium increases the nonlinear penetration effect; the nonrelativistic plasma frequency ω_p required to reflect a strong wave is $\omega_p^2 \approx (\omega L/c)^{1/2}(eE_i\omega/mc)$ where L is the density gradient scale length and E_i the electric field in the absence of a plasma.

We consider the propagation of strong, linearly polarized waves[1-4] in a cold, uniform plasma with no magnetic field, retaining all self-consistent fields. The solution for circularly polarized waves is already available.[1,2] Akhiezer and Polovin[1] have derived the nonlinear equations governing linearly polarized waves propagating in the z-direction:

$$\frac{d^2\rho_x}{d\zeta^2} + \left(\frac{1}{\beta^2 - 1}\right)\frac{\beta\rho_x}{\beta(1+\rho^2)^{1/2} - \rho_z} = 0 \tag{1}$$

$$\frac{d^2}{d\zeta^2}\left[\beta\rho_z - (1+\rho^2)^{1/2}\right] + \frac{\rho_z}{\beta(1+\rho^2)^{1/2} - \rho_z} = 0\ . \tag{2}$$

Here $\rho = p/mc$ is the dimensionless electron momentum, βc is the phase velocity, and all spatial and temporal dependence occurs in the combination $\zeta = \omega_p c^{-1}(z - \beta ct)$, where ω_p is the nonrelativistic plasma frequency. The formulas for the electric field, etc., in terms of ρ are in the paper of Akhiezer and Polovin.[1] Our solution, which is valid in the limit $\beta \gg 1$, and where $|\rho_x| \gg 1$ throughout most of a period yields

$$\rho_x = \rho_o - \frac{(\zeta - \frac{1}{4}P)^2}{2(\beta^2 - 1)} \qquad 0 \leqq \zeta \leqq \frac{P}{2}$$

$$= -\rho_o + \frac{(\zeta - \frac{3}{4}P)^2}{2(\beta^2 - 1)} \qquad \frac{P}{2} \leqq \zeta \leqq P \tag{3}$$

where $\rho_o = (\frac{1}{4}P)^2 [2(\beta^2 - 1)]^{-1} \gg 1$. The solution[5)] for ρ_z is $\rho_z = \rho_o R/\beta$ where

$$R = A(1-\eta^2) - \frac{2}{3}\eta^2 + \frac{1}{3}(1-\eta^2)\,\ell n(1-\eta^2) \tag{4}$$

and

$$A = \frac{1}{6}\,\ell n \frac{36\beta^2\rho_o^2}{4\rho_o^2 + 9\beta^2} \tag{5}$$

$$\eta = (4\zeta/P) - 1 . \tag{6}$$

Equation (4) is not valid in a small boundary layer near $\eta = 1$ where special considerations must be used.

Figure 1 shows the wave form of the linearly polarized wave.

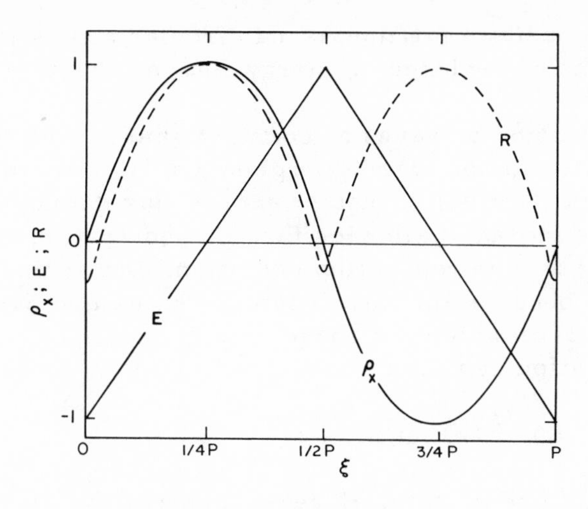

Fig. 1

Strong linearly polarized electromagnetic waves. The curves show transverse ρ_x and longitudinal R components of electron momentum [see Eq. (4)] , as well as the transverse electric field. All quantities have been normalized to their maximum value. The period is P.

The results for linearly and circularly polarized waves are given in the following table.

Table I

Polarization	Propagation Condition	Energy Flux
Linear	$1-\frac{\pi}{2}\frac{\omega_p^2}{\omega^2\nu}>0$	$\frac{cE^2}{72\pi\beta}\left[\log\left(\frac{576\beta^2\rho_o^2}{4\rho_o^2+9\beta^2}\right)+\frac{2}{3}\right]$
Circular	$1-\frac{\omega_p^2}{\omega^2\nu}>0$	$\frac{cE^2}{4\pi\beta}$

The propagation condition for linear and circular waves is almost identical, as is the dependence of energy flux on phase velocity β.

The propagation of waves at constant energy flux allows a "bootstrap" penetration of overdense plasmas: As a wave nears its reflection point ($\beta \rightarrow \infty$), constant energy flux demands that the electric fields increase, allowing further penetration of the plasma according to the propagation condition. But eventually reflections occur because the wavelength becomes comparable with the distance over which the phase velocity changes. This yields the propagation condition

$$\omega_p^2 < (\omega L/c)^{1/2} (eE_i \omega/mc) \tag{7}$$

to within factors of order unity. Here E_i denotes the electric field strength in vacuum and L is the density gradient scale length.

DISCUSSION

The reason why strong electromagnetic waves penetrate plasmas more readily than small-amplitude waves is that the plasma current is limited to the value nec, instead of increasing with E as $ne^2E/m\omega$. A wave will be reflected only if the plasma current is large enough to cancel the displacement current. Relativistic effects thus diminish the ability of the plasma to act as a dielectric. The model of fixed ions limits our calculation to values of $\nu = eE/mc\omega$ in the range $1 < \nu < M_i/m_e$. Qualitatively one expects that when ions become relativistic, the current they generate will be subject to the same limitation and will not cause major changes in the propagation condition. In linearly polarized waves with phase velocity large compared to the velocity of light, both the longitudinal motion of electrons and their density perturbations are small, because of the self-consistent electrostatic fields. Nevertheless the longitudinal motion is important, since particle energy flux exceeds field energy flux in linearly polarized waves. One additional caution must be given: the solutions presented here may not be stable; indeed strong ac fields lead to instabilities in nonrelativistic plasmas.[6]

The strong magnetic dipole radiation believed to occur in pulsars[3,7] is the principal astrophysical application of our results. Figure 2 shows plasma density required to reflect the dipole radiation according to (7), with the additional assumption that the plasma density scale length L is of the order of the distance R from the pulsar.

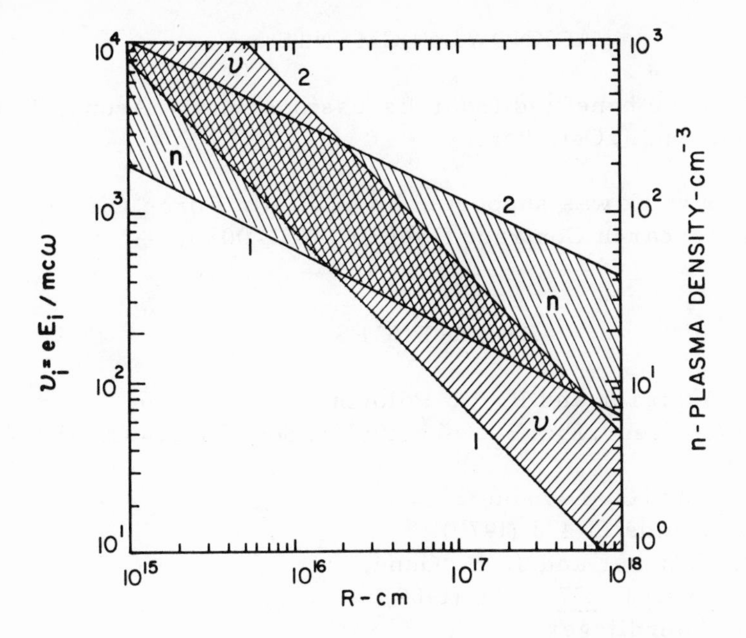

Fig. 2

Vacuum strength parameter $\nu_i = eE_i/mc\omega$, and plasma density [see Eq. (7)] required to reflect electromagnetic waves, vs radius R. The curves are based on magnetic dipole radiation from the Crab pulsar (cf. Ref. 3 and 7) with surface fields in the range from (1) $6 \cdot 10^{11}$ to (2) $4 \cdot 10^{12}$ gauss. In evaluating (7), we assume L = R.

The term involving L in (7) is an important feature not found in previous theories.[2,3,7] One can conclude that it may be possible for pulsar radiation to penetrate to large distances in the medium of the Crab, thus allowing the possibility of an energy source in these large volumes. On the other hand, penetration of the filaments with their higher densities and shorter scale lengths appears doubtful. Thus the results presented here would support the hypothesis that the "amorphous mass" region of the Crab is filled with magnetic dipole radiation, since the electron density in this region is less than 1 cm^{-3}.[8]

ACKNOWLEDGMENTS

We have benefited from discussions with J. Arons, P. Kaw, R. Kulsrud, and J. Ostriker.

This work was supported by U.S. Air Force Office of Scientific Research Contract F 44620-70-C-0033.

REFERENCES

1. A. I. Akhiezer and R. V. Polovin
Zh. Eksp. Teor. Fiz. 30, 915 (1956); [Sov. Phys.-JETP 3, 696 (1956)]
2. P. Kaw and J. Dawson
Phys. Fluids 13 472 (1970)
3. J. P. Ostriker and J. E. Gunn,
Astrophys. J. 157, 1395 (1969)
P. D. Noerdlinger
Phys. Fluids 14, 999 (1971);
J. E. Gunn and J. P. Ostriker
Astrophys. J. 165, 523 (1971)
4. H. R. Jory and A. W. Trivelpiece
J. Appl. Phys. 39, 3053 (1968)
5. E. Kamke, Differentialgleichungen: Lösungsmethoden und Lösungen I. (Akademische Verlagsgesellschaft Geest und Portig K.-G., Leipzig, 1961), p. 453, Eq. 2.231.
6. P. K. Kaw and J. M. Dawson
Phys. Fluids 12, 2586 (1969)
D. F. Dubois and M. V. Goldman
Phys. Rev. Letters 14, 544 (1965)
7. F. Pacini
Nature 219, 145 (1968)
J. P. Ostriker and J. E. Gunn
Astrophys. J. (Letters) 164, 195 (1971).
8. I. S. Shklovsky,
Supernovae (Wiley, New York, 1968), p. 261.

A THREE-DIMENSIONAL RELATIVISTIC COMPUTATION FOR THE PULSAR MAGNETOSPHERE*

G Kuo-Petravic†, M Petravic†, and K V Roberts$^{\phi}$

†Dept of Engineering Science, Oxford University

$^{\phi}$Culham Laboratory, UKAEA, Abingdon, Berks.

1. INTRODUCTION

Astronomical objects for which gravitational forces are dominant are usually spherical in form, and many computer calculations have been successfully carried out in this 1-dimensional regime. Examples include stellar structure and evolution [1], stellar pulsation [2] and collapse [3], and the dynamics of star clusters [4]. When rotation is important the geometry typically becomes axially symmetric (or 2-dimensional) as in a disc-shaped galaxy, and if a magnetic field is added one reaches the full 3-dimensional case. There are many situations in cosmical physics and astrophysics where it would be useful to be able to obtain detailed computer solutions of sets of coupled time-dependent 3-dimensional magnetofluid equations. Depending on the problem these equations should include a variety of physical effects, e.g. gravitation, special or general relativity, ionization, radiation or pair production processes and so on.

To facilitate calculations of this kind the authors have developed a general-purpose program generator called ZODIAC, to be briefly described in §4. Essentially, the ZODIAC generator enables the physicist to write down sets of equations in a concise symbolic vector form, equivalent to the notation of mathematical physics. These equations are then automatically converted into efficient target code for any chosen type of computer. The advantage is that new physical effects can quickly be added, so that ZODIAC provides

---

* Work carried out in co-operation with the Max Planck Institut für Plasmaphysik, Garching, F.R. of Germany.

a flexible research tool for many different astrophysical calculations.

This paper describes one example of current interest, namely a numerical study of the behaviour of the magnetosphere surrounding a rapidly rotating neutron star. It is believed [5] that this provides the underlying structure for pulsars although the detailed radiation mechanism is not yet clear. We explain how the model is set up on the computer and give some preliminary results. The calculations carried out so far have been scale-free but where appropriate the parameters corresponding to the Crab pulsar CP 0532 will be quoted.

In accordance with the usually accepted model [5] we assume a perfectly conducting star of radius $R_s \simeq 10^6$ cm with a dipole magnetic field of strength $B_s \simeq 10^{12}$ gauss at the poles, rotating with angular frequency ω. An important parameter is the light radius $R_L = C/\omega$. (For the Crab pulsar $\omega \simeq 60\,\pi$ sec^{-1} and $R_L \simeq 1.6 \times 10^8$ cm.) Two limiting cases have mainly been discussed in the literature. In Case 1 the magnetic dipole is oriented parallel to the rotation axis. This system cannot radiate energy _in vacuo_, but Goldreich and Julian [6] and Sturrock [7] have argued that because of the high vacuum electric field $E_{\parallel}$, parallel to the magnetic field at the stellar surface, a plasma must be generated, and the inertia of this plasma will then twist the magnetic field lines into a spiral form, causing a braking torque on the star. Case 2 assumes that the axes are orthogonal. Here one obtains twisted field lines, 30Hz radiation and a braking torque even _in vacuo_ [8], but the high $E_{\parallel}$ again suggests that a plasma must be generated.

Rotation can supply the power source, and explain the pulsed emission [8] provided that the axes are non-parallel, but it is not clear where the X-ray, optical and radio emissions are generated, nor how far the low frequency 30Hz waves will propagate. Lerche [9], for example, requires the 30Hz waves to propagate many light radii from the star in order to explain the continuum emission from the Crab nebula, while Endean and Allen [10] argue that the pulsed emissions are generated at $r \simeq 2R_L$. Gold [5] originally proposed $r \simeq R_L$, and Sturrock [7] has recently suggested that they are generated very close to the star itself, in the vicinity of the magnetic poles. In this situation it would clearly be useful to have a family of self-consistent computer calculations, carried out for a range of plasma densities, various orientations of the dipole axes, and with different physical effects and boundary conditions at the stellar surface included. We have performed one such preliminary calculation so far, assuming orthogonal axes and other parameters defined in §6, and describe this as an example of the type of work for which ZODIAC can be used.

2. METHOD OF CALCULATION

To an observer rotating with the star the solution should appear steady, apart from fluctuations which are responsible for the fine structure of the individual pulses [11]. In a fixed frame all functions therefore take the form $f(r,\theta,\varphi + \omega t)$. Our method is to transform to the moving coordinate system by a purely Galilean transformation which leaves the variables ρ, V_r, V_θ, V_φ, H_r, H_θ, H_φ, E_r, E_θ, E_φ unaltered except for the elimination of a phase factor $\exp(i\omega t)$. This avoids any artificial complications due to the fact that at large radii the coordinate frame is rotating with a speed greater than c; notice in particular that V_φ is not changed, and that there is no transformation between the electric and magnetic fields. Maxwell's equations in the moving frame then become for example

$$\frac{\partial \underline{\underline{H}}}{\partial ct} = - \text{Curl } \underline{E} - \frac{\partial \underline{\underline{H}}}{\partial \varphi}, \quad \frac{\partial \underline{\underline{E}}}{\partial ct} = \text{Curl } \underline{H} - \underline{j} - \frac{\partial \underline{\underline{E}}}{\partial \varphi} \qquad (1)$$

(the other two being unaltered), and a similar advective term $\partial/\partial\varphi$ is introduced wherever $\partial/\partial ct$ appears.

The calculation begins with arbitrary but self-consistent initial values; e.g. a static magnetic dipole field, $\underline{E} = 0$, charge neutrality, a prescribed radially symmetric plasma density, and zero plasma velocity. Physical boundary conditions are then imposed at the surface of the rotating star, and the correct solution should then propagate outwards, a steady state ultimately being reached. This method enables standard time-dependent computational techniques to be used to solve a steady state problem. It is also instructive to watch how the solution sets itself up. If the true solution exhibits turbulent or regular fluctuations [7], one would expect to obtain the mean behaviour in this way and then to study the fluctuations by a separate calculation.

Some complications are introduced by the space mesh and by the radial boundary conditions. In explicit time-dependent calculations the maximum stable timestep Δt is determined by the minimum spacestep, which leads to problems at small r and also near the poles of the coordinate system (rotation axis) since the mesh intervals are $(\Delta r, r\Delta\theta, r\text{Sin}\theta\Delta\varphi)$. In the first runs we have avoided these problems by working with an artificially increased stellar radius $R_{min} = 0.2\ R_L$ or $0.5\ R_L$, and by cutting off the calculation at an 'arctic circle' $\theta = 30^o$. If it turns out that the excluded regions are physically important they can however be taken into account.

Physical boundary conditions are required at $r = R_{min}$ and

R_{max}. The inner boundary condition is critical and will be discussed in §§ 5 & 6. It is difficult to know what to do at the outer boundary, where the calculation is artificially broken off, but no great penalty in computer time is involved if R_{max} is made sufficiently large so that effects do not propagate back during the time of interest, and this is the approach that has so far been most successful.

Symmetry enables the vacuum field calculation to be restricted to ¼ of the total solid angle. When plasma is present it is necessary to compute ½ of the total, unless the electron and ion masses are equal when again ¼ is sufficient. So far we have only studied the equal-mass case.

3. THE ENDEAN-ALLEN ANALYTIC SOLUTION

An analytic solution for a dipole orthogonal to the axis of rotation has been given by Endean and Allen [10] :

$$E_\theta = \frac{M\omega}{4\pi r^2}(kr-i)e^{i(\varphi+\omega t-kr)}\ , \ E_r = 0,\ E_\varphi = i\text{Cos}\theta\ E_\theta \tag{2}$$

$$H_\varphi = \frac{M}{4\pi r^3}((kr)^2-ikr-1)e^{i(\varphi+\omega t-kr)}\ , \tag{3}$$

$$H_r = \frac{2\text{Sin}\theta}{\omega r}E_\theta\ ,\ H_\theta = -i\text{Cos}\theta\ H_\varphi \tag{4}$$

where M is the dipole moment, $\omega = ck$, and the real part of the solution is understood. Since it is convenient to use this solution as a partial check on our computing technique we have plotted the H-field in the equatorial plane which was sketched by hand in ref.10. This is shown in Fig.1. It clearly exhibits the twisting of the field lines into a spiral structure at radii $r \gtrsim R_L$. Far out, each field line must move outwards one wavelength $\lambda = 2\pi R_L$ during each revolution, and the rotation of such a field structure appears like a plane wave in any particular direction. Notice how the 'returning field lines' appear to coalesce. This does not contradict the condition div $\underline{H} = 0$ since we are plotting only a 2D section of a 3D field.

4. THE ZODIAC GENERATOR

The program generator enables a physical equation such as the first Maxwell equation (1) to be programmed in the form [12,13]

```
EQUATE(H,DIFF(H,MULT(DCT,SUM(CURL(E),DPHI(H)))));        (5)
```

which is independent of the coordinate system, the number of dimensions, the difference scheme used, and of the computer on which the optimized target code is to run. Formula (5) is then interpreted as an Algol 60 statement, which when executed automatically generates the required code and punches it out on cards. This deck of cards is then used as a subroutine in a master Fortran program which contains the necessary housekeeping and control modules, printing and graphical routines and so on. The target code may be in any language but so far IBM 360 assembler code and Fortran have been used for production runs. To include a new physical term it is simply necessary to repunch one or more statements such as (5) and to return the generator, taking of order 10 seconds on the IBM 360/91. To change the coordinate system or the difference scheme, the appropriate modules in ZODIAC are 'unplugged' and replaced by others. Further details are given in refs. 12 & 13.

5. VACUUM CALCULATION

As a test of the computational method, Maxwell's equations were solved in the region of calculation ($R_{min} \leqslant r \leqslant R_{max}$, $\pi/6 \leqslant \theta \leqslant \pi/2$, $0 \leqslant \varphi \leqslant \pi$) using the leapfrog difference scheme. Ideally the leapfrog difference equations take a very simple form, pointed out by Buneman [14], in which the six quantities $\underline{E}$, $\underline{H}$ are all defined on separate lattices. However, since in a full plasma calculation we do need all the variables defined at every mesh point in order to evaluate the total stress tensor, this general scheme was used for the test case also.

The initial field was that of a static dipole, with $\underline{E} = 0$. The time-dependent equations (1) were then solved with H_θ , H_φ held constant on the two innermost spherical shells $r = R_{min}$, $R_{min} + \Delta r$. The results showed that a good approximation to the analytic solution is obtained if $R_{min} \simeq 0.2$ and $R_{max} \simeq 8.5$. Notice that with this difference scheme it is necessary to specify the boundary conditions on two shells in order to avoid decoupling of the lattices. Only two components H_θ and H_φ need however be specified, and the equations div $\underline{H} = 0$, div $\underline{E} = 0$ are guaranteed by the initial conditions and by the conservative form of the difference equations.

6. PLASMA CALCULATION

In formulating the plasma equations it is necessary to make a number of decisions of a physical nature. We have endeavoured to choose the simplest model that is capable of giving meaningful results. Once this has been understood it should be a straightforward matter to try out various alternative assumptions. The equations used are :

$$\frac{\partial n}{\partial ct} = - \operatorname{div}(n_o \underline{u}) + \frac{\partial n}{\partial \varphi} , \tag{6}$$

$$\frac{\partial (n\underline{u})}{\partial ct} = - \operatorname{div}(n_o \underline{u}\underline{u}) + \frac{e}{mc^2} n_o (\gamma \underline{E} + \underline{u} \times \underline{H}) + \frac{\partial (n\underline{u})}{\partial \varphi} \tag{7}$$

(separate equations for positive and negative particles), Maxwell's equations (1), and

$$\underline{j} = 4\pi e \, (n_{o+} \underline{u}_+ - n_{o-} \underline{u}_-) \quad , \tag{8}$$

where

$$\underline{u} = \gamma \frac{\underline{v}}{c} , \ \gamma = \frac{1}{\sqrt{1 - v^2/c^2}} = \sqrt{1+u^2} \ , \ n = \gamma n_o \ . \tag{9}$$

It is necessary to use special but not general relativity, and gravitational forces have been neglected compared to electromagnetic effects. Since there may well be regions in which one sign of charge dominates, the 2-fluid equations have been used. We have however set the masses of the positively and negatively charged particles equal in order to simplify the calculation and to reduce the region of calculation as mentioned in §2.

Viscosity, resistivity and heat conduction introduce complications in a relativistic calculation, and since the dominant effects are believed to be inertial they have been omitted. For the same reason the plasma pressure has been assumed to be zero.

The most serious computational problem is likely to be the large numerical coefficient which occurs as a factor in the third term of (7), which is effectively the ratio of the gyrofrequency $eH/\gamma mc$ to the angular frequency ω of the star. This has the value 5×10^{13} for zero-energy protons at the stellar surface, but decreases as $1/r^3$ in the dipole field and $1/r$ in the wave zone. Some alleviation will occur if $\gamma \gg 1$ and $\underline{E} + \underline{v} \times \underline{H} \approx 0$, but computers cannot deal readily with frequency ratios of this magnitude. The standard theoretical technique would be to replace the transverse components of the equations of motion (7) by guiding-centre equations, thus averaging over the gyromotion, but it is unclear how to proceed when $\gamma \gg 1$ and $E/H \simeq c$, and the equations would become complicated to solve numerically. To obtain preliminary results we have therefore arbitrarily increased the particle mass m to about 10^7 proton masses, a device which is perhaps equivalent to the Von Neumann technique [15] of artificially increasing the viscosity in order to compute shocks. Once a steady-state solution has been reached this factor can be varied in order to determine the errors that it introduces.

The main physical uncertainty concerns the plasma boundary conditions at $r = R_{min}$, which together with the fixed boundary conditions on $\underline{H}$ should uniquely determine the solution. As a start, it has been assumed that the total masses of both the positive and the negative ions are held constant on each of the two innermost shells, but that the densities are free to adjust themselves in the θ and φ directions. Thus the inner boundary can act as both a source and sink of particles.

7. RESULTS

Fig.2 shows the velocity fields at a particular angle at time step 300, when the 'switch-on' wave has reached just beyond twice the light cylinder. As expected, the velocity component u_φ is comparable with the co-rotation velocity but smaller. Fig.3 shows how the wave moves on between steps 100 and 300. The angular positions for the two figures are $\theta = 75^{o}$ and $\varphi = 52.5^{o}$.

8. ACKNOWLEDGEMENT

The authors would like to thank the Institut für Plasmaphysik, Garching, Germany for making available their excellent IBM 360/91 computing facilities.

REFERENCES

1. L G Henyey and R D Levee, Methods of Computational Physics 4 (1965); R. Kippenhahn, A Weigert and E Hofmeister loc cit 7 129 (1968).

2. R F Christy, Methods of Computational Physics 7 191 (1968).

3. M M May and R H White, Methods of Computational Physics 7 219 (1968); S A Colgate and R H White, Ap.J. 143 626 (1966); J R Wilson Ap.J. 163 209 (1971).

4. L H Spitzer Jr. and M H Hart, Ap.J. 164 399 (1971), 166 483 (1971).

5. T Gold, Nature 221 25 (1969).

6. P Goldreich and W H Julian, Ap.J. 157 869 (1969).

7. P A Sturrock, Ap.J. 164 529 (1971).

8. J P Ostriker and J E Gunn, Ap.J. 157 1395 (1969).

9. I Lerche, The Astrophysical Journal, 159 229, (1970).

10. V G Endean and J E Allen, Nature 228 348 (1970).

11. F G Smith, M.N.R.A.S. 149 1 (1970).

12. K V Roberts and J P Boris, 'The Solution of Partial Differential Equations using a Symbolic Style of Algol', Journal of Computational Physics, in the press.

13. M Petravic, G Kuo-Petravic and K V Roberts, 'Automatic Optimization of Symbolic Algol Programs. I.General Principles' Culham Laboratory Preprint CLM-P274 (1971).

14. O Buneman, in 'Relativistic Plasma', ed. O Buneman and W B Pardo, P.205, Benjamin, New York, 1968.

15. R D Richtmyer and K W Morton, 'Difference Methods for Initial Value Problems', Interscience, New York, 2nd Ed. 1967.

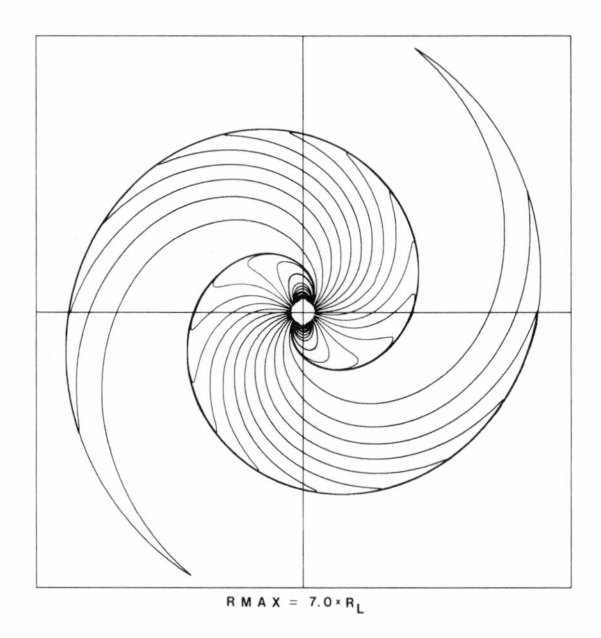

Fig.1 Magnetic field lines in the equatorial plane for the Endean-Allen analytic solution.

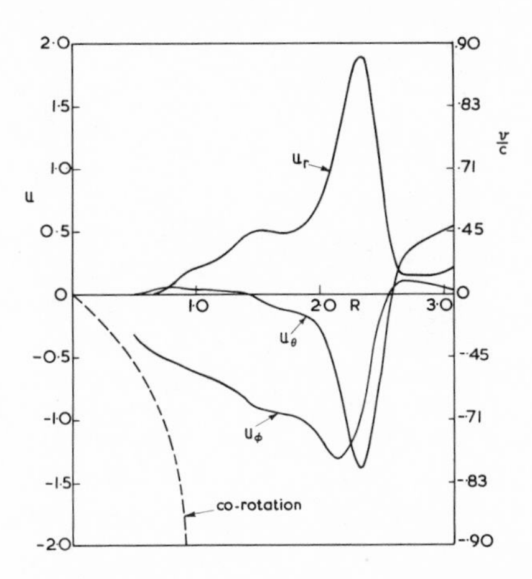

Fig.2 The three components of velocity $u(= \frac{v}{c}/\sqrt{1 - \frac{v^2}{c^2}})$ as a function of radial distance from star measured in units of light cylinder radius. The dashed line shows the value of u_φ which the particles would have if they were co-rotating with the star.

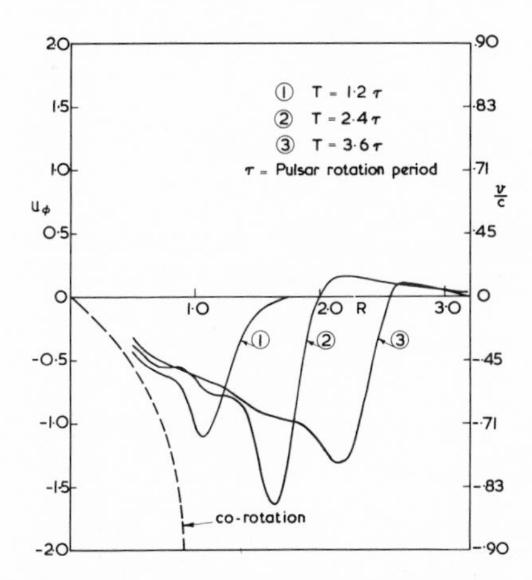

Fig.3 The variation of the azimuthal component of velocity $u_\varphi(= \frac{v}{c}/\sqrt{1 - \frac{v_\varphi^2}{c^2}})$ as a function of radial distance from star measured in units of light cylinder radius. This figure shows the time development of u_φ. At T = 0 the rotation of the star is 'switched-on'. τ is the star rotation period.

ON THE ORIGIN OF PULSAR RADIATION

V.V. Zheleznyakov

Gor'kii Radiophysics Institute

I. INTRODUCTION

The pulsar theory involves three main questions: 1) type and constitution of the central body (of the star); 2) structure of the magnetosphere; 3) mechanisms of radio emission from pulsars. A successive solution of these problems is a desirable way to develop the theory. It is very complicated to carry out the program especially if we take into account that the criterion of validity of the theory is at the very end of the scheme and involves the comparison of theoretical peculiarities of radiation with the observed ones. This lecture deals with the more limited problem of the choice of those conditions in the magnetosphere of the star (the position and velocity of the source, the magnetic field strength, the spectrum of charged particles, etc.) which provide generation of radiation with observed characteristics.

II. FORMATION OF RADIATION DIAGRAM OF PULSARS

At present it is generally accepted that a pulsar is a rotating dense (neutron) star. The rotation period defines the period of pulse repetition. Pulses are formed by beaming radiation of the pulsar. Since the beaming appears in all kinds of radiation ranging from radio emission to gamma rays it is clear that this is the result of one common effect independent of the concrete mechanism of generation. The formation of the narrow diagram of pulsar radiation is asserted to take place due to the source motion round the neutron star with a velocity close to the velocity of light (Smith 1969, 1970). The effect of relativistic beaming enables one to understand the whole series of important properties of the observed pulsar radiation (Zheleznyakov, 1971).

1) This effect automatically explains the pulse character of radiation connecting the beaming directly with rotation of the star. Interrupting pulses occur if the velocity v_o of the source corotating with the star is close to the speed of light c. For example, for PSR 0833 (the period $P=9\times10^{-2}$sec, pulse duration $\Delta P/P\simeq 1/45$, the index of radio emission $\alpha\simeq1$) we obtain $v_o=0.8c$; the beam width $\Delta\theta=40°$ and the radius of the source orbit $r=1.3\times10^8$cm. The ratio $\Delta P/P$ appears to be small even at comparatively wide diagrams of radiation (because of shortening of the pulse radiating along $\vec{v}_o$). Due to this it is possible to register the radio emission from a considerable part (up to 1/3) of the existing pulsars despite their "pencil" beam.

2) The pulse duration does not depend on the frequency if the relativistic source possesses a power frequency spectrum. This explains that the pulse duration of radio emission does not depend (or depends rather weakly) on the frequency since the power radio emission spectrum is typical of pulsars.

3) A complicated profile of the pulses may have several reasons.

a) A few local sources in the magnetosphere of the pulsar. Examples are NP 0532, NP 0527 and CP 0950 in which we observe interpulses approximately in the middle between the main pulses. To explain these interpulses in the frame of the pulsar model with the "pencil" beam oriented along the source velocity, it is necessary to assume the existence of the second source which is localized almost diametrically opposite the first (see Fig.1). The sources of optical and radio emission in each pulse are shown in the figure lying in the same plane passing through the rotation axis of the star (for example, above the poles of the dipole magnetic field of the star). This provides sending radio and optical pulses to the Earth simultaneously with an accuracy of $\Delta t<1$msec as pointed out by Rankin et al. (1970).

b) Anisotropy of diagrams in the frame of reference connected with a source. Such anisotropy may cause the

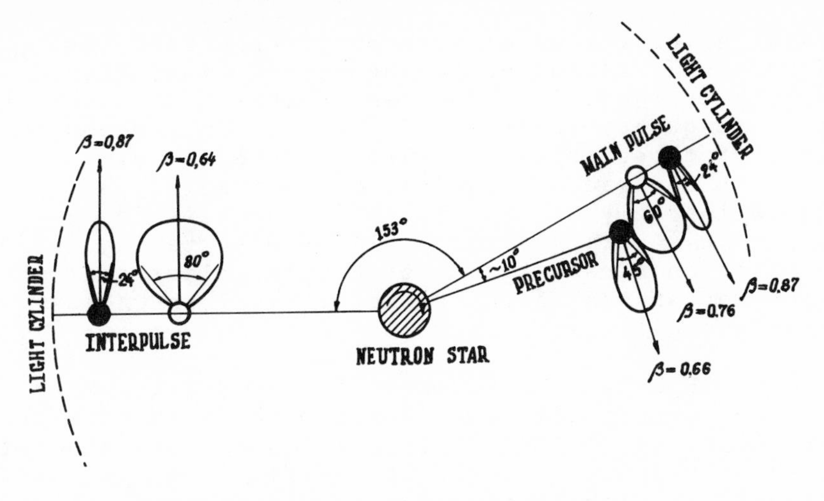

Fig. 1. Positions of optical (black circles) and radio emission (light circles) sources projected into the equatorial plane and radiation patterns providing the observed pulse duration of the Crab pulsar.

double-peaked pulses typical of many pulsars (CP 1919 a.o.).

c) Time variation of the radiation power of the source. Such a modulation may be considered as the simplest reason for occurrence of the second period in pulsars ("marching" subpulses). Because of Doppler effect the modulation period fixed by the observer on the Earth is $P_2 = P_2'(1-\beta^2)^{-1/2}(1-\beta\cos\theta)$, where P_2' is the modulation period in the frame of reference accompanying the source; θ is the angle between the velocity $\vec{v}_0$ and the direction to the Earth; $\beta = v_0/c$. At $\beta \simeq 1$ the subpulse duration at the edges of the pulse (where $\theta \simeq (1-\beta^2)^{1/2}$) will become two times longer than in the middle according to the observational data obtained by Backer (1970) for 1919. Note that the account of relativistic aberration in more complicated variants of the origin of marching subpulses leads to analogous results relative to the change of subpulse duration over the pulse.

III. MECHANISMS OF OPTICAL, X-RAY AND γ-RADIATION

When solving the problem of radiation mechanisms of pulsars one should bear in mind essential differences in the character of radio emission on the one hand and optical, X-ray and γ-radiation on the other. They are referred both to the brightness temperature and to the behavior of frequency spectra. If we take into account the existence of three types of pulsars: type I - with emission only in the radio range (most pulsars); type II - with radiation only at the higher frequencies in X-rays (probably pulsar CEN X-3); type III - with radio, optical and X-radiation (Crab pulsar) it becomes clear that the radiation mechanisms in the radio, optical and X-ray range must be different.

In the RF range the registered flux density from Crab pulsar corresponds best to the high brightness temperature T_b (up to 10^{28}°K) which indicates the coherent mechanism of radiation (see section IV). For optical radiation and X-rays T_b is considerably smaller (about 10^{11}°K at $\lambda \approx 0.5\mu$) proving that the attempt to associate this radiation with the action of incoherent mechanisms is correct.

If the radiation in the infrared, optical and X-ray ranges is produced by the incoherent synchrotron mechanism (Bertotti et al. 1969; Ginzburg and Zheleznyakov 1969, 1970; Shklovsky 1970) and γ-radiation arises from the inverse Compton effect, one may select the energetic spectrum of relativistic electrons and other parameters of the source such that its radiation corresponds to the observed level of emission from the Crab pulsar (Zheleznyakov and Shaposhnikov).

The frequency spectrum of the radiation has a maximum possibly in the optical range and decreases towards infrared and X-rays (see Fig. 2). The γ-ray spectrum is unknown; however, the measured flux of γ-radiation with the energy $h\nu > 0.6$MeV per photon (Hillier et al., 1970) corresponds to the frequency spectrum with the index $\alpha = 0.2$ lying considerably higher than the extension of the spectrum from the X-ray region. The detailed character of the spectrum between optical radiation and X-rays is also not clear. We shall consider that in this interval the spectrum is also power law type with the index $\alpha = 0.2$ providing a continuous transition from the observed flux in optics to that in X-rays at 1 keV. The observed power spectrum with the index $\alpha = 1.2$ in the X-ray region will be provided if the reabsorption is inessential and the energetic spectrum of electrons radiating over this interval is also a power spectrum with the index $\gamma = 2\alpha + 1 \approx 3.4$ (for electrons with the energy E E^{*}). The power energy spectrum of electrons with the index $\gamma = 2\alpha + 1 \approx 1.4$ for particles

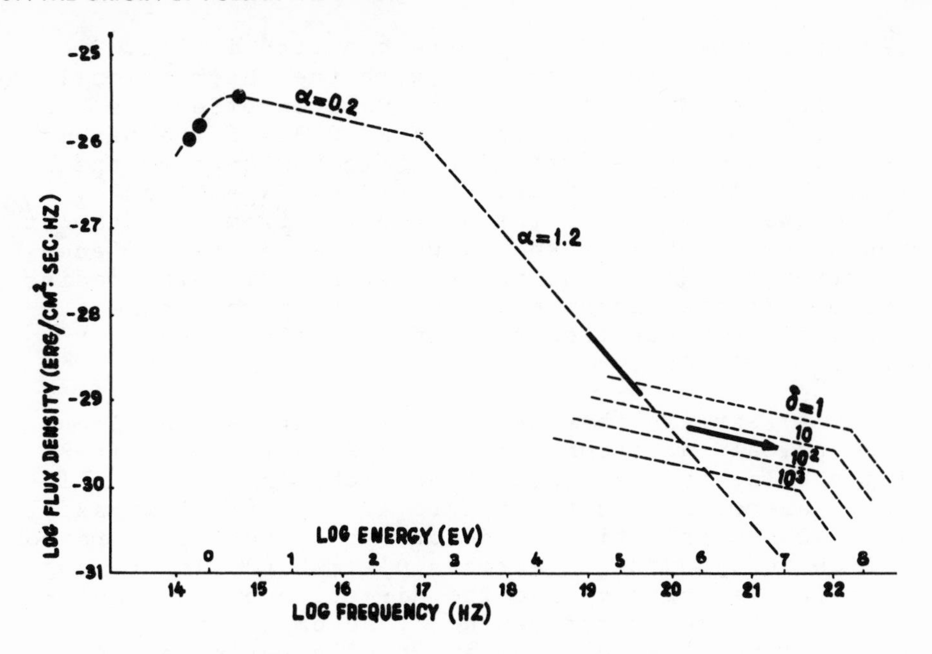

Fig. 2. The frequency spectrum of NP 0532 (the main pulse). Solid lines and points are the observed values. Dashed lines are the theoretical values.

whose energy does not exceed E^{*} corresponds to the power spectrum with $\alpha \simeq 0.2$ in the interval of "optics-X-ray up to 1 keV". The cut-off in the frequency spectrum when passing from optical to infrared rays may be associated with the synchrotron reabsorption in the infrared. As a result we have that the linear source dimension is $L \sim 5\times10^{7}\delta^{1/17}$cm; the magnetic field is $H\sim6\times10^{4}\delta^{4/17}$oe; $E^{*}\sim 3\times10^{8}\delta^{-2/17}$ev; density of electrons radiating in the range from infrared to X-rays is $N_e\sim6\times10^{11}\delta^{-7/17}$ electrons cm^{-3}. Here all values are expressed through the parameter δ - the ratio of the energy density of the magnetic field to that of electrons N_e. For the synchrotron mechanism the parameter δ is to be determined (for the stability of the radiating system it must not become less than unity). System parameters depend weakly on the concrete value δ which may be estimated according to the observed γ-rays, having assumed that they are produced by the inverse Compton effect.

The flux of γ-radiation from electrons giving also the synchrotron radiation ranging from the infrared to X-rays at the different values of the parameter δ is characterized by a set of spectra depicted in Fig. 2 (right). From their comparison with the observational data it is clear that the inverse Compton effect is enough to create γ-radiation from NP 0532 if the parameter $\delta \sim 30$. Finally we determine the parameters of the source: $L \simeq 6\times10^{7}$cm; $H \simeq 10^{5}$oe; $E^{*} \simeq 2\times10^{8}$ev; $N_e \simeq 10^{11}$cm^{-3}. If the magnetic field in the source far from the star at distance $r \simeq 1.3\times10^{8}$cm is known, we can estimate surface field strengths H_o of the neutron star having the radius $r_o \simeq 10^{6}$cm. If the magnetic field in the magnetosphere does not differ much from the dipole field, then $H_o \sim Hr^3/r_o^3 \sim 2\times10^{11}$oe.

We may explain the turn of the polarization plane of the optical radiation from NP 0532 by the angle140° during a pulse by the synchrotron mechanism if the relativistic aberration is taken into account (Zheleznyakov, 1971). The polarization plane of the optical synchrotron radiation in opposite directions $\theta' = \pm\pi/2$ (θ' is the angle between the velocity $\vec{v}_o$ and the direction of the radiation in the frame of reference A', accompanying the source) will be the same for any configuration of the magnetic field if the densities of relativistic electrons radiating in these directions coincide. The directions $\theta' = \pm\pi/2$ correspond to the edges of the radiation diagram in the frame of reference connected with the observer and to the beginning and end of the pulse of radiation. In the system A', the polarization planes coincide at $\theta' = \pm\pi/2$; the aberration effect when passing into the system A leads to the difference in the position angles at the edges of the radiation pattern (Fig. 3). The character and direction of the turning of the polarization plane over the pulse is determined by the concrete structure of the magnetic fields in the source.

IV. MECHANISMS OF RADIO EMISSION FROM PULSARS

Mechanisms of radio emission from pulsars must be coherent, i.e. subject to the condition that the radiation power of an object $W' > aL^3$ (a is the emissivity per unit volume of the source which is composed of spontaneous radiation of separate particles). Note that for the incoherent mechanism $W' \lesssim aL^3$. Radiation of particles with the equilibrium velocity distribution (the thermal radiation) is a particular case of such a mechanism. In this case $T_b \leq T \lesssim \bar{E}/\kappa$ where T is the kinetic temperature, $\bar{E}$ is the mean energy of particles, κ is the Boltzmann constant. As a rough estimate the condition $T_b \lesssim \bar{E}/\kappa$ is also valid for other incoherent mechanisms. In case of pul-

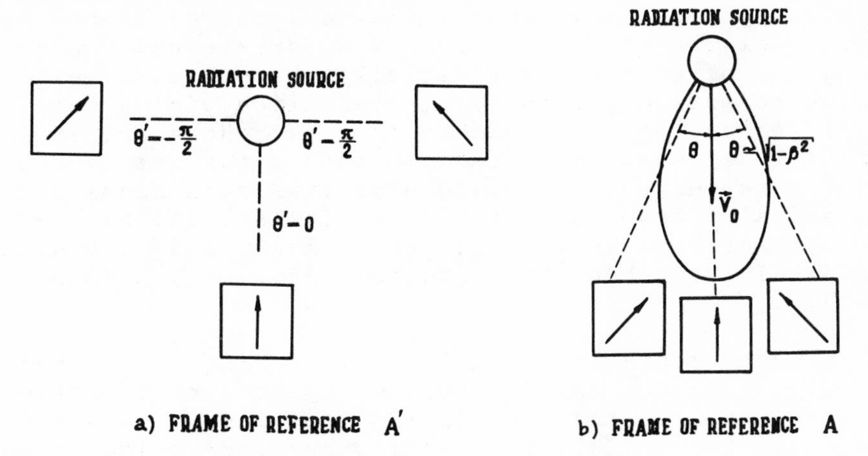

Fig. 3. Explanation of the turn of the polarization plane of optical radiation from Crab pulsar.

sars $\bar{E}\gtrsim 10^{24}$ev! Therefore, one should prefer coherent mechanisms which raise the radiation flux up to the observed level without the assumption of the very high energy of radiating particles.

In principle two variants of coherent mechanisms are possible:

a) The "antenna" mechanism in which the criterion $W'>aL^3$ is provided due to preliminary phasing (bunching) of particles. The phase of the spontaneous radiation of separate particles is not of random character and when summing the fields the radiation power W' is obtained greater than aL^3;

b) The maser mechanism does not need such initial phasing of particles. There is primarily produced (and supported) only such regular particle distribution over momenta and energies as provides the inverse population of energetic states. In such a system the reabsorption coefficient $\mu<0$ due to which the amplification of the spontaneous radiation propagating in the system takes place. Then again $W'>aL^3$; this means that the particle radiation is again phased but under the action of the amplifying radiation (auto-phasing). Over a certain stage the maser mechanism is reduced to the antenna one: in the steady-state amplification regime the electron motion in the system is equivalent to a complex system of electric currents whose phase varies considerably at distances of the order of the wavelength λ. If we create and support this system of currents artificially, namely using external electro-motive forces, their radiation yields the same effect as the maser mechanism does. However, it is practically impossible to realize such a current system without maser effect. Therefore antenna mechanisms of radio emission from pulsars (in particular, the mechanisms by Eastlund 1968, 1969; Lerche 1970, and others) are scarcely probable (see in detail Ginzburg, Zheleznyakov 1970).

In maser mechanisms the intensity I increases exponentially along the ray due to the negative reabsorption in the source. In the case $|\mu L| \ll 1$ the maser mechanism is noneffective: $I \simeq aL$, i.e. it is the same as the incoherent mechanism. If $|\mu L| \gg 1$ the intensity increases sharply by a factor of $\exp|\mu L|/|\mu L|$ as compared with the value aL (at μ=const.).

Coherent maser mechanisms may be divided into two groups. In group I the negative reabsorption (amplification) takes place directly at the radio waves. In group II the enhancement occurs at plasma waves which cannot escape from dense plasma into the interstellar medium. The radio emission appears as a result of conversion (transformation) of plasma waves into electromagnetic ones. An example of group I mechanisms is the coherent synchrotron mechanism acting in the system of relativistic electrons in the magnetic field (Zheleznyakov 1966, Zheleznyakov and Suvorov 1971). An example of group II mechanisms is the plasma wave amplification in the stream-plasma system (two-stream instability) and their conversion into electromagnetic ones due to scattering on plasma particles or the other plasma waves.

In group I mechanisms the whole electromagnetic energy comes freely from the source (without conversion relaxing in general the efficiency of the generation mechanism). In group II mechanisms the radiation is weakened by conversion. However, in pulsars - the sources of small dimensions with the high brightness temperature - the conversion must be rather strong. The energy comparable with that of excited plasma waves transforms into electromagnetic radiation (due to induced scattering of plasma waves; see in detail Ginzburg, Zheleznyakov, Zaitzev 1969).

At present none of the known models of the origin of radio emission (Sturrock 1970; Chiu, Canuto 1971;

Ginzburg, Zheleznyakov 1970b; Kaplan, Tsytovich 1969; and others) is still developed such that one can be sure of its agreement with real sources of radio emission from pulsars. Note that the sources of radio emission must act independently of optical and X-ray sources (the existence of three types of pulsars and other reasons point to this circumstance). However the comparative proximity of radio and optical sources in the magnetosphere of NP 0532 permits us to conclude that the order of magnitude of the magnetic field and possibly the particle density in both sources are the same ($H \sim 10^5$oe, $N_e \sim 10^{11}$el/cm^3). Consequently, the generation of radio emission from pulsars (at least in Crab) occurs under conditions differing considerably from the solar ones. If in the solar corona the frequency of radio emission $\nu \gtrsim \nu_L$, ν_H then in the source of radio emission NP 0532 $\nu \ll \nu_L$, ν_H (ν_L, ν_H are the nonrelativistic values of the plasma frequency and electron gyrofrequency). It is not out of the question that the radiation at the lower frequencies is produced by ions rather than electrons. The problem of escaping of the radio emission from the source into the interstellar medium becomes very important.

Besides the high intensity and power frequency spectrum, the generation mechanism must provide the polarization of radio emission from pulsars. There are two possibilities of explaining the polarization characteristics. The first is to associate these characteristics completely with the conditions in the source and the effect of relativistic orbiting of the source (Smith 1970); the second to associate them with the conditions of transition of radio emission from the dense magnetosphere into the interstellar medium, the source having to generate the radiation containing one type of waves (ordinary or extraordinary) (Ginzburg, Zheleznyakov, Zaitsev 1969; Swarup et al. 1969; Zheleznyakov 1970). The realization of either this or that possibility depends on concrete conditions of escape of the radio emission. The first possibility more probably takes place when the polarization of radiation coming from the source does not vary up to emanating into interstellar medium. The linear polarization of most pulsars appears if the ordinary and extraordinary wave polarization in the source is also linear. Then, with the relativistic motion of the source, the orientation of the polarization plane remains unchanged (as in the case of the radio emission from the pulsar in Crab) when the polarization plane coincides with the orbital plane or is orthogonal to it (when observing along the direction perpendicular to the rotation axis of the star). If the polarization plane is inclined to that of the source orbit then the polarization turn takes place (PSR 0833-45 et al.) as in the case of optical radiation of NP 0532 (see Fig. 3). A peculiar change in the direction of polarization inside the pulse of the pulsar CP 0328 is explained provided that the character of normal wave polarization is circular along $\theta'=\pm\pi/2$ and linear at $\theta'=0$. Then the polarization is circular (opposite signs) at the edges of the pattern of the radiation pulse and is linear in the middle of the pulse.

V. CONCLUSION

1) The whole series of singularities of radiation from pulsars is due to the fact that the formation of the radiation diagram takes place owing to relativistic motion of the source round the star.

2) Assuming an incoherent synchrotron mechanism in infrared, optical, X-ray ranges and the Compton mechanism for the γ-radiation from the Crab pulsar gives an explanation of the frequency spectrum of the radiation observed, and permits a determination of the basic parameters of the radiating object and an estimate of the dipole surface magnetic field of the star.

3) The concrete mechanism of radio emission remains unclear. However it must undoubtedly be a coherent emission process (at least in NP 0532) in a dense plasma with a strong magnetic field ($H\sim10^5$oe).

REFERENCES

Backer, D.C., 1970, Nature 227, 692.

Bertotti, B., A. Cavaliere, F. Pacini, 1969, Nature 223, 1351.

Chiu, H.Y., V. Canuto, 1971, Ap. J., 163, 577.

Eastlund, B.J., 1968, Nature 220, 1293; 1969, Nature 225, 430.

Ginzburg, V.L., V.V. Zheleznyakov, V.V. Zaitzev, 1969, Astrophys. Space Sci. 4, 464.

Ginzburg, V.L., V.V. Zheleznyakov, 1969, UFN, 99, 514-524.

Ginzburg, V.L., V.V. Zheleznyakov, 1970, Comments Astrophys. Space Phys. 2, 167-197.

Hillier, R.R., W.R. Jacksen, A. Murray, R.M. Redfern, R.G. Sale, 1970, Ap. J. 162, L177.

Kaplan, S.A., V.N. Tsytovich, 1969, Symposium on Pulsars and High Energy Activity in Supernovae Remnants, Rome, Dec. 18-20, 1969.

Lerche, I., 1970, Nature 159, 229.

Rankin, J.M., J.M. Comella, H.D. Craft, D.W. Richards, D.B. Campbell, C.C. Counselman, 1970, Radio pulsar NP 0532 (preprint).

Shklovsky, I.S., 1970, Nature 225, 251.

Smith, F.G., 1969, Nature 223, 934; Smith F.G., 1970, Monthly Notices Roy. Astron. Soc. 149, No. 1, 1-15.

Sturrock, P.A., 1970, Nature 227, 465.

Swarup, G., S.M. Chitze, R.P. Sinha, 1969, Symposium on Pulsars and High Energy Activity in Supernovae Remnants, Rome, Dec. 18-20, 1969.

Zheleznyakov, V.V., 1971, Astrophys. Space Sci. (in press).

Zheleznyakov, V.V., 1966, JETP 51, 570.

Zheleznyakov, V.V., E.V. Suvorov, 1971, Astrophys. Space Sci. (in press).

Zheleznyakov, V.V., 1970, Radiofizika 13, 1842.

STRONG MAGNETIC FIELD EFFECTS IN THE PULSAR CRUSTS AND ATMOSPHERES

G. Kalman*†, P. Bakshi*† and R. Covert†

*Physics Department, Boston College

†Astrophysics Institute, Brandeis University

1. INTRODUCTION

Strong magnetic fields of the order $10^{12} \sim 10^{13}$ gauss are believed[1] to exist in the vicinity of the surface of neutron-stars. The existence of such high fields brings about important changes in the behavior of the crust and the surrounding atmosphere: some of these effects have been discussed previously by various authors[2,3]; in the present paper we discuss some further physical processes which can play an important role in the behavior of pulsars.

The problems we deal with are classified as follows:

1) Equilibrium and equation of state of an electron gas in a strong magnetic field; gravitational stability of such a system.

2) Magnetic polarizability of an electron gas; the problem of magnetic phase transitions.

3) Polarization matrix of a relativistic electron gas; dispersion relation for electromagnetic waves.

4) Polarization matrix for the vacuum; some consequences.

2. EQUILIBRIUM AND EQUATION OF STATE OF AN ELECTRON GAS IN A STRONG MAGNETIC FIELD

The physical effect responsible for the peculiar equilibrium behavior of the electron gas in a magnetic field is the existence of quantized LANDAU-levels and the interaction of the associated electronic magnetic moments. To describe this situation we construct a phenomenological Hamiltonian,

$$\mathcal{H} = \mathcal{H}^{o} - \sum_{i} \int_{o}^{\underset{\sim}{B}_1} \underset{\sim}{\mathcal{M}}_i (\underset{\sim}{B}) \cdot d\underset{\sim}{B} + \frac{1}{2} \sum_{ij} \mathcal{M}_i^{\mu} \psi^{\mu\nu} (\underset{\sim}{r}_{ij}) \mathcal{M}_j^{\nu} \qquad (1)$$

where $\mathcal{H}^o$ is the magnetic field-free Hamiltonian, M_i is the magnetic moment of the i-th particle and $\psi^{\mu\nu}$ is the dipole-dipole tensor potential. Such a Hamiltonian has all the desired properties (in particular, $\mathcal{M}^\mu \equiv \Sigma \mathcal{M}_i^\mu = -(d\mathcal{H}/dH^\mu)$ although $\mathcal{H} \neq \Sigma \mathcal{H}_i$) and is easily amenable to HARTREE and higher order equilibrium calculations. The second quantized version of (1) is

$$\mathcal{H} = \sum_\alpha \varepsilon_\alpha(B) a_\alpha^\dagger a_\alpha + \frac{1}{4} \Sigma \{\mathcal{m}_\alpha \mathcal{m}_\beta + \mathcal{m}_\gamma \mathcal{m}_\delta\} \langle\alpha\beta|\psi|\gamma\delta\rangle a_\alpha^\dagger a_\beta^\dagger a_\gamma a_\delta \qquad (2)$$

where $\mathcal{m}$ now is the effective magnetic moment, including the magnetic moment resulting from the magnetic field dependence of the Fermi-energy; α, etc. are labels of one-particle states in the magnetic field and $\varepsilon_\alpha(B)$ is the corresponding one-particle energy.

In the HARTREE approximation that we consider, the total energy is

$$U = \langle\mathcal{H}\rangle = E + \frac{4\pi}{2V}\mathcal{M}^2, \qquad E = \Sigma n_\alpha \varepsilon_\alpha, \ \mathcal{M} = \Sigma n_\alpha \mathcal{m}_\alpha = -\frac{\partial E}{\partial B} \qquad (3)$$

and the equation of state is determined by the density (n) dependence of the pressure P:

$$P = \frac{1}{V}\{n\frac{\partial E}{\partial n} - \frac{4\pi}{2V} M^2\} \simeq \frac{2}{3}\frac{1}{V}\{E + MB\} \quad . \qquad (4)$$

Our main concern is the situation where only a few LANDAU-levels are occupied. All the thermodynamic quantities can be evaluated for this case exactly and an illustration is given in Fig. 1. The discontinuous behaviour at the points where new LANDAU-levels are getting occupied, should be noted. Of special interest is the behavior of the logarithmic derivative

$$\gamma = d(\log P)/d(\log n) \quad . \qquad (5)$$

It oscillates between a maximum and 0, where the maximum is

$$\gamma_{max} = \frac{5}{3}\left\{1 + \frac{4}{15}\left(k + \frac{5}{2}\,\frac{2S_1 - T_2}{2S_o - k^{1/2}} - \left[1 - \frac{5}{2}\,\frac{T_1}{2S_o - k^{1/2}}\right]\frac{T_1 - 2S_o + k^{1/2}}{T_o - (1/2)\,k^{-1/2}}\right)\frac{1}{k - \dfrac{2S_1}{2S_o - k^{1/2}}}\right\} \qquad (6)$$

with

k − 1 = the number of LANDAU-levels occupied,

$$S_n = \sum_{j=o}^{k-1} (k-j)^{1/2}\, j^n \ , \quad T_n = \sum_{j=o}^{k-1} (k-j)^{-1/2}\, j^n \quad .$$

This function is given in Fig. 2. The implications of this result are discussed below.

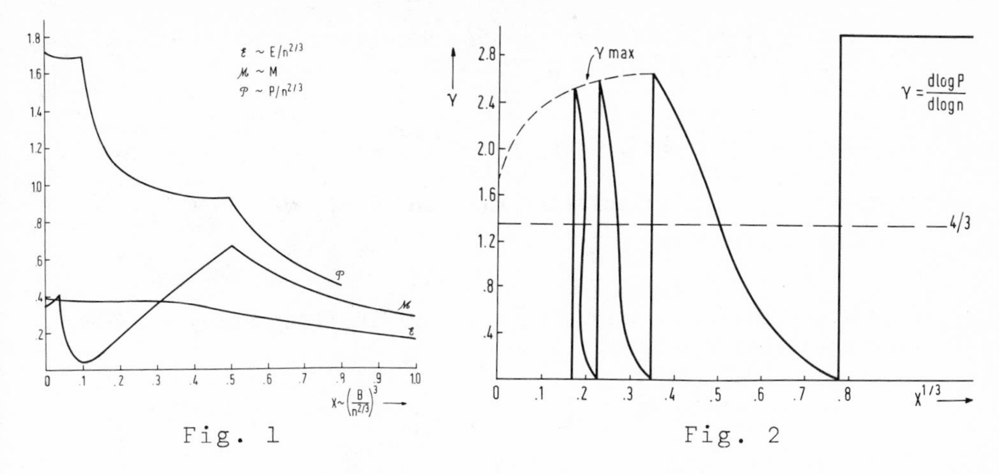

Fig. 1 Fig. 2

3. GRAVITATIONAL STABILITY

For a star of uniform density the condition for gravitational stability is $\gamma > 4/3$. Should the crust be situated in a uniform magnetic field and possess a uniform density, its ability to withstand the gravitational pressure would survive for certain values of the magnetic field only. A temporal decay of the magnetic field would bring an initially stable situation into an unstable one, and the crust would collapse to such a higher density that could render the electron gas stable for the new magnetic field.

The lack of uniformity makes the stability criterion more delicate. There are three important factors in order of decreasing importance: (i) the electron-crust extends over a shell of finite width only; (ii) the density, pressure and γ are highly non-uniform within the shell; (iii) the magnetic field is also non-uniform both because it is an explicit function of the position and because the magnetization varies with the density. In the following we describe a model which probably accounts with a reasonable accuracy for (i) and (ii); the effect of the spatial non-uniformity of B we ignore.

Adopting a trial displacement function

$$\frac{\delta r}{r} \equiv \xi = \xi(r) \tag{7}$$

an effective trial $\bar{\gamma}$ can be calculated[4] which sets an upper bound on the actual $\bar{\gamma}$ determining the stability. Let R be the inner radius and $R_o = \rho R$ the outer radius of the crust. Then

$$\xi(r) = 1 - \frac{R}{r} \tag{8}$$

is compatible with an incompressible core. With this choice

$$\bar{\gamma} = \frac{\int W_1(x)\ P\{n(x)\}\ \gamma\ \{n(x)\}\ dx}{\int W_o(x)\ P\ \{n(x)\}\ dx} \tag{9}$$

$$W_1 = 4\text{-}12x + 9x^2,\ W_o(x) = 3\text{-}12x + 9x^2,\ x = \frac{r}{R}\ .$$

For the purpose of calculating n(x) and P(x), we adopt

$$P = An^{\Gamma} \tag{10}$$

where Γ is some suitable average of the rapidly fluctuating γ. This yields the equilibrium distribution

$$n = n_o\ (\frac{1}{x} - \frac{1}{\rho})^{1/\Gamma-1} \tag{11}$$

The actual n-dependence of Γ we approximate by a sawtooth-like behavior,

$$\gamma^{(k)}(n) = \gamma^{(k)}_{max}\ \frac{n - n^{(k-1)}}{n^{(k)} - n^{(k-1)}} \tag{12}$$

valid between the two points in space where the filling of the k-th LANDAU-level has just begun and where it is completed; $\gamma^{(k)}_{max}$ is given by (6) and

$$n^{(k)} = (2^{1/2}/\pi^2)\ (S_o - \frac{1}{2}\ k^{1/2})\ (\hbar eB/c)^{3/2}$$

$$\rightarrow (2^{1/2}/3\pi^2)\ (\hbar eB/c)^{3/2}\ k^{3/2} \qquad k>>1 \tag{13a}$$

or

$$n^{(k)}/10^{28}\ cm^{-3} \rightarrow (0.133)(B/10^{12}Gs)^{3/2}k^{3/2}\ . \tag{13b}$$

A computer program for the calculation of $\bar{\gamma}$, based on the above model, has been worked out. Since $3 > \gamma\ max > 5/3$, the convenient Γ=2 has been chosen. The other input parameters are ρ, the magnetic field and the density at the core. These determine k via (11) and (13a) and $\bar{\gamma}^k$ can be calculated: it is a monotonically decreasing function of k. If $\bar{\gamma}^{(k)}\ (B,\rho) < 4/3$, an instability follows, leading to a collapse, whose extent Δr can be estimated as follows.

If the value of $\bar{\gamma}$ is marginal, instability ensues when the "local" maximum drops below the critical 4/3. Then the density has to increase to such an extent that a "local" minimum of $\bar{\gamma}$ is passed and the subsequent higher local maximum is reached (see Fig. 3). Thus

$$\Delta r = (5.76)(R_o/(\rho -1))(B/10^{12}Gs)(n_{core}/10^{28}\ cm^{-3})^{-2/3}\ . \tag{14}$$

The computer calculation for the accurate determination of $\bar{\gamma}$ is in progress.

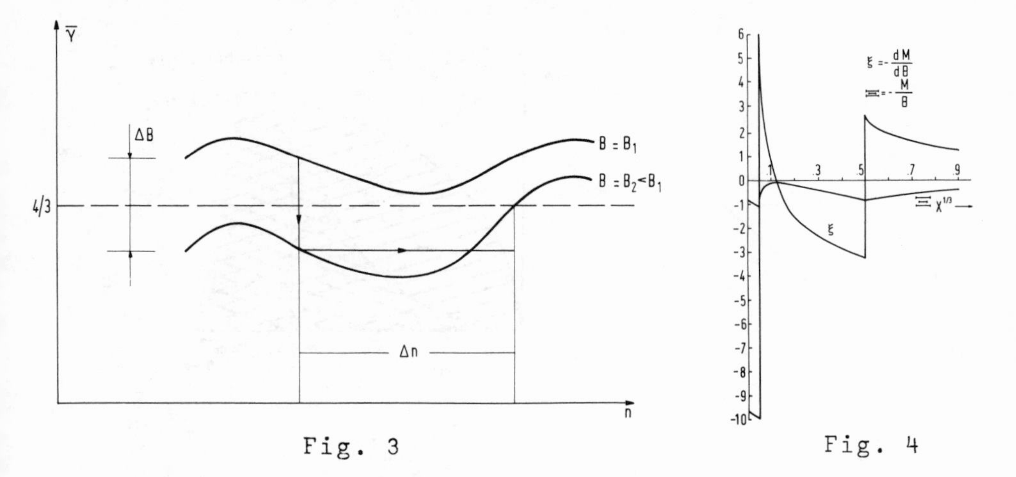

Fig. 3 Fig. 4

4. MAGNETIC POLARIZABILITIES AND MAGNETIC PHASE TRANSITIONS

The magnetic polarizability tensor $\xi_{\mu\nu} = -\partial M_\mu(\underset{\sim}{\kappa})/\partial B_\nu(\underset{\sim}{\kappa})$ describes the response of the system to κ-dependent (or uniform if $\kappa \to o$) perturbations. In the coordinate system where B = (OOB), and in the limit $\kappa \to o$, only the diagonal elements survive [5],

$$\xi_{\mu\nu}\,(o) = \begin{pmatrix} \Xi & & \\ & \Xi & \\ & & \xi \end{pmatrix} \cdot \tag{15}$$

The dispersion relation

$$\det\left|1 + 4\pi\underline{\underline{\xi}}(\underset{\sim}{\kappa})\right| = o \tag{16}$$

if satisfied for some real $\underset{\sim}{\kappa}$ indicates the instability of the system with respect to a phase transition[6] into a periodic structure characterized by the particular $\underset{\sim}{\kappa}$-value. The elements of $\underline{\underline{\xi}}$ are monotonically increasing functions of κ in the vicinity of $\kappa \to o$. Therefore, for (16) to be satisfied, $\xi_{\mu\nu}(o) < o$ is required. Thus

$$1 + 4\pi\;\Xi = o, \qquad\qquad 1 + 4\pi\;\xi = o \tag{17}$$

can be identified as marginal stability criteria. Since $\Xi = -(M/B)$ also[7], (17) is identical to the condition for a ferromagnetic phase transition, first identified by Chiu and Canuto [1].

The calculation of Ξ and ξ is based on

$$\Xi = \frac{1}{B}\;\frac{\partial E}{\partial B}\,, \qquad\qquad \xi = \frac{\partial^2 E}{\partial B^2}\;\cdot \tag{18}$$

For the case when only a few LANDAU-levels are occupied, the Ξ and ξ can be calculated exactly (Fig. 4). The latter is discontinuous at each LANDAU-level. In order to ascertain at what parameter values the dispersion relations can be satisfied, the

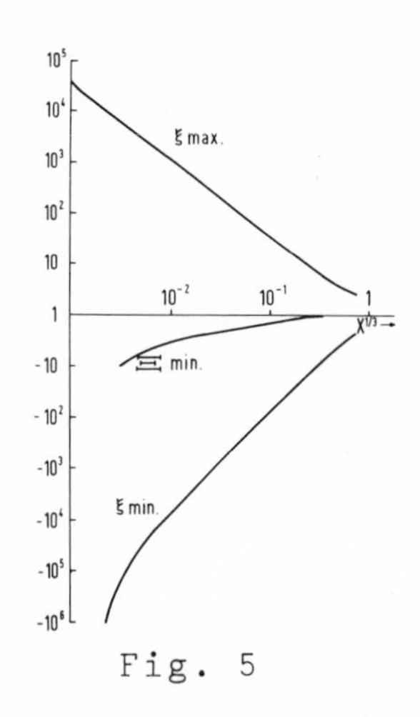

Fig. 5

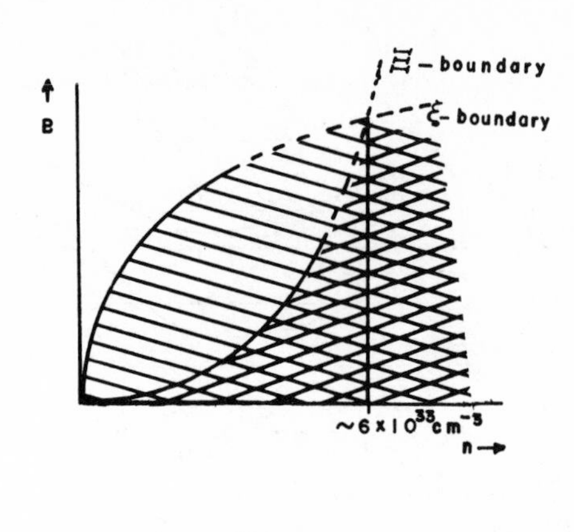

Fig. 6

envelopes of Ξ_{min}, ξ_{min} ξ_{max} have been calculated (Fig. 5):

$$\Xi_{min} = -(4/3)\eta\ (S_o - \frac{1}{2}k^{1/2})^{2/3}\ \{k - \frac{5}{2}\ \frac{S_1}{S_o - \frac{1}{2}k^{1/2}}\}$$

$$\rightarrow -\ (1/2)(2/3)^{2/3}\ (1 + 6C_o)\ \eta k^{1/2},\quad C_o = 0.04127$$

$$\xi_{min} = -5\eta\ (S_o - \frac{1}{2}k^{1/2})^{2/3}\quad \{k - \frac{S_1}{S_o - \frac{1}{2}k^{1/2}}\} \tag{19}$$

$$\rightarrow -\ 2(3/2)^{1/3}\ \eta k^2$$

$$\xi_{max} = (2/3)\eta\ \{2k - 2\ \frac{T_1 - 2S_o + k^{1/2}}{T_o - \frac{1}{2}k^{-1/2}}\left(1 - \frac{5}{4}\ \frac{T_1}{S_o - \frac{1}{2}k^{1/2}}\right) + \frac{5}{2}\ \frac{2S_1 - T_2}{S_o - \frac{1}{2}k^{1/2}}\}$$

$$\eta = (2\pi^2)^{-2/3}\ (e^2/\hbar c)\ (\hbar/mc)\ n^{1/3}\ .$$

The arrow refers to the asymptotic behavior as level index $k\rightarrow\infty$.

The resulting phase transition boundaries are (Fig. 6)

$$B \sim n^{4/3}\ :\ (\Xi),\qquad B \sim n^{5/6}\ :\ (\xi)\ . \tag{20}$$

When the ξ-instability supersedes the Ξ instability, it makes questionable the existence of the ferromagnetic state.

5. POLARIZABILITY OF AN ELECTRON GAS: DISPERSION RELATION

Wave propagation and dynamical response characteristics of the electron gas in the, and near the, crust are conveniently described by the polarization tensor $\pi_{ij}(\underset{\sim}{\kappa},\omega)$ which can be calculated with the aid of particle GREEN functions along the line previously employed for the magnetic field free situation[8,10].

The electric polarizability is determined by $\alpha = -(\pi/\omega^2)$ and leads to the dispersion relation $1 + 4\pi\alpha = c^2\kappa^2/\omega^2$. In order to study the penetration into and excitation of waves in the electron gas, we calculated $\pi_{\pm} = \pi_{11} \pm i\pi_{12}$ in the long wave length limit $(\kappa\to 0)$. The results indicate that the cyclotron resonance is replaced by two series of resonances, due to combined relativistic and quantum effects

$$\omega_n^2 = 2B\ \{\lambda_n - (\lambda_n^2 - 1)^{1/2}\}$$
$$\lambda_n^{(1)} = (m^2/B) + 2n+1,\quad \lambda_n^{(2)} = \{(m^2+p_f^2(n))/B\} + 2n+1 \tag{21}$$

where $p_f(n)$ is Fermi momentum for level n. For $B \ll m^2$ or $\omega_o \ll m$, these reduce to

$$\omega_n^{(1)} = \omega_o - (n+\tfrac{1}{2})\ (\omega_o^2/m)$$
$$\omega_n^{(2)} = \{m\omega_o/(m^2+p_f^2)^{1/2}\} - \{(n+\tfrac{1}{2})\omega_o^2/(m^2+p_f^2)^{1/2}\}\ . \tag{22}$$

A detailed analysis of the ensuing dispersion relation and its implications will be given elsewhere.

6. VACUUM POLARIZATION: SOME CONSEQUENCES

As a by-product of the calculation of $\pi_{ij}(\kappa\omega)$ referred to before, one obtains the dielectric and magnetic polarizability of the vacuum. In the $B_o \ll B_{crit}$ limit, previously obtained results[9] can be recovered. The modification of a uniform (or a quasiuniform) field appears in this context as a result of the difference in the renormalization when $B_o = 0$ or $B_o \neq 0$. In particular, one finds that the electric field is modified to

$$\underset{\sim}{E} = \underset{\sim}{E}_{bare}\ (1 + \frac{2}{45}\ (B_o^2/B_c^2)) - \frac{7}{45}\ \underset{\sim}{B}_o\ (\underset{\sim}{B}_o \cdot \underset{\sim}{E}_{bare})/B_c^2\ . \tag{23}$$

The first factor is a mere change of scale of the interaction strength, whereas the more interesting anisotropic effect contained in the second factor leads to a fanning out of the electric field lines, away from the magnetic field lines. The modified Coulomb field due to a point charge is given by

$$\underset{\sim}{E} = \frac{1}{r^2}\ \{(1 + \frac{2}{45}\ b^2)\ \hat{r} - (\frac{7}{45}\ b^2 \cos\theta)B_o\},$$
$$= \frac{1}{r^2}\{(1 + \frac{2}{45}b^2 - \frac{7}{45}b^2\ \cos^2\theta)\hat{r} + (\frac{7}{90}\ b^2\ \sin 2\theta)\hat{\theta}\},\ b=(B_o/B_c). \tag{24}$$

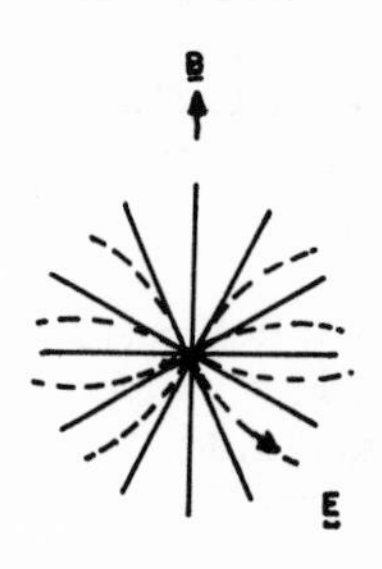

Fig. 7

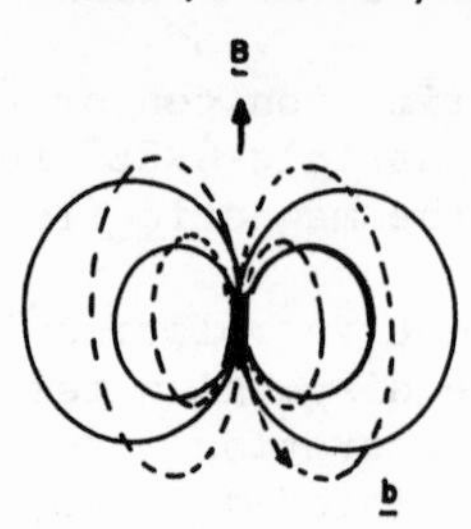

Fig. 8

This field has the feature of pulling like charges towards the equatorial plane (see Fig. 7).

A static, quasiuniform magnetic field is modified to

$$\underset{\sim}{B} = \underset{\sim}{B}_{bare} \left(1 + \frac{2}{45} (B_o/B_c)^2\right) + \frac{4}{45} \underset{\sim}{B}_o \, (\underset{\sim}{B}_o \cdot \underset{\sim}{B}_{bare})/B_c^2 \qquad (25)$$

The scale factor is identical for electric or magnetic effects. However the change of sign implies an opposite tendency as regards the anisotropic effects. The magnetic lines tend to align themselves closer to the external field direction, for example, leading to a transverse contraction of the dipole field pattern (Fig. 8).

This work was partly supported by Contract No. AF 19628-69C-0074.

REFERENCES

1. See, e.g., A.G.W. Cameron: Neutron Stars, Ann.Rev.Astr. Astrophys. 8, 179 (1970).
2. V. Canuto and H.Y. Chiu: Phys.Rev. 173, 1210,1220,1229 (1968); H.Y. Chiu and V. Canuto: Phys.Rev. Letters, 21, 2,110 (1968).
3. P. Goldreich & W.H. Julian, Ap.J. 157, 869 (1969); P.A. Sturrock, SUIPR Report No. 365, 1970.
4. P. Ledoux, Handbuch der Physik, 51, 605 (1958); ibid 353 (1958).
5. K.I. Golden and G. Kalman: J.Stat.Phys. 1, 415 (1969).
6. J.J. Quinn: Phys.Rev. Letters 16, 731 (1966). H. Lee, M. Greene and J. Quinn: Phys. Rev. Letters, 19, 428 (1967). M.Ya. Azbel: Soviet Phys. JETP, 26, 1003 (1968).
7. J.J. Quinn: J.Phys. Chem. Solids, 24, 933 (1963).
8. G. Kalman and B. Prasad: Proc. AAS Meeting, Austin (1969).
9. R. Braier and P. Breitenlohner: Nuovo Cimento, XVII, 261 (1967).
10. V.N. Tsytovitch: Soviet Phys. JETP, 13, 1249 (1961).

COSMIC RAY SPECTRUM AND PLASMA TURBULENCE

V.N. Tsytovich

P.N. Lebedev Physics Institute

U.S.S.R. Academy of Sciences, Moscow

ABSTRACT

1. The main features of the observed cosmic ray electron and ion spectra are considered as fundamental properties of cosmic rays. It is asked whether it is possible to obtain from these properties any information about the physical parameters in the cosmic ray sources. The modern theory of plasma turbulence is applied to this question. It is shown that the observed power law spectra of cosmic rays $f_\varepsilon \sim 1/\varepsilon^\gamma$ (f_ε - the distribution function in energies ε; $\int f_\varepsilon d_\varepsilon = n_*$ - the density of cosmic rays) with average γ near to 2.7 give very rigid restrictions for possible mechanisms of acceleration and of energy losses of cosmic rays. Both must be very effective, be correlated with each other, and essentially exceed the usual energy losses of cosmic rays in a quiet plasma. The ratio of the diffusion coefficient $D(\varepsilon)$ in energy space that describes the acceleration, to the energy losses of a single particle per unit time $A(\varepsilon)$, must be, with good accuracy, proportional to the energy $D(\varepsilon)/A(\varepsilon) = \alpha\varepsilon$, with the constant coefficient α of order unity.

2. The analyses show that these restrictions cannot be satisfied for collisionless magnetohydrodynamic turbulence, which produces an acceleration similar to that of Fermi acceleration, but are easily satisfied for Langmuir turbulence. In this case the anomalous energy losses of cosmic rays are due to inverse Compton effect on plasma waves. The correlation between the accelera-

tion and energy losses is simply due to the fact that they both are caused by the same physical mechanism: the interaction of cosmic ray particles with turbulent pulsations (spontaneous and induced). The coefficient α derived in this case is a function of γ, of order unity; the equation for γ has the form $\gamma + 2 = 1/\alpha(\gamma)$.

3. The problem of cosmic ray electron spectra is discussed. Two possibilities for the conditions in the cosmic ray electron source are considered: 1) relativistic ($\varepsilon \gg m_e c^2$) electrons in a "cold" nonrelativistic turbulent plasma; 2) turbulent relativistic electron plasma (the cold plasma is absent). It is shown that in both cases the distribution function of relativistic electrons f_ε is $f_\varepsilon \sim \varepsilon^2$ for $\varepsilon \ll \varepsilon_*$ and $f_\varepsilon \sim 1/\varepsilon^\gamma$ for $\varepsilon \gg \varepsilon_*$. In the first case ε_* plays the role of an injection energy and depends on the energy density of turbulence W and n, the "cold" plasma density: $\varepsilon_*/m_e c^2 = \sqrt{nm_e c^2/W}$. In the second case the ε_* can be considered as an energy of the order of the average particle energy ("temperature") of the relativistic turbulent electron plasma (the difference with the case of quiet plasma is that the distribution function for $\varepsilon \gg \varepsilon_*$ is of the power law type instead of exponential):

$$\frac{\varepsilon_*}{m_e c^2} \approx \left(\frac{W}{n_* m_e c^2}\right)^{2/7} (4\pi n_* r_e^3)^{-1/7} \quad ; \; r_e = \frac{e^2}{m_e c^2} \quad .$$

In the absence of an external magnetic field the equation for γ has only one solution independent of W, which gives $\gamma = 3$. In the weak magnetic field case, $\xi_e = \dfrac{eH}{m_e c\omega_{pe}} \ll 1$, the value of γ depends on both ξ_e and W/H^2 and lies in the interval $0.9 < \gamma < 3$. The ω_{pe} is the plasma frequency, i.e. $\omega_{pe} = \sqrt{4\pi n_e{}^2/m_e}$ for a "cold" plasma and $\omega_{pe} = \sqrt{8\pi n_* e^2/3\varepsilon_*}$ for a relativistic plasma. The possible direction of development of this theory is discussed.

4. The problem of cosmic ray ion spectra is considered also for two possible conditions: 1) cosmic ray ions in a "cold" turbulent plasma; 2) relativistic turbulent ion plasma. It is shown that in a "cold" plasma the cosmic ray ion spectrum can be of the power law type only for $\varepsilon > \varepsilon_{*i}$ with $\dfrac{\varepsilon_{*i}}{m_i c^2} = \dfrac{m_i}{m_e}\sqrt{\dfrac{nm_e c^2}{W}} \gg \dfrac{m_i}{m_e}$, where

m_i is the ion mass. But in a relativistic turbulent plasma $\varepsilon_{*i}/m_i c^2$ can be of order unity, i.e.

$$\frac{\varepsilon_{*i}}{m_i c^2} \approx \left(\frac{W}{n_{*i} m_i c^2}\right)^{2/7} (4\pi n_{*i} r_i^3)^{-1/7} \; ; \; r_i = \frac{e^2 Z^2}{m_i c^2} \; , \; Z$$

is the atomic number. Thus the cosmic ray ion population must be of two kinds. This result may possibly lead to a new interpretation of the observed change in the spectra of cosmic ray ions at 10^{15}-10^{17} ev. For both populations, when $\varepsilon > \varepsilon_{*i}$, the theory predicts a power law spectrum with γ near 2.7 as one of the most probable. This appears if $\xi_i = eH/m_i c\omega_{pe}$ is assumed to be very small (if $H^2/8\pi = nT_e$ in a cold plasma, $\xi_i \approx \frac{v_{Te}}{c}\frac{m_e}{m_i} \lll 1$). Then it appears that $\gamma \approx \gamma_* = 2.7$ is a singular point of the equation which determines γ. More precisely, a change in ξ_i and W/H^2 of many orders of magnitude does not essentially change γ. The derived value of γ_* is close to that observed in cosmic rays near the Earth. It is also shown that one cosmic ray source cannot at the same time be the source of cosmic ray ions and cosmic ray electrons. But in the case of a cold plasma the production of cosmic ray electrons in the presence of the process of production of very high energy cosmic ray ions is possible only up to electron energies of the order $\varepsilon_* \frac{m_i}{m_e}$ (thus $f_\varepsilon \sim 1/\varepsilon^\gamma$ in $\varepsilon_* < \varepsilon < \varepsilon_* \frac{m_i}{m_e}$).

The time needed for spectrum formation diminishes with the particle energy. Thus it is determined by the characteristic acceleration time at $\varepsilon \simeq \varepsilon_*$. For the case of cosmic ray ions in a cold plasma

$$\frac{1}{\tau_i} \simeq \frac{4\pi n e^4 Z^4}{m_i^2 c^3} \sqrt{\frac{W}{n m_e c^2}} \, .$$

The preferential acceleration of heavy multicharged ions is predicted. Even for protons with $W/nm_e c^2 \sim 10^{-2}$ and $n \sim 10^{13}$ cm^{-3}, the acceleration time is of the order of a few years.

5. For the case of infinite turbulent plasma in the frame of the process considered, there are no limits on the maximum energies of cosmic ray particles. The hope exists that near collapsed objects, or objects that are close to the stage of gravitational collapse, the plasma can be considered as approximately infinite. The estimate of possible maximum energies of cosmic ray particles for known objects (of finite dimensions), such as quasars, the first stages of supernova explosion, pulsars, and so on, gives a value of the order $\varepsilon_{max}/mc^2 \sim 10^4$-$10^6$.

THE PROPERTIES OF MAGNETIC NEUTRAL SHEET SYSTEMS

S.W.H. Cowley

Physics Department

Imperial College, London

Much work has been devoted in the past two decades to the investigation of the properties of the flow of a highly conducting fluid near X-type neutral lines in the magnetic field. This work was initiated by Dungey (1953) who pointed out that it was only in such regions that large currents could be produced in a plasma without being opposed by the electromagnetic force $\underline{j} \wedge \underline{B}/c$. Indeed, he showed that such a region is unstable with respect to the growth of the current.

The equations governing the motion in these models are Maxwell's equations; the continuity equation, the hydromagnetic equation of motion and Ohm's law in the form

$$\underline{E} + \frac{\underline{u} \wedge \underline{B}}{c} = \underline{j}/\sigma \tag{1}$$

Far away from the neutral sheet or line, where the magnetic field has the value B_o, the fluid flows towards it with velocity u_o and the currents are very small. Then from Ohm's law the electric field is given by $E_o = u_o B_o / c$ and is directed along the neutral line. In the steady state this electric field is uniform, hence the current density at the neutral sheet is $j_o = \sigma E_o$. The thickness of the sheet in which these currents flow is such that the total current just matches the change in the magnetic field. Hence

$$\ell \simeq \frac{c^2}{4\pi \sigma u_o} \tag{2}$$

where ℓ is the half-thickness of the current sheet.

As an example we consider (with Yeh and Axford (1970)) the case where the fluid flows with velocity u_o all the way to the neutral sheet from either side, postulating the existence of a plasma sink at the neutral sheet. The induction equation is

$$\frac{\partial B_y}{\partial t} = \pm u_o \frac{\partial B_y}{\partial x} + \frac{c^2}{4\pi\sigma}\frac{\partial^2 B_y}{\partial x^2} \quad \text{for } x \gtrless 0 \qquad (3)$$

Considering $x > 0$ (and $B_y > 0$ for $x > 0$) the first term in (3) represents a rise in the magnetic field due to the inward convection of the magnetic field with the plasma (influx of Poynting electromagnetic energy), while the second term gives a decay of the field due to 'resistive annihilation' (joule heating of the plasma). In the steady state the source and sink of the magnetic field balances and the solution to (3) is

$$B_y = \pm B_o\,(1 - \exp(\mp 4\pi\sigma\, u_o x/c^2)) \quad \text{for } x \gtrless 0 \qquad (4)$$

The scale length of the field change is as given above by equation (2). Although this equation appears to determine ℓ for a given u_o, Parker (1957, 1963) attempted to determine ℓ by considering the physical nature of the postulated plasma sink at the neutral sheet. He assumed that the fluid flows out of the system along the field lines (in the $\pm$ y direction); the length of the system in this direction is 2L (see Fig. 1). For the incompressible case we have, for continuity

$$u_o L = v\,\ell \qquad (5)$$

and it is shown that the fluid is accelerated out of the system by the pressure gradient and emerges at approximately the Alfvén velocity. Thus

$$\ell \simeq \frac{u_o L}{V_A} \qquad (6)$$

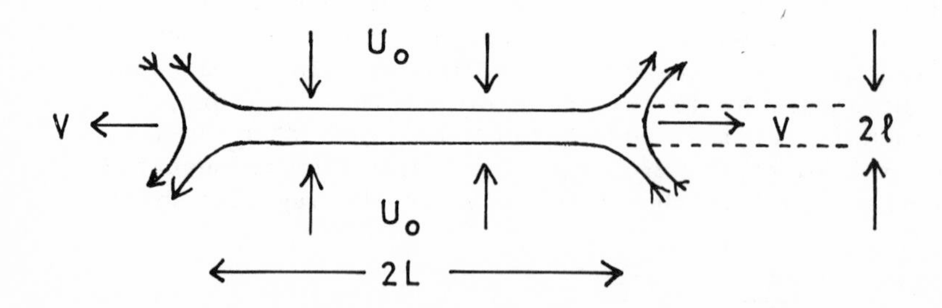

Figure 1: Field and flow configuration envisaged in the analysis of Parker (1963).

and hence $u_o \simeq V_A\left(\frac{c^2}{4\pi\sigma V_A L}\right)^{\frac{1}{2}}$ and $\ell = L\left(\frac{c^2}{4\pi\sigma V_A L}\right)^{\frac{1}{2}}$ (7a,b)

The quantities B_o, N_o (particle number density), σ and L are assumed known; L is taken to be a typical dimension of the system. However, Petschek (1964) first suggested that, like ℓ, L should be determined from (6) once u_o is given. Further away from the neutral line he suggested that magnetic field energy can be converted to fluid energy by the presence of standing MHD shocks. Such solutions for the flow have been obtained by Yeh and Axford (1970) who showed that a pair of shocks are, in fact traversed by the streamlines (Fig. (2)). Separate solutions were obtained for infinitely-conducting flow external to the neutral line (involving the shocks) and for the region near the line where finite conductivity must be taken into account. While it is not rigourously shown how the two solutions match, it is clear they are qualitatively similar in form. On the basis of the fluid theories, therefore, it appears that the reconnection rate (given by u_o) may occur at essentially any speed consistent with the boundary conditions. The higher the speed, the smaller the spatial extent of the 'diffusion region'.

For many neutral sheet and line problems in nature (e.g. the Earth's magnetic tail, solar wind and possibly various astrophysical objects) the fluid is collisionless. The above MHD theories are then open to the following criticisms.

(a) The 'diffusion regions' are typically of very small spatial extent, even smaller values than typical particle gyroradii are possible. MHD assumptions then break down (e.g. the pressure tensor $\underline{\underline{P}} \simeq p\,\underline{\underline{1}}$), and we call into question the validity of treating with fluid equations the properties of a collision-free plasma near a neutral point.

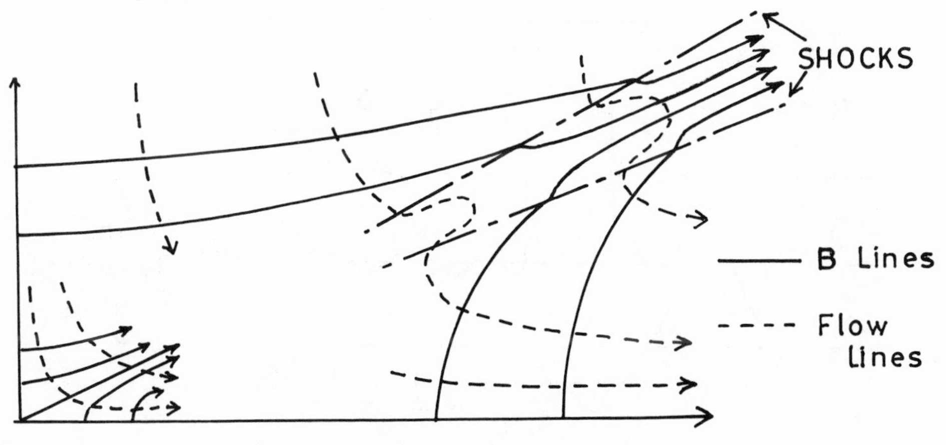

Figure 2: Flow and field for the solutions obtained by Yeh and Axford (1970).

(b) It is always assumed that the conductivity is homogeneous and isotropic. For a collision-free system the collision mean-free-time may be replaced in the theory of conductivity by the time the particle remains in the system, and this depends on the field geometry (see for example Speiser (1970)). The conductivity is then by no means homogeneous or isotropic, and currents should be determined by studies of particle trajectories, rather than by simply giving a value to σ .
(c) Dungey (1953) and Speiser (1965, 1968) have investigated the motion of particles near a neutral line in the presence of a 'reconnection' electric field. They showed that particles drift into the field reversal region from both sides under the action of the electric field, and then oscillate about the neutral sheet, becoming accelerated along it by the magnetic field. (Fig. 3). It is clearly these particles which provide the current in the field reversal region in the collisionless models. Inclusion of a weak field component normal to the sheet in the 'x' configuration causes them to turn away from the neutral line as they accelerate until they move out of the sheet along the magnetic field lines, as envisaged in the fluid theories. However, 'charge separation' occurs at the sheet and leads to important variations in the 'third' dimension (i.e. along the neutral line), particularly if the normal component of the field is very small.

The first description of the properties of a neutral sheet model for a collision-free plasma was given by Alfvén (1968). In view of the above comments it is perhaps not surprising that the considerations involved are rather different from those of the fluid theories. He considered a neutral sheet configuration of finite width d in the third dimension (along the electric field), the

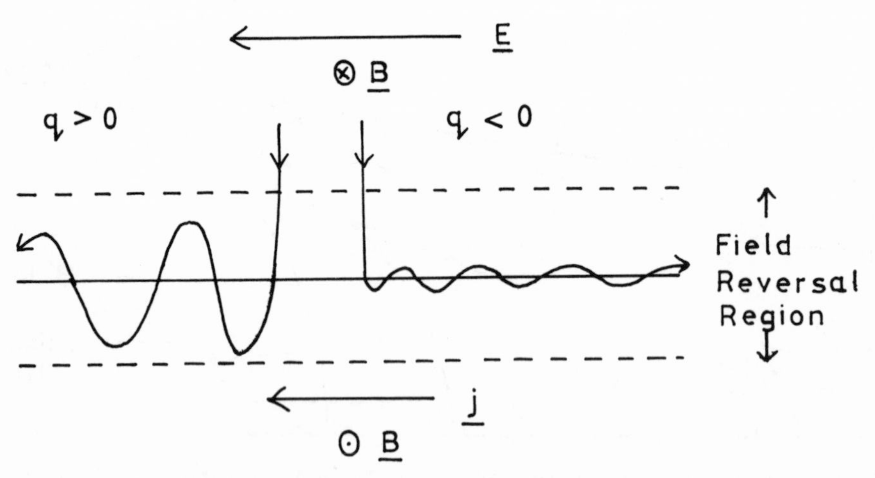

Figure 3: Particle motions near a neutral sheet, following Speiser (1965).

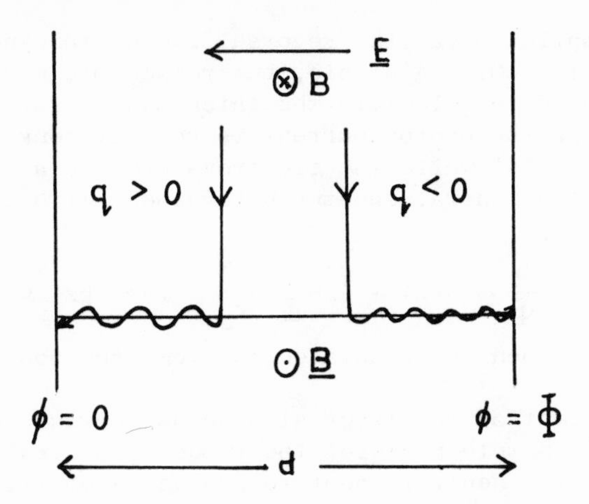

Figure 4: Field geometry of the system discussed by Alfvén (1968).

system being bounded by 'condenser plates', which represent equipotential boundaries and a sink for charged particles (Fig. 4). The magnetic field outside the field reversal region is B_o and the density of positive particles N_o (with equal numbers of electrons). Following Speiser's analysis of the particle motions we assume that all incoming particles contribute to the current. The flux of positive or negative particles flowing into the sheet gives the total current flowing in the field reversal layer. For self-consistency between the current and magnetic field the total potential is found to be

$$\Phi = \frac{B_o^2}{4\pi N_o e} \tag{8}$$

The average velocity of a positive particle on leaving the system is then the Alfvén velocity of the external medium. The philosophy of this calculation is simply that from a knowledge of the particle trajectories a self-consistent incoming flux of plasma can be computed to produce the current required by the change in the magnetic field. No knowledge of the detailed structure of the sheet is required, and none is obtained from the calculation.

Cowley (1971b) also considered conservation of energy and momentum for the system. It was first of all shown that the above self-consistent potential gives conservation of energy, the incoming Poynting flux of electromagnetic energy being directly converted into kinetic energy of particles as they are accelerated along the sheet. Momentum conservation needs to be considered in the direction parallel to the sheet because we have positive particles emerging from one boundary with the same energy spectrum as electrons from the other, but carrying a factor $(mp/me)^{\frac{1}{2}}$ times more momentum. This implies a varying reversal region thickness, it being thickest at $\phi = 0$. A simple, but reasonable magnetic field model can then be used to calculate the thickness. For instance, if it is assumed that the proton current is roughly constant over a region of thickness 'a' while the electrons provide a very thin current layer near the neutral sheet (in agreement with trajectory studies) then

$$B_x(z) \simeq B_o\left\{ \frac{\phi}{\Phi}\,\mathrm{sgn}(z) + \left(1-\frac{\phi}{\Phi}\right)\frac{z}{a}\right\} \quad \text{for} \quad |z| \leqslant a. \tag{9}$$

The value of 'a' can then be calculated for momentum conservation.

The charge content of the layer also needs careful consideration, since, in the simple picture, the incoming neutral plasma charge separates at the neutral sheet to produce a positively charged beam accelerating towards $\phi = 0$ and a negatively charged beam accelerating towards $\phi = \Phi$. We expect the sheet to be largely positively charged because the protons spend much longer times in the sheet than electrons, due to their much larger mass. Cowley (1971a) showed that the effect of this positive charge is to localize the Alfven potential drop near the $\phi = 0$ boundary, which by increasing the accelerating electric field and decreasing the distance travelled by protons reduces the time they spend in the sheet and hence the charge content. For quasi-neutrality in the sheet the proton charge must be balanced, in the main, by electrons drifting through the field reversal region towards the neutral sheet, and this, too, implies a value for the sheet thickness (Cowley (1971b)). Using the same model magnetic field as was used

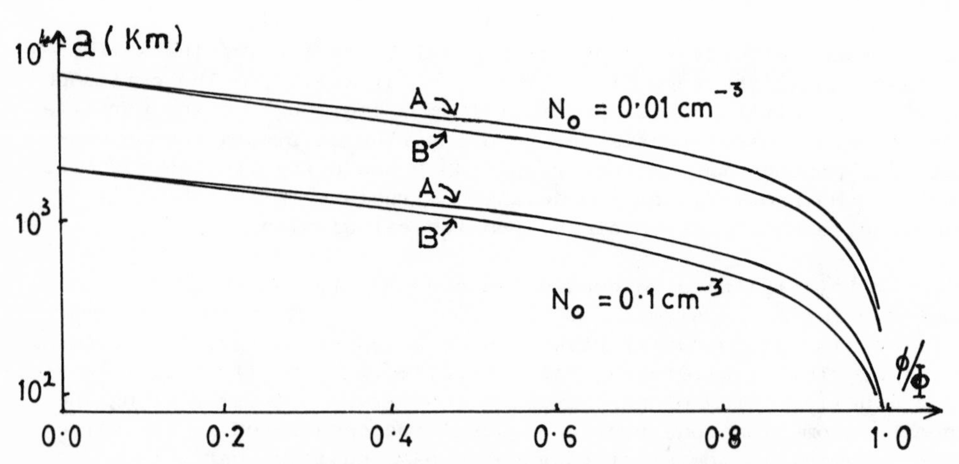

Figure 5: Sheet thickness parameter 'a' versus ϕ/Φ calculated for B_o=10 γ and $N_o \simeq 0.1$, 0.01 for (a) momentum conservation; (b) charge neutrality.

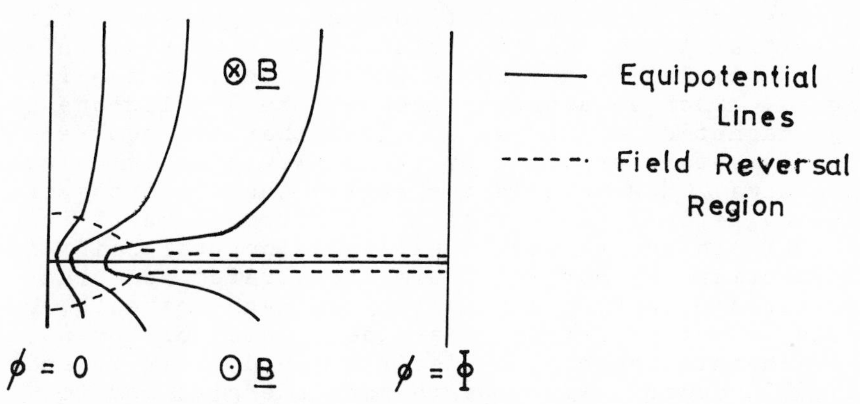

Figure 6: Neutral sheet structure as indicated by Cowley (1971b)

in the momentum calculation very good agreement with the latter is found (Fig. 5). Near $\Phi = 0$ the momentum and charge calculation give

$$a \simeq \frac{c}{e}\left(\frac{2\ m_p}{N_o}\right)^{\frac{1}{2}} \tag{10}$$

and if we take, very roughly $a \simeq v_p/\Omega$ where $v_p \simeq cE/_B$ then we obtain

$$E \simeq \frac{B^2}{c}\left(\frac{2}{\pi\ m_p\ N_o}\right)^{\frac{1}{2}} .$$

Hence the potential drop across the system is localized near the $\Phi = 0$ boundary over a distance

$$\mathcal{L} \simeq \frac{\Phi}{E} \simeq \frac{c}{e}\left(\frac{m_p}{N_o}\right)^{\frac{1}{2}}$$

which is comparable with the proton plasma wavelength and the sheet thickness given by equation (10). Thus charge neutrality in the sheet is achieved by enhanding the electric field near $\Phi = 0$ which reduces the proton charge content as above described, and also by widening the sheet to make the adiabatic electron charge content significant. If the electric field were uniform then we would have, for consistency between the sheet thickness and the oscillation amplitude of the Speiser particles

$$a \simeq \frac{v_p}{\Omega} = \frac{m_p c^2 E_o}{cB_o}$$

which gives, for tail parameters, $a \simeq 2 \rightarrow 20$ Km for $N_o \simeq 0.1 \rightarrow 0.01$ cm^{-3}, values much smaller than those given in Fig. (5). Charge neutrality could not be satisfied for such field configurations, nor momentum conserved.

It should be noted that the theory presented here assumes that the plasma is cold i.e. the thermal energy of the particles is unimportant compared with the flow energies. Such an assumption should be a valid one for the geomagnetic tail if we consider that the neutral sheet current is provided by the inflow of cold polar wind plasma. However, in the region where the component of the magnetic field normal to the current sheet is not negligible in the three-dimensional X-neutral point field configuration the particles are accelerated onto the field lines rather than out into the magnetosheath, resulting in the formation of a plasma sheet of hot particles. In this region the electric-field drift velocities are small compared with the thermal energies and to zeroth order may be neglected in investigating the gross structure of the current sheet, as has been done by Schindler (1971).

ACKNOWLEDGMENT

The author wishes to thank Professor J.W. Dungey for introducing him to this subject and for valuable discussions. Financial support was provided by the Science Research Council of Great Britain.

REFERENCES

Alfvén, H. 1968, J. Geophys. Res. 73, 4379-4381.
Cowley, S.W.H. 1971(a), Cosmic Electrodyn. 2, 90-104.
Cowley, S.W.H. 1971(b). To be published in "Magnetosphere-Ionosphere Interactions", proc. April 1971 Advanced Study Inst., Norway; University of Oslo Press; also Imperial College Scientific Report, June 1971.
Dungey, J.W. 1953, Phil. Mag. 44, 725-738.
Parker, E.N. 1957, J. Geophys. Res. 62, 509-520.
Parker, E.N. 1963, Astrophys. J. Suppl. Ser. 77, 8, 177-211.
Petschek, H.E. 1964, AAS-NASA Symposium on the Physics of Solar Flares (Ed: W.N. Hess), NASA SP-50, 425-439.
Schindler, K. 1971. Ninth ESRO Summer School and Summer Advanced Institute, Cortina, 1971 (to be published in the proceedings).
Speiser, T.W. 1965, J. Geophys. Res. 70, 4219-4226.
Speiser, T.W. 1968, J. Geophys. Res. 73, 1112-1113.
Speiser, T.W. 1970, Planet. Sp. Sci. 18, 613-622.
Yeh, T. and W.I. Axford. 1970, J. Plasma Phys. 4, Part 2, 207-299.

References

Alfvén, H. 1968. Some properties of magnetospheric neutral surfaces. J. Geophys. Res. 73, 4379-4381.

Cowley, S.W.H. 1971(a), The adiabatic flow model of a neutral sheet. Cosmic Electrodyn. 2, 90-104.

Cowley, S.W.H. 1971(b), Some properties of magnetic neutral sheet systems. To be published in 'Magnetosphere-Ionosphere Interactions', proceedings of the April 1971 Advanced Study Inst., Norway; University of Oslo Press.
Also Imperial College Scientific Report, June 1971.

Dungey, J.W. 1953. Conditions for the occurrance of electrical discharges in astrophysical systems. Phil. Mag. 44, 725-738.

Parker, E.N. 1957. Sweet's mechanism for merging magnetic fields in conducting fluids. J. Geophys. Res. 62, 509-520.

Parker, E.N. 1963. The solar flare phenomenon and the theory of reconnection and annihilation of magnetic fields. Astrophys. J. Suppl. Ser. 77, 8, 177-211.

Petschek, H.E. 1964. Magnetic field annihilation. AAS-NASA Symposium on the Physics of Solar Flares (ed: W.N. Hess), NASA SP-50, 425-439.

Speiser, T.W. 1965. Particle trajectories in model current sheets, 1: Analytical solutions. J. Geophys. Res. 70, 4219-4226.

Speiser, T.W. 1968. On the uncoupling of parallel and perpendicular motion in a neutral sheet. J. Geophys. Res. 73, 1112-1113.

Speiser, T.W. 1970. Conductivity without collisions or noise. Planet. Sp. Sci. 18, 613-622.

Yeh, T. and W.I. Axford. 1970. On the re-connexion of magnetic field lines in conducting fluids. J. Plasma Phys. 4, part 2, 207-229.

FIELD LINE MOTION IN THE PRESENCE OF FINITE CONDUCTIVITY

Thomas J. Birmingham

Laboratory For Space Physics

NASA Goddard Space Flight Center, Greenbelt, Maryland

I. INTRODUCTION

Magnetic field lines are convected with the motion of a perfectly conducting magnetohydrodynamic fluid.[1] Application of this frozen-in field line property is made in diverse areas of cosmic plasma physics; for example, theorists postulate the co-rotation of the magnetospheric plasma of pulsars[2] and stars[3], the effusion of coronal magnetic field lines with the expanding solar wind[4], and the solar-wind-driven convection of low energy terrestrial magnetospheric plasma.[5-7]

In this paper we relate plasma motion to field line motion when the conductivity is imperfect. The imperfect conductivity may result from collisions between plasma particles and neutrals as in the case of the earth's ionosphere, from the classical interaction of charged particles with one another, or from the scattering of charged particles by the enhanced field fluctuations which characterize a turbulent plasma.[8]

We assume that the magnetic field **B** is a given, known function of space and time. We further assume that all field lines intersect an ideally conducting surface S and identify them by their points of intersection.[9, 10] This is a valid identification regardless of whether the region outside S is filled with a perfectly conducting plasma or an imperfectly conducting one. For many problems, it is also a useful identification (cf. Ref. 11 and references cited therein).

II. EULER POTENTIALS

Assume that $\mathbf{B}(\mathbf{r}, t)$ is sufficiently regular that it can be described by Euler potentials [12] α and β

$$\mathbf{A} = \alpha(\mathbf{r}, t)\nabla\beta(\mathbf{r}, t)$$
$$\mathbf{B} = \nabla\times\mathbf{A} = \nabla\alpha\times\nabla\beta \tag{1}$$

The Euler potentials are constant on $\mathbf{B}$-lines ($\mathbf{B}\cdot\nabla\alpha = \mathbf{B}\cdot\nabla\beta = 0$). Together with $s(\mathbf{r}, t)$, arc length along $\mathbf{B}$-lines, α and β can be used as a spatial coordinate system, albeit one which varies with time.

The electric field $\mathbf{E}(\mathbf{r}, t)$ is

$$\mathbf{E} = -\frac{1}{c}\frac{\partial \mathbf{A}}{\partial t} - \nabla\Phi$$
$$= -\frac{\mathbf{w}\times\mathbf{B}}{c} - \nabla(\Phi + \Psi) \tag{2}$$

where

$$\Psi = \frac{\alpha}{c}\frac{\partial\beta}{\partial t} \tag{3}$$

and

$$\mathbf{w} = \left(\frac{\partial\beta}{\partial t}\nabla\alpha - \frac{\partial\alpha}{\partial t}\nabla\beta\right)\times\frac{\mathbf{B}}{B^2} \tag{4}$$

It can be verified directly that $\mathbf{w}$ satisfies the equation $\nabla\times(\mathbf{E} + \mathbf{w}\times\mathbf{B}/c) = 0$ and is therefore a valid flux-preserving field line velocity [13, 14]. Further, the velocity $\mathbf{w}$ preserves the α, β labels of a field line. [14]

The gauge freedom is eliminated by specifying α and β on S. Thus $\mathbf{A}, \alpha, \beta, \Phi, \Psi$ and $\mathbf{w}$ become uniquely defined. Jones and Birmingham [10] validate this method of choosing a gauge.

We restrict S here to be a rigid, stationary conductor. (The extension to a moving, non-rigid conductor is direct.[10]) Our gauge choice is $\partial\alpha/\partial t = \partial\beta/\partial t = 0$ on S. It follows then that Ψ and $\mathbf{w}$ also both vanish on S. It further follows from Eq. (2) that Φ = const = 0 on S, since $\mathbf{E}_{tan} = 0$ at a stationary, rigid, ideally conducting surface. From Eqs. (2) and (4) we find that $\mathbf{w}$ differs from $\mathbf{v} = c\mathbf{E} \times \mathbf{B}/B^2$, the particle $\mathbf{E} \times \mathbf{B}$ drift velocity, by

$$\mathbf{v} - \mathbf{w} = \frac{c\,\mathbf{B} \times \nabla(\Phi + \Psi)}{B^2} \tag{5}$$

Note from Eq. 2 that $\Phi + \Psi$ is a potential for the electric field component parallel to $\mathbf{B}$: $E_{\parallel} = -\partial(\Phi + \Psi)/\partial s$. If the plasma has infinite parallel conductivity σ_0, $E_{\parallel} \equiv 0$ at all times. Hence $\Phi + \Psi$ is constant along each $\mathbf{B}$-line. Since each $\mathbf{B}$-line intersects S where $\Phi + \Psi$ is zero, we conclude that when $\sigma_0 = \infty$, $\Phi + \Psi = 0$ and $\mathbf{v} \equiv \mathbf{w}$ everywhere outside S. In the infinite conductivity limit, particles drift with the field line motion.

Our conclusion that $\mathbf{v} \equiv \mathbf{w}$ when $\sigma_0 = \infty$ follows from a particle picture of plasma motion rather than a fluid description, the one more customarily used in discussing field line motion. Note also that we have specified nothing about the perpendicular conductivity.

III. PARALLEL RESISTIVITY

Let a given, known electric field $\mathbf{E}_0(t)$ be imposed on the plasma at $t = 0$ and suppose that $\mathbf{E}_0$ has a component parallel to $\mathbf{B}$. The source of $\mathbf{E}_0$ might be a time-varying magnetic field $\partial\mathbf{B}/\partial t$ or charge accumulations outside the region of space being considered. (We assume $\nabla \cdot \mathbf{E}_0 = 0$ throughout the spatial domain of interest.)

The plasma reacts in a way tending to make $\Phi + \Psi$ constant in space. The time it takes to eliminate variations in $\Phi + \Psi$ depends on the freedom with which plasma particles can react to $\mathbf{E}_0$. This freedom in turn depends importantly on the collision rate of charged particles. So long as variations in $\Phi + \Psi$ exist between adjacent points on adjacent field lines $\mathbf{B} \times \nabla(\Phi + \Psi) \neq 0$ and we conclude from Eq. 5 that $\mathbf{v} \neq \mathbf{w}$.

We assume that the plasma responds electrostatically, setting up an electric field $-\nabla\Phi_1$ by charge separation but having negligible magnetic effect. ($\mathbf{E}_0$ may contribute a Φ_0 to the total potential.) The total electric field is $\mathbf{E} = \mathbf{E}_0 - \nabla\Phi_1$. We further assume in this Section that the motion of plasma is strictly along $\mathbf{B}$. Our theory is linear in the smallness of $\mathbf{E}$ so that the time varying part of $\mathbf{B}$ is regarded as small.

We can determine $\Phi + \Psi$ by integrating

$$\int_{s0}^{s} d\,s'\, E_{\|}$$

outward along each field line from the conducting surface at $s = s_0(\alpha, \beta)$ where $\Phi + \Psi = 0$. To do this we must, however, know Φ_1.

Φ_1 is determined from Poisson's Equation

$$\nabla^2 \Phi_1 (\mathbf{r}, t) = -4\pi\rho = 4\pi B \int_0^t d\,t' \frac{\partial}{\partial s}\left(\frac{j_{\|}}{B}\right) \tag{6}$$

the latter form following from charge continuity.

The current $j_{\|}$ (p), Laplace transformed in time (p is the Laplace variable), is related to the electric field

$$j_{\|} (p) = K(p, s) \left(E_{\|o} - \frac{\partial \Phi_1}{\partial s}\right) \tag{7}$$

by solving the MHD equations for electrons and a single species of ions, neglecting pressure gradients but including the effect of collisions between charged particles and a fixed background of scatterers. The proportionality factor is

$$K(p, s) = \frac{\omega_e^2/4\pi\,[p + \nu_i + \delta\,\nu_e]}{p^2 + (\nu_i + \nu_e)\,p + \nu_i\,\nu_e} \tag{8}$$

where ω_e(s) is the electron plasma frequency, ν_i (s) and ν_e (s) are the ion and electron collision frequencies respectively, and δ is the mass density ratio $N_e\, m_e/N_i\, m_i$. K includes inertial effects which become negligible in regions which are collision dominated $\nu_e, \nu_i >> \omega_e$. In such regions K becomes the direct conductivity $\sigma_0 = \omega_e^{\,2}\,(\nu_e^{-1} + \delta\,\nu_i^{-1})/4\pi$.

We are thus able to integrate Poisson's Equation outward from S along each field line

$$\Phi_1(s, p) = \int_{s0}^{s} ds' \left\{ E_{\|o} - \frac{B(s')}{B(s_0)} \frac{K(s_0)}{K(s')} \left[\frac{A'}{p + 4\pi K(s_0)\cos^2\theta} \right] \right.$$

$$- \frac{p}{4\pi} \left[\frac{\partial}{\partial \alpha} \left(\frac{\nabla \Phi_1 \cdot \nabla \alpha}{B} \right) + \frac{\partial}{\partial \beta} \left(\frac{\nabla \Phi_1 \cdot \nabla \beta}{B} \right) \right] \left. \int_{s'}^{s} ds'' \frac{B(s'')}{K(s'')} \right\} \qquad (9)$$

$$+ \frac{p}{4\pi} \int_{s0}^{s} ds' \frac{B(s')}{K(s')} \left[\frac{\nabla \Phi_1 \cdot \nabla s}{B} \bigg|_{s0} - \frac{\nabla \Phi_1 \cdot \nabla s'}{B(s')} \right]$$

Here $A' = E_{\|o}(s_0, t = 0)$ and θ is the angle which the field line makes at s_0 with the surface normal. Two boundary conditions have been used in obtaining Eq. (9): $\Phi_1(s_0, p) = 0$ and $\partial \Phi_1(s, p)/\partial s|_{s0} = E_{\|o}(s_0) - A'[p + 4\pi K(s_0) \cos^2\theta]^{-1}$, the latter following from charge continuity at S and the assumption that electric field lines intersecting S terminate because of the presence of a surface charge density.

While formidable, Eq. (9) is rigorous and general. It is intended that relevant cosmic processes can be modeled by reasonable variations in **B**, $\mathbf{E}_0$, and plasma properties for which this equation simplifies. Let us now, for example, consider a two-dimensional model which illustrates some of the physics involved in the reaction of the earth-magnetosphere system to an imposed $\mathbf{E}_0$.

We consider the region $x \geq 0$ above an infinitely extended, ideally conducting plane at $x = 0$. The zeroth order magnetic field $\mathbf{B} = -B_0(\hat{e}_x \cos\theta + \hat{e}_z \sin\theta)$ is homogeneous, everywhere inclined at the angle $\pi - \theta$ with respect to the x-axis. (The z-axis lies in the conducting plane.) The imposed electric field $E_0 = -\hat{e}_x \tilde{A}$ is constant in time and normal to the conducting plane. The plasma properties vary spatially so that $K = K(x, z)$.

If we further assume that the z-dependence is weak so that $\partial/\partial z$ is small, Eq. (9) can be solved for this model. The quantity $\Phi + \Psi$, of importance in Eq. (5), is to lowest order in x/L (L is the scale length of the variation of K in z)

$$\Phi(x, z, t) + \Psi(x, z, t) = -\frac{\tilde{A}}{2\pi i} \int_0^x dx' \int_{-i\infty+\Gamma}^{i\infty+\Gamma} dp$$

$$\exp pt \frac{1}{p + 4\pi K(x', z, p)\cos^2\theta} = -\frac{\tilde{A}}{2\pi i} \int_0^x dx' F(x') \qquad (10)$$

the p integration, being, as usual, to the right of all singularities in the complex p-plane.

In regions where $\omega_e >> \nu_e, \nu_i$, F behaves dominantly as exp $- \nu_e t$ $\cos[(\omega_e \cos\theta)t]$. (In the presence of plasma instabilities, such as the ion-acoustic instability, ν_e may be some appreciable fraction of ω_i, the ion plasma frequency[8, 15, 16].) Since ω_e varies spatially, this response is not, strictly speaking, a plasma wave. Nevertheless it has a rapidly time-fluctuating character and is negligible if one time averages over the interval $(\bar{\omega}_e \cos\theta)^{-1}$, the period of plasma waves which would exist if the entire region had the mean density. We shall henceforth assume that at any time the entire variation in $\Phi + \Psi$ along a field line occurs over the collision dominated region $\nu_e, \nu_i >> \omega_e$. We shall assume further that $\nu_{e,i} = 0$ and σ_0 is infinite elsewhere.

We consequently obtain from Eq. 10

$$\Phi + \Psi = -\tilde{A} \int_0^x dx' \exp - [4\pi\sigma_0(x') t \cos^2\theta] \tag{11}$$

Note that the integrand in Eq. 15 relaxes most slowly in low conductivity regions. The contribution to $\Phi + \Psi$ from such regions is, however, in proportion to their spatial dimension.

The result obtained for this model by adding to Eq. (11) the solution to Eq. 10 of first order in x/L and plugging that expression into Eq. 5 is

$$\begin{aligned} \mathbf{v} - \mathbf{w} = \frac{c\tilde{A}}{B_0} \hat{e}_y \Bigg\{ & \sin\theta \exp - f(x) + \frac{f(x)\cos\theta}{\sigma_0(x)} \int_0^x dx' \frac{\partial \sigma_0(x', z)}{\partial z} \\ & \exp - f(x') + \sin\theta \tan\theta \frac{\partial \sigma_0}{\partial z} \int_0^x dx' \frac{\sigma_0(x)}{[\sigma_0(x) - \sigma_0(x')]^2} \\ & \exp - f(x) \left(1 - \left[1 - f(x')\left(1 - \frac{\sigma_0(x')}{\sigma_0(x)}\right)\right] \exp - [f(x') - f(x)\right) \end{aligned} \tag{12}$$

where $f(x, z, t) = 4\pi\sigma_0(x, z) t \cos^2\theta$.

Equation 12 illustrates the fact that although one may be at a place where the conductivity is good and $f(x) >> 0$ so that variations in $\Phi + \Psi$ have locally disappeared, $\mathbf{v} - \mathbf{w}$ can be non-zero if (1) the conductivity is anomalously low, $f(x') \simeq 0$, anywhere on the field line connecting the observation point and $x = 0$ and (2) there is a z-variation in the conductivity. In this case a gradient in $\Phi + \Psi$ perpendicular to $\mathbf{B}$ exists even in the highly conducting plasma region. Similar conclusions also follow when $\mathbf{E}_0$ has a z-variation as would be expected in more realistic models.

At a point x_0 such that the entire poorly conducting region lies between $x = 0$ and $x = x_0$, particles drift a distance of order

$$y = \frac{c\tilde{A}}{B_0}\frac{d}{L}\frac{1}{\sigma_{min}} \tag{13}$$

away from field lines in this model. Here d is the thickness in x of the low conductivity region and σ_{min} is the minimum value of the direct conductivity on the field line passing through the observation point.

The earth's atmosphere is a low conductivity region which under certain conditions allows the long term drift of magnetospheric plasma relative to field lines. Under other conditions, however, $\Phi + \Psi$ may be equalized between field lines by perpendicular charge displacements. We now discuss this aspect of the problem briefly.

IV. PERPENDICULAR CONDUCTIVITY

Consider that sufficient time has elapsed that a quasi-equilibrium in $\Phi + \Psi$ along field lines has been established. Since an $\mathbf{E}_\perp$ exists, perpendicular charge displacements occur in collision-dominated regions where the transverse conductivity is non-zero.

Our further calculations indicate that $\Phi + \Psi$ differences between field lines relax to zero via such perpendicular charge displacements in a characteristic time $(4\pi\sigma_1)^{-1}$, where σ_1 is the transverse conductivity in the collisional plasma. Furthermore, if

$$\gamma = \frac{\sigma_1 B_0^2}{\nu_e \rho c^2} << 1 \quad (\rho = \text{mass density})$$

the plasma in the collisional region is massive in the sense that in the discharge time $\mathbf{j} \times \mathbf{B}$ forces are insufficient to overcome inertia and start the plasma into bulk motion. For the earth's ionosphere $\gamma \sim \mathcal{O}(10^{-6})$, an estimate based on parameter values taken from Maeda and Kato.[17]

If the $\Phi + \Psi$ difference between field lines in the conducting region can be eliminated via this transverse charge displacement mechanism without the need for charge motion through regions of anomalously reduced σ_0, then σ_{min} in Eq. 13 should be interpreted as the minimum conductivity along the discharge path, quite probably the transverse conductivity σ_1.

We emphasize that the discussion in this Section assumes that the potential difference between field lines is not externally maintained. In the convection of the earth's magnetospheric plasma[5-7] the solar wind is the source of a constant emf. When an externally maintained emf exists particles may drift indefinitely with respect to field lines if there is an insulating layer between the plasma and S. Indeed, in such situations the identification of field lines as rooted in S is not a practical one.

V. DISCUSSION

The relative displacement $\Delta r_1 = \int dt\ \mathbf{v} - \mathbf{w}$ between drifting particles and moving field lines has significance only when compared with some other pertinent length Δr_2, for instance the total particle drift distance $\int dt\ \mathbf{v}$. When $\Delta r_2 >> \Delta r_i$ it makes sense to follow particles by tracing field lines. Each problem and model must, however, be examined individually for this criterion. In many instances, and the case of terrestrial magnetospheric impulses is exemplary here, the time scale over which E_0 is imposed is so long compared with $(\sigma_{min})^{-1}$ that Δr_2 is enormous compared with Δr_1.

ACKNOWLEDGMENT

I thank Dr. Frank C. Jones of GSFC for numerous illuminating discussions.

REFERENCES

1. Cowling, T. G., Magnetohydrodynamics, Interscience Publishers, New York, 1957.

2. Gold, T., Rotating Neutron Stars as the Origin of the Pulsating Radio Sources, Nature, 218, 731, 1968.

3. Davis, L., Jr., Stellar Electromagnetic Fields, Phys. Rev., 72, 632, 1947.

4. E. N. Parker, Interplanetary Dynamical Processes, Interscience Publishers, New York, 1963.

5. Axford, W. I. and C. O. Hines, A Unifying Theory of High-Latitude Geophysical Phenomena and Geomagnetic Storms, Can. J. Physics, 39, 1433, 1961.

6. Dungey, J. W., Interplanetary Magnetic Field and the Auroral Zones, Phys. Rev. Letters, 6, 47, 1961.

7. Brice, Neil M., Bulk Motion of the Magnetosphere, J. Geophys. Res., 72, 5193, 1967.

8. Sagdeev, R. Z., On Ohm's Law Resulting from Instability, Proceedings of Symposia in Applied Mathematics, Vol. 18, American Mathematical Society, Providence, Rhode Island, 1967.

9. Birmingham, T. J and F. C. Jones, Identification of Moving Magnetic Field Lines, J. Geophys. Res., 73, 5505, 1968.

10. Jones, F. C. and T. J. Birmingham, Identification of Moving Magnetic Field Lines 2. Application to a Moving Non-rigid Conductor, J. Geophys. Res., 76, 1849, 1971.

11. Conrath, Barney J., Radial Diffusion of Trapped Particles with Arbitrary Pitch Angle, J. Geophys. Res., 72, 6069, 1967.

12. Stern, David, Geomagnetic Euler Potentials, J. Geophys. Res., 72, 3995, 1967.

13. Newcomb, W. A., Motion of Magnetic Lines of Force, Ann. Phys., 3, 347, 1958.

14. Northrop, T. G., The Adiabatic Motion of Charged Particles, John Wiley & Sons (Interscience), New York, 1963.

15. Sellen, J. M., W. Bernstein, B. D. Fried, and C. F. Kennel, Anomalous Resistivity Due to Ion Acoustic Turbulence, B.A.P.S., 15, 1427, 1970.

16. Gary, S. Peter, and J. W. M. Paul, Anomalous Resistivity Due to Electrostatic Turbulence, Phys. Rev. Letters, 26, 1097, 1971.

17. Maeda, K. and S. Kato, Space Science Reviews, 5, 57, 1966.

2. Dungey, J. [illegible], [illegible] Electrodynamics [illegible], Phys. Rev., 79, [illegible], 1950.

3. [illegible] Processes, [illegible] Publishers, [illegible], 1963.

4. [illegible] W. [illegible] A [illegible] Theory of [illegible] [illegible] and [illegible], Can. J. Phys., [illegible]

5. [illegible] Fields and [illegible] Applied [illegible] [illegible], 1961.

6. [illegible] [illegible], [illegible] 1961.

7. [illegible] Instability [illegible] [illegible] Measurements [illegible] [illegible], 1961.

8. [illegible] and [illegible] of Moving Magnetic [illegible], 1968, [illegible]

9. [illegible] Detection of Moving Magnetic Field Lines [illegible] Conductors, [illegible] Res. [illegible], 1971.

10. [illegible] Charged Particles with [illegible] [illegible], 1961.

11. [illegible] Res., [illegible] 1967.

12. [illegible] Magnetic Lines of Force, Am. J. Phys., [illegible]

13. [illegible] Motion of Charged Particles, John Wiley & Sons [illegible], New York, [illegible]

14. [illegible] and [illegible] A.P.S., 15, [illegible], 1970.

15. [illegible] S. Peter, [illegible] W. M. Fairbank, Applications [illegible] Phys. Rev. Letters, 26, 1097, 1971.

16. [illegible], Phys. [illegible], 390.

COLLISIONLESS SHOCKS

J.W.M. Paul

U.K.A.E.A. Research Group

Culham Laboratory, Abingdon, Berkshire

1. INTRODUCTION

Collisionless shock waves appear with increasing frequency in the literature of space and astrophysics. The adjective collisionless is deceptive. The classical definition of a collision involves a series of small angled deflections by the THERMAL ELECTRIC FIELD FLUCTUATIONS, $\langle E^2 \rangle$, in the plasma. However many plasmas are far from thermal equilibrium (e.g. severe gradients in a shock) and the free energy can usually couple into the COLLECTIVE (i.e. wave) degrees of freedom of the plasma. This coupling is a plasma instability and leads to a SUPRA-THERMAL level of $\langle E^2 \rangle$.

If many such degrees of freedom are excited with random phases, the plasma is said to be TURBULENT. Non-linear wave-wave mode coupling can generate this 'wave chaos' in analogy with the particle-particle origin of 'molecular chaos'. There are interesting conceptual questions here which we must leave.

We restrict our discussion to electrostatic turbulence with
(i) scale $\ll L_s$ = shock width (i.e. microturbulence),
(ii) fluctuating potential $\phi \ll (\kappa T_e/e)$ so that large angle deflection and trapping are not important, (i.e. weak turbulence).
Under these conditions the turbulent $\langle E^2 \rangle$ results in more rapid deflection of the particles than for a thermal plasma. There is an effective or turbulent collision frequency i.e. $\nu^* > \nu_{thermal}$. The plasma can be described reasonably well by a FLUID MODEL with ENHANCED TRANSPORT coefficients derived from ν^*. The dependence of these transport coefficients is determined by the instability and the non-linear processes. The dissipation, i.e. entropy increase, arises through the randomness of the turbulence.

Experiments[1,2] have demonstrated that 'collisionless shocks' exist in which an enhanced collision rate arises from microturbulence and that the shock can be described by an MHD fluid model with enhanced transport coefficients.

2. COSMICAL COLLISIONLESS SHOCKS

We shall briefly mention some of the circumstances in which collisionless shocks occur in the literature of space and astrophysics, recognizing that in the latter case it is often conjecture.

(i) Space: Satellites have observed both the earth's bow shock[3,4] and interplanetary shocks in the solar wind.
(ii) Solar: Optical[5] and radio observations[6,7] demonstrate the emission of shocks from solar flares. Some models of flares invoke an internal shock as well.
(iii) Stellar: Flare stars and supernovae may involve shocks.
(iv) Galactic: Galactic 'jets' and 'explosions' may involve shocks as may the expanding 'plasma blobs' of double radio sources.

The appeal of a collisionless shock in situations (ii) to (iv) is that it can convert kinetic energy, through the mediation of plasma turbulence, into the observed NON-THERMAL ELECTROMAGNETIC EMISSION. Turbulence can stochastically accelerate particles to sufficient energy for synchrotron emission and can also directly emit at ω_{pe} and its harmonics. Both of these emission processes are observed.

It should be noted that the kinetic energy of the streaming plasma can generate the required turbulence even if no shock wave forms ahead of it. Both 'piston' and shock can be turbulent.

3. LABORATORY EXPERIMENTS

Just as laboratory spectroscopy and atomic physics have contributed to the understanding of classical (i.e. optical) astrophysics so we expect laboratory plasma physics to contribute to the understanding of modern (e.g. radio) astrophysics. Phenomena are more readily understood when they and the theories involved can be studied in the laboratory under the 'microscope', rather than the telescope.

Laboratory experiments on collisionless shocks can be divided into three classes depending on the nature of the flow and piston.

Flow experiments; A quasi-steady supersonic plasma flow can be produced by an arc or a nozzle. When the flow impinges on an obstacle, such as a magnetic dipole, a bow shock is produced.

Plasma pistons: A more dense plasma can be made to compress a less dense one and produce a shock. The more dense plasma can be produced within a plasma by (i) increasing the degree of ionization using pulsed UV light or by (ii) ionizing a solid target using a powerful pulse of laser light. Alternatively, plasmas of different density, separated by a negatively biassed grid, can be pulsed into contact. If the two plasmas interact sufficiently strongly a shock should be produced.

Magnetic piston: A rising magnetic field acts on a highly conducting plasma like a piston. Three cylindrical configurations are used; (i) Z-pinch with axial current, (ii) the theta-pinch with azimuthal current, both of which give radial compression, and (iii) the annular shock tube with radial current and axial compression.

The time and space scales must be right for piston and shock formation. In small apparatus, for example, the field may diffuse before any compression occurs or the driver plasma may not have time to interact with and hence compress the ambient plasma.

4. SHOCK DISCONTINUITIES

Assuming a shock forms, its nature depends on the piston and the compression wave which it can generate. In the absence of a magnetic field, there is only the one sound speed $c_o^2 = \gamma p/\rho$. However in a magnetized plasma there are three anisotropic sound speeds, slow (c_s), intermediate (c_i) and fast (c_f). All of these, except the intermediate wave, should theoretically steepen to form shocks. For MHD stability, the change of flow speed across the shock, in its frame, must jump only one sound speed (i.e. only one wave is trapped and steepens).

Regarding the shock as a discontinuous jump from state 1 to state 2, the conservation relations and Maxwell's equations determine the shock jumps (i.e. Rankine-Hugoniot or de Hoffman-Teller relations). These are uniquely defined by four parameters.

(i) Ratio of initial plasma to magnetic pressure $\beta = 2\mu_o p/B^2$

(ii) Ratio of shock to sound or Alfven speed (C_A), i.e. Mach Nos. $M = V_s/C_o$; $M_A = V_s/C_A = \sqrt{\mu_o\rho}\, V_s/B$ (S.I. units)

(iii) The angle between shock velocity vector $\bar{V}_s$ and $\bar{B}$

(iv) The ratio of specific heats γ.

For both fast and slow shocks $\bar{V}_s$, $\bar{B}_1$, $\bar{B}_2$ are coplanar. For fast shocks $B_2 > B_1$ and the ratio $R = \kappa T_2/\frac{1}{2}MV_s^2$ tends to limit $1/\gamma$ as $M_A \to \infty$. For slow shocks $B_2 < B_1$ and R is not limited.

5. LIMITATION OF STEEPENING

The internal structure of the shock will be determined by the first process which can limit the steepening and satisfy the jump relations.

Dispersive limitations: Consider a collisionless plasma without instability. Through the dispersion relation, $D(\omega,k) = 0$, the phase velocity of a wave is related to its wave number, $k = 2\pi/\lambda$. When a dispersive effect decreases $d\omega/dk$ at a certain k_o, steepening results in a slower phase velocity and a trailing wave train; e.g. (i) non-magnetic ion sound waves at $k_o \sim 2\pi/\lambda_D$, (ii) fast waves with $\bar{V}_s \perp \bar{B}$, $\beta < 1$, at $k_o \sim 2\pi(\omega_{pe}/c)$. When $d\omega/dk$ increases, steepening will produce a forward wave train, but some damping is essential because it cannot extend indefinitely forward; e.g. fast wave propagating obliquely to B at $k_o \sim 2\pi(\omega_{pi}/c)$. Both these dispersive processes limit steepening and produce a large amplitude wave. Neither produces the required entropy increase for a shock without some additional process.

Dissipative Limitations: The gradients produced by simple steepening or the above large amplitude waves have free energy which can be dissipated through thermal collisions or instability driven turbulence. Both of these processes can be described by a transport coefficient and for a strong shock only viscosity (μ) and/or resistivity (η) need be considered. In a viscous shock dissipation through ∇v is dominant while in a resistive shock ∇B (i.e. current J) is dominant.

Particle Effects: The above gradients have associated electric fields and these can reflect some of the incident ion distribution function. In the absence of a magnetic field these particles are lost, and no steady state is possible. The steepening can still be limited by the energy loss. In the presence of a magnetic field the reflected particles gyrate, gain energy and pass back through the shock. Theories exist which predict that the phase mixing of these gyrating particles behind the shock gives rise to the required entropy change for a shock to form. Trapping of particles within wave trains can also produce an entropy increase.

The structure of a shock will depend on which of these various processes limits the steepening.

6. SCALING LAWS

In relating large scale natural phenomena ($L \sim 10^5 - 10^8$ m) to small scale laboratory experiments ($L \sim 10^{-3} - 10^0$ m) it is important to recognize a simple physical process and attempt to simulate it in the laboratory by scaling the relevant parameters.

For shock waves the important parameters are:

(i) Jump parameters: β, M_A, θ, γ
(ii) Plasma parameters: n_e, T_e, T_i, ion mass and charge

(iii) Flow parameters: V_s
(iv) Fields: B, E, ϕ (potential)
(v) Microscopic parameters: distribution $f^n s$ $f_e(v)$, $f_i(v)$,
lengths λ_D, λ_{coll}, r_{ce}, r_{ci} and $\alpha = \left(\frac{\omega_{ce}}{\omega_{pe}}\right)^2 = \left(\frac{r_{ce}}{\lambda_D}\right)^2$.

Collisionless plasma effects should obey the Vlasov equation for which the exact scaling laws are (ref. 8)

$$t \propto L;\ n \propto L^{-2};\ B \propto L^{-1};\ E \propto L^{-1}\ (\lambda_D \propto L,\ r_c \propto L)$$
$$\alpha, \beta, \gamma, V_s, M_A, M, T, \phi, f_e, f_i, \text{ independent.}$$

Clearly this scaling does not include thermal fluctuation effects (e.g. λ_{coll}, N_D) or radiation effects.

For collisionless shocks we shall find that the shock width depends on (c/ω_{pe}) or (c/ω_{pi}) for magnetic shocks and λ_D for non-magnetic. All of these follow the Vlasov scaling $L \propto n^{-\frac{1}{2}}$ and fortunately the scaling curve also passes through the region of laboratory plasmas. For example

	L(m)	n(m^{-3})	B(T)
Bow shock	10^5	10^6	10^{-9}
Laboratory	10^{-2}	10^{20}	10^{-2}

Qualitative scaling of thermal effects $L_s < \lambda_{coll}$ can also be obtained in the laboratory.

7. ELECTROSTATIC SHOCKS

In a collisionless unmagnetized plasma ion inertia produces dispersion at ω_{pi}. This gives rise to large amplitude trailing waves with $\lambda \sim \lambda_D$ and these have been observed in the laboratory[9,10].

At high Mach number $M > 1.6$, the E-field of the wave reflects particles forward and these are lost. The wave itself take up the reaction and becomes irregular, and non-steady.

At much higher M it is not even possible to produce a piston because the interstreaming velocity is too great for instability[1c]. There is, as yet, no evidence of ion-ion streaming instability producing a viscous shock.

8. MAGNETIC SHOCKS : MACROSTRUCTURE

We shall restrict our discussion to fast shocks because these are observed in the laboratory. We shall classify by the jump parameters M_A, β and θ.

There are several critical Mach numbers in shock physics. The most important, M_A^*, occurs when the plasma is heated sufficiently for the flow behind a magnetic shock to become sub-sonic as well as sub-magnetosonic. Above M_A^* non-magnetic sound waves can steepen, on a scale length for which the magnetic field and plasma are resistively decoupled, and form a sub-shock. Below M_A^* we expect a resistive magnetic shock and above M_A^* an additional viscous electrostatic sub-shock.

8.1 $\theta = 90^o$ (i.e. $V_s \perp B$), $\beta_1 \ll 1$, $M_A < M_A^* \sim 3$

Weak shocks with structures dominated by classical resistivity are observed[11]. These change to a broader structure with $L_s \sim 10(c/\omega_{pe})$ when the electron drift velocity (v_d) within the shock is sufficient to drive the ion wave instability (i.e. $v_d > C_o$).

Large amplitude trailing waves are observed at low densities[9,11]. These result from the dispersive effect of electron inertia and have $L_s \sim (c/\omega_{pe})$. For $\alpha \geq 1$, $v_d \to c$ and relativistic limitation yields $L_s = \sqrt{\alpha}(c/\omega_{pe}) = v_A/\omega_{pi}$, which is observed[9]. At higher density there is time for the two-stream instability driven by $v_d \sim v_{eth}$ to grow within the shock, and then a non-oscillating structure with $L_s \sim 10(c/\omega_{pe})$ appears[9].

The characteristic non-classical (i.e. collisionless) shock for this θ, β, M_A, has $L_s \sim 10(c/\omega_{pe})$ and has been studied in many laboratories[1,2,9,11-14]. The 'collisionless' nature of these shocks is demonstrated by the inadequacy of classical transport coefficients to explain the observed electron heating which requires $\eta^* \sim 100\eta_{sp}$(classical)[13b]. Also in some cases $\tau_s (= L_s/V_s)$ is much shorter than the classical collision time (τ_{ei})[11]. As $r_{ci} \gg L_s \gg r_{ce}$, the ions are unmagnetized while the electrons experience drift motiond due to ∇B, ∇n, ∇T and $\nabla \phi$ (i.e. E_L). This latter is dominant in most experiment. E_L arises from ∇p_e and the Hall effect and adiabatically slows down the ions. The electrons are irreversibly heated and satisfy the conservation relations.

The observed drift velocity exceeds the critical value for electrostatic instability. As the electrons are heated usually $T_e > T_i$, and so ion wave (I.W.) or I.W. coupled to electron cyclotron wave (E.C.W.) instability can occur[15]. There is experimental evidence for I.W. turbulence[1] and the η^* derived from the macrostructure scales[11] in agreement with the predictions of the Kadomtsev-Sagdeev theory for I.W. turbulence.

8.2 $\theta = 90^o$, $\beta_1 \ll 1$, $M_A > M_A^* \sim 3$

For $6 > M_A > M_A^* \sim 3$ the resistive shock with $L_s = L_R \sim 10(c/\omega_{pe})$ has a broad 'foot' in front with $L_F \sim 2(c/\omega_{pi}) \sim 8\ L_R$[13]. The sharp rise L_R is the same as for $M_A < M_A^*$ and is dominated by I.W.

turbulence. The foot appears to be formed by the gyration of ions ($L_R \sim r_{ci}$) in front of the resistive structure. However the observed electric potential ϕ_R is not sufficient to reflect the required number of ions.

Theory predicts that above M_A^* there should be a viscous sub-shock at the rear of the resistive shock and that this sub-shock should heat ions. The observed electron heating is inadequate to satisfy the conservation relations and so ion heating is assumed to occur within a sub-shock. There are two suggested mechanisms for reflections of ions within the sub-shock. Firstly, the ion heating makes reflection more probable. Secondly the sub-shock can consist of damped ion inertia waves and the overshoot will reflect adequately.

The fraction of the jump ΔB across L_F increases until for $M_A \sim 6$ there is no L_R[13]. This structure has oscillations behind and is unsteady.

8.3 $\theta = 90^o$, $\beta_1 \gg 1$

There are only two experiments[2,11,12,14] in this regime. The most striking difference from $\beta \ll 1$ is that there is no structural change at M_A^* and that $L_s \sim (c/\omega_{pi})$ for $3 < M_A < 9$. There is a clearly observed change from electron heating below, to increasing ion heating above M_A^*, with the observed $T_e + T_i$ fitting the conservation relations. There is also direct evidence for microturbulence as for $\beta \ll 1$.

8.4 Oblique Shocks

For oblique propagation, Whistler dispersion produces a forward wave train which is observed in piston experiments[16]. Non-classical damping of the oscillations corresponds to an η^* similar to that observed for perpendicular propagation. For $M_A > M_A^*$, there is no 'foot' or other evidence of reflected ions, but the observed structures require a viscous sub-shock at the rear. Under certain conditions, which are not fully understood, the Whistler oscillations develop high frequency components as the shock propagates and these eventually destroy the regular structure.

In the steady flow experiments[17] there is no evidence of a forward wave train. Both in and behind the shock, $L_s \sim c/\omega_{pi}$, there is macroscopic electromagnetic turbulence. In these experiments there appears to be sufficient time for the high frequency instability, mentioned above, to convert the steady oscillations to turbulence.

These observations emphasise the problem of time scales in simulation experiments.

8.5 Parallel Propagation

For parallel propagations there are two classes of shock: the non-magnetic shock and the 'switch-on' shock. The latter generates a transverse component of B, and occurs in a limited region of parameter space defined by

$$\beta_1 \leq 2/\gamma \quad \text{and} \quad 1 \leq M_A \leq \hat{M}_A \;; \quad \hat{M}_A = [\gamma(1 + \beta_1) + 1]^{\frac{1}{2}}/(\gamma - 1)^{\frac{1}{2}} .$$

Such shocks have been observed[1c] in the laboratory but not studied in detail.

9. MICROSTRUCTURE OF SHOCKS

The inadequacy of thermal transport leads directly to a search for collective effects and microturbulence. The micro-turbulent electric fields, $\langle E^2 \rangle$, will scatter and heat particles while the corresponding density fluctuations $\langle \delta n_e^2 \rangle$ can scatter photons and allow a direct measurement of the level and spectrum of the turbulence in terms of the Fourier transform $\langle \delta n_e^2 (\omega, k) \rangle$. The nature of the fluctuations can be deduced and hence $\langle E^2 \rangle$ derived from the measured $\langle \delta n_e^2 \rangle$. This technique has been used on two perpendicular shock experiments.

The results from the first, TARANTULA[1], with $\beta \ll 1$, $M_A < M_A^*$, $T_e > T_i$, are summarized. The fluctuations are SUPRA-THERMAL by more than two orders of magnitude. For a given $\bar{k}$ there is a dominant mode with frequency ω such that (ω, k) fits and scales as for ION WAVES. The mode, however has a short coherence time $\tau \sim 2\pi/\omega$ demonstrating a high degree of RANDOMNESS OF PHASE. This turbulence is grossly ANISOTROPIC being confined to within 50° from the direction of the electron current within the shock. At present these measurements are restricted to the plane perpendicular to B. However, if the fluctuations are ion waves, as seems probable, this anisotropy should form a cone about the driving electron current. The wave number spectrum has the form $\langle \delta n_e^2 (\omega, k) \rangle \propto (1/k^3) \ln (1/k\lambda_D)$ in agreement with the predictions of non-linear theory[18,19].

The non-linear theory of ion wave turbulence is discussed in terms of a balance of linear growth at $k \sim 1/\lambda_D$ against non-linear diffusion to lower k. This diffusion results from the 3-wave process in the form of resonant wave decay[19] or non-resonant wave scattering on particles[18]. Both processes give the observed k-spectrum. The decay process is only possible because of the short coherence time but, when possible, as the resonant process it should dominate. However as yet there is no clear agreement between experiment and non-linear theory.

The measured level of turbulent energy ($\leq 2\%$ thermal), potential fluctuations ($e\phi \leq 1\% \, \kappa T_e$) and randomness of phase are used to justify a STOCHASTIC treatment of the electron heating by the turbulence. This yields a mean effective resistivity within the shock which is a half that required experimentally.

The second experiment[2] involving photon scattering has $\beta > 1$, $M_A \gtrsim M_A^*$ and $T_e < T_i$. The measurements are similar to the above although no scaling with ion plasma frequency is reported. This similarity is surprising because for $Te < Ti$ ion wave turbulence is not expected and the E.C.W. appears necessary[20].

10. COMPUTER SIMULATION OF MICROSTRUCTURE AND TURBULENCE

Paricle simulation computations, usually one-dimensional, have followed the development of turbulence driven by a current across a magnetic field. However the conflicting results require more discussion that is possible here[21,22].

11. COLLISIONLESS SHOCKS OUTSIDE THE LABORATORY

Finally we return to the cosmical scene to consider two examples of collisionless shocks.

(i) The Earth's Bow Shock

Measurements from spacecraft show clearly the existence of a collisionless bow shock but equally clearly it is NOT STEADY in position or structure. It is, in general, a fast oblique shock with $M_A \sim 8 > M_A^*$, $\beta \gtrsim 1$, $T_{e1} \gtrsim T_{i1}$, $T_{e2} < T_{i2}$. Unfortunately the shock moves in position with a velocity comparable to or greater than that of the space craft. This results in multiple crossings and ambiguous length scales. The length scales for the shock transition vary considerably for different parameters and from crossing to crossing. Detailed results from OGO V[3] and VELA 4[4] show magnetic field changes in distance $L_s \sim 10\ c/\omega_{pe}$ and less frequently c/ω_{pi}, while temperature changes are over $L \sim 10\ c/\omega_{pe}$ for ions and $30\ \lambda_D$ for electrons (rare observation). In one crossing a reversible wave train with $\lambda \sim c/\omega_{pe}$ (i.e. dispersive wave train) was observed. Some of these results are surprising for $\beta \geq 1$ and $M_A > M_A^*$.

High frequency FLUCTUATING ELECTRIC FIELDS are observed in regions of high magnetic field gradient. However the ambiguity of velocity makes the k and ω scales uncertain. The frequency spectrum appears as discrete modes which tend to broaden and merge towards the rear. There is some similarity here with the current driven turbulence observed in laboratory shocks[1,2].

(ii) Solar Flare Shocks

A shock-like disturbance has been observed, by both optical[5] and radio[6] emission, to emanate from the sudden release of energy in a solar flare on the disc of the sun. The radio emission at ω_{pe}

and $2\ \omega_{pe}$ (Type II) must arise from some COLLECTIVE EFFECT within the shock front.

A limb flare has been observed[7] to give rise to type IV radio emission propagating outwards from a solar flare like a shock wave. Analysis suggests that this synchrotron emission commences when the disturbance forms a collisionless shock through two-stream instability and that it disappears when the shock broadens at $M_A = M_A^*$. The emission is thought to be compatible with synchrotron emission from a TURBULENT PLASMA with a few electrons STOCHASTICALLY accelerated to a few MeV. This calculation assumed current driven ion wave turbulence of the Kadomtsev form[18].

12. CONCLUSIONS

There is now clear evidence for the turbulent nature of collisionless resistive ($M_A < M_A^*$) shocks in the laboratory. A reasonable degree of self-consistency and agreement with theory has been obtained. Some of these features are also observed in space and solar shocks.

Future laboratory effort should move on to particle effects, including acceleration, in both electrostatic and high M_A magnetic shocks and, if possible, also onto electromagnetic emission from shocks.

The task of understanding natural phenomena is intrinsically difficult because of irreproducibility and uniqueness. Also physical processes can not be isolated as in the laboratory. Fortunately there is no need to understand cosmic plasmas in great detail. If the basic processes involved can be understood as a result of theory and experiment, then an adequate model can be constructed. While not accurate in detail it should then have a high degree of plausibility.

REVIEWS

PAUL, J.W.M., Physics of Hot Plasmas, Ed. Rye and Taylor, Oliver & Boyd, pp.302-345, 1968.

Proceedings Conference Collision-free Shocks in the Laboratory and Space, ESRO - SP-51, 1969.

CHU, C.K. and GROSS, R.A. Advances in Plasma Physics , Vol. 2, Ed. Simon and Thompson, Academic Press, pp.139-201, 1969.

HINTZ, E., Methods of Experimental Physics, Vol. 9A, Ed. Griem and Lovberg, Wiley, pp.213-274, 1970.

TIDMAN, D.A. and KRALL, N.A., Shock Waves in Collisionless Plasmas, Wiley, 1971.

REFERENCES

1. PAUL, J.W.M. et al., Nature, 223, 822 (1969); Phys. Rev. Lett., 25, 497 (1970); IAEA Conference, Madison, J9, 1971.
2. KEILHACKER, M. et al., Phys. Rev. Lett., 26, 694 (1971); IAEA Conf., Madison, J10, 1971.
3. FREDRICKS, R.W. et al., Phys. Rev. Lett., 21, 1761 (1968), and 24, 994 (1970).
4. MONTGOMERY, M.D. et al., J. Geophys. Res., 75, 1217 (1970).
5. MORETON, G.E., Astron. J., 69, 145 (1964).
6. WILD, J.P. et al., Nature 218, 536 (1968).
7. LACOMBE, C. et al., Astron. and Astrophys., 1, 325 (1969).
8. SCHINDLER, K., Rev. Geophys., 7, 51 (1969).
9. KURTMULLAEV, R.Kh. et al, IAEA Conference, Novosibirsk, A1, 1968.
10. TAYLOR, R.J. et al., Phys. Rev. Lett., 24, 206 (1970).
11. HINTZ, E. et al., IAEA Conference, Madison, J11, 1971.
12. HINTZ, E. et al., IAEA Conference, Novosibirsk, A2, 1968.
13. PAUL, J.W.M. et al., Nature, 208, 133 (1965); 216, 363 (1967); ESRO Report No. SP-51.
14. KEILHACKER, M. et al., IAEA Conference, Novosibirsk, A3, 1968; Z. Physik, 223, 385 (1969).
15. GARY, S.P. et al., J. Plasma Phys., 4, 739, 753 (1970).
16. ROBSON, A.E. et al., IAEA Conference, Novosibirsk, A6, 1968.
17. PATRICK, R.M. et al., Phys. Fluids, 12, 366 (1969).
18. KADOMTSEV, B.B. Plasma Turbulence, Academic Press, 1965.
19. TSYTOVICH, V.N., Culham Laboratory Preprint CLM-P 244 (1970).
20. LASHMORE-DAVIES, C.N., J. Phys. A, 3, L40 (1970).
21. LAMPE, M. et al., Phys. Rev. Lett., 26, 1221 (1971).
22. FORSLUND, D. et al., IAEA Conference, Madison, E18, 1971.

NON-LINEAR EVOLUTION OF FIREHOSE-UNSTABLE ALFVEN WAVES

K. Elsässer and H. Schamel

Max-Planck-Institut für Physik und Astrophysik

München

SOLAR WIND ANISOTROPIES

Due to the rarefaction of the expanding solar wind, plasma particles will become more or less collisionless. For protons this may happen at solar distances smaller than 1 AU (1). Beyond that critical distance one could try to describe the particles by considering adiabatic invariants. Assuming for simplicity a radial magnetic field $B_o(r) \sim 1/r^2$ (r = solar distance) conservation of magnetic moment would imply $T_\perp \sim B_o \sim 1/r^2$, where $T_\perp$ is the kinetic proton temperature associated with the velocity components perpendicular to B_o. On the other hand the parallel temperature $T_\parallel$ is not expected to change appreciably, so a temperature anisotropy $T_\parallel - T_\perp > 0$ is expected to develop in the collisionless regime (2). Indeed proton anisotropies of $T_\parallel / T_\perp \gtrsim 2$ are typical values measured in the "quiet" solar wind at the earth's orbit, but the above argumentation would eventually lead to much higher values (1). Thus a collisionless mechanism is required which keeps the anisotropies low.

ANISOTROPY INSTABILITIES

If $T_\parallel - T_\perp$ becomes large enough the plasma may be destabilized. The classical candidate for the required collisionless dissipation is the firehose instability which evolves if

$$\sum_j (\beta_\parallel^j - \beta_\perp^j) - 2 > 0$$

where $\beta_{\parallel}^{j}$ ($\beta_{\perp}^{j}$) is the ratio of the parallel (perpendicular) pressure of the particle species j to the magnetic field energy density. This instability has recently been discussed in connection with collisionless shocks (3) and the solar wind (4). In the latter paper a resonant ion cyclotron instability was also considered. Both instabilities are favoured by large electron anisotropies. But at the earth's orbit the instability conditions can be fulfilled only by transient, not "quiet", configurations. Nevertheless it remains an interesting problem to determine the equilibrium spectrum associated with these instabilities, since, on the one hand, magnetic fluctuations have been observed in the solar wind over a large frequency range, and, on the other hand, other observations like the high temperature of the α particles (1) or the relatively fast corotation of the solar wind due to pressure anisotropy (7) could perhaps be related to these fluctuations.

PARAMETERS FOR FLUID-LIKE BEHAVIOUR

We have calculated the relaxation of the (non-resonant) firehose instability (5). A fluid-like behaviour of this instability can be expected if the highest wave vector k_B of the unstable modes is $\lesssim 1/2 \times R_+^{-1}$, where R_+ is the ion Larmor radius. From the dispersion relation (evaluated for adiabatic, not resonant particles) we have (4)

$$k_B = \left[\Delta/(\tfrac{1}{4}\, B^2 - P)\right]^{1/2} R_+^{-1}$$

with

$$\Delta = \{\sum_j (\beta_{\parallel}^{j} - \beta_{\perp}^{j}) - 2\} / \beta_{\parallel}^{+}$$

$$B = 1 + 2 \times (\beta_{\parallel}^{+} - \beta_{\perp}^{+})/\beta_{\parallel}^{+}$$

$$P = (\overline{v_{\parallel}^4} - \tfrac{3}{2}\, \overline{v_{\parallel}^2 v_{\perp}^2})_+ / (\overline{v_{\parallel}^2})_+^2$$

The index + refers to the protons; a bar means averaging with the velocity distribution function. The condition of a short wavelength stabilization and fluid-like behaviour requires $P<(1/4)B^2$ and $0<\Delta<<1$. For a negligible electron anisotropy and a bi-Maxwellian proton distribution this would imply $\beta_{\parallel}^{+}>15$, a value which is too high for the solar wind at the earth's orbit, by an order of magnitude. But we do not need to insist on a bi-Maxwellian proton distribution; a truncated Maxwellian distribution near the sun (6), for instance, could help to lower the value of P. In any case, we may claim that at least for the disturbed solar wind adiabatic wave-particle interactions may be important, and for simplicity we restrict our discussion to this case.

NONLINEAR AMPLITUDE EQUATIONS

If the spectrum equation holds during the whole quasilinear stabilization process (8), nonlinear terms should usually be included (9). Therefore we start with a nonlinear amplitude equation for the unstable magnetic perturbations and try to make it consistent with both the Vlasov and Maxwell equations. This procedure has been outlined in more detail for electrostatic plasma turbulence (10). Circularly polarized Alfvén waves are characterized by a wave vector k parallel to $\pm B_o$, and by a polarization index $\sigma = \pm 1$ which discriminates between right- and left-hand polarization (if k is given). A three wave interaction among these waves is not possible because the polarization condition $\sigma_1=\sigma_2+\sigma_3$ cannot be fulfilled. Therefore we expect a third order amplitude equation of the following type:

$$\frac{\partial}{\partial t} B_1 = v_{12'3'4'} \exp \{i\int_o^t d\tau(\omega_1-\omega_{2'}-\omega_{3'}-\omega_{4'})\}B_{2'}B_{3'}B_{4'}$$

$$+ \delta\gamma_1 B_1$$

The first term on the right hand side (the slashes indicate the summation with respect to (k,σ) e.g. $2' \to \sum_{k_2,\sigma_2}$) describes the direct four-wave coupling - $B_2 \exp(-i\int_o^t d\tau\omega_2)$ is, for instance, the magnetic field perturbation associated with a partial wave (k_2, σ_2) -, while the second term takes account of the fact that all perturbations are defined with respect to a slowly varying background. In (5) we have evaluated the coefficients for adiabatic particles with the following results:

$$v_{1234} = \delta(k_1-k_2-k_3-k_4)\delta(\sigma_2,-\sigma_1)\delta(\sigma_3,\sigma_1)\delta(\sigma_4,\sigma_1)(2+\beta_\parallel^+\times\Delta)/\alpha_1$$

$$\delta\gamma_1 = -i\ \frac{\partial\gamma_1}{\partial t}\Big/(k_1^2 v_A^2 \alpha_1)$$

with

$$\alpha_1 = \frac{\omega_1}{k_1^2 v_A^2} + \frac{\sigma_1}{\Omega_+} (\beta_{\shortparallel}^+ - \beta_{\perp}^+) \tag{1}$$

ω_1 is the complex frequency of the wave (k_1, σ_1); $\partial\gamma_1/\partial t$ is the time derivative of the growth rate which was obtained from the dispersion relation and the quasilinear diffusion equation. v_A is the Alfvén velocity of the protons.

SPECTRUM EQUATION

The amplitude equation is only a formal starting point. The goal of any description of turbulence is a closed equation for the energy density of the waves, i.e. the spectrum equation. Due to the specific nature of the nonlinear interaction we may use a Gaussian closure of the hierarchy of wave correlations (i.e. discarding the four-wave correlations). Using the functional formalism of (11) we derive in (5) the following spectrum equation

$$\frac{\partial}{\partial t} I_1 = 2\Gamma_1 I_1 \tag{2}$$

with

$$\Gamma_1 = \gamma_1 \left[1 + \left\{ (1 + \tfrac{1}{2}\beta_{\shortparallel}^+ \cdot \Delta) W + \frac{-\partial\gamma_1/\partial t}{k_1^2 v_A^2} \right\} \Big/ (k_1^2 v_A^2 |\alpha_1|^2) \right]$$

$$W = \int dk_1 I_1 \quad ; \quad I_1 = |B_1|^2 \exp(2\int_0^t d\tau \gamma_1(\tau))/B_o^2$$

and the following equation for the truncation error δI_1 of I_1:

$$\frac{\partial}{\partial t} \delta I_1 = 2\gamma_1 \delta I_1 + W^2 I_1/(|\omega_1||\alpha_1|^2) \tag{3}$$

The latter is meant only for the initial phase where typically γ_1 is not small compared to $\text{Re}(\omega_1)$. Later on the quasilinear relaxation $\gamma_1 \to 0$, $\text{Re}(\omega_1) \to \neq 0$ leads to the typical weak turbulence situation where four-wave correlations play no role in the spectrum equation (11). The main exception occurs at $k_1 << k_B(0)$ where $\text{Re}(\omega_1) \gtrsim 0$ from the beginning, and where the relaxation time is long;

here we need a cut-off in order to obtain stabilization in a finite time.

NUMERICAL EXAMPLE AND DISCUSSION

According to the spectrum equation (2) equilibrium is reached if (and only if) $\gamma_1=0$ for all values of k_1. That means the weak turbulence equations do not introduce any process leading to equilibrium which would compete with the quasilinear relaxation process $\gamma_1 \to 0$. They modify this process only in one specific way: Since $\Gamma_1 > \gamma_1$ a higher equilibrium level of wave energy will be reached within a shorter time, as compared with the pure quasilinear case. This is particularly pronounced at small k's where the quantity $k_1 \cdot |\alpha_1|$ (equation (1)) becomes small. This feature has been shown by a numerical example, where equation (2) was integrated simultaneously with the equations of velocity moments (5). In the figure we show the time development of the spectrum for the following initial conditions: $\beta^+_{\parallel}=20$; $\beta^+_{\perp}=18$; $\beta^-_{\parallel}=20.114$; $\beta^-_{\perp}=19.8$; $\Delta=1.57\times10^{-2}$; $W=4\times10^{-5}$; $P=3\times(1-\beta^+_{\perp}/\beta^+_{\parallel})$. This choice corresponds to a high-β plasma with bi-Maxwellian proton distribution. Due to the low value of W the boundary of stabilization ($k_B \approx 4.1\ R_+^{-1}\Delta^{1/2}$ for t=0) does not move to the right any more for t<0, so we have included all

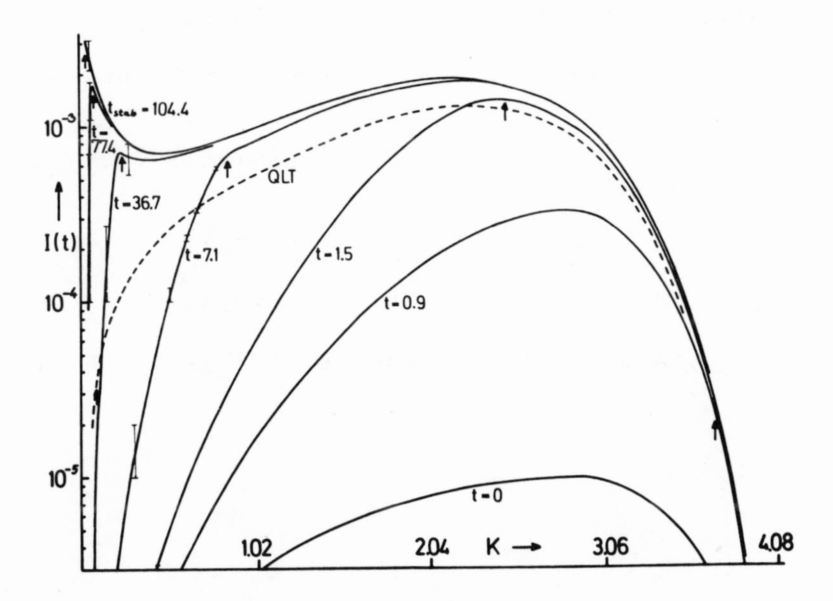

The time development of the spectrum

unstable waves at $t \to -\infty$. The arrows at each curve indicate k_B. The dashed line is the quasilinear equilibrium spectrum. The bars indicate the truncation error. Thus we have confirmed the overall picture of quasilinear theory for the case of firehose unstable Alfvén waves; if they are present in the solar wind, they should be even more detectable than predicted by quasilinear theory: They grow higher and faster, and feed more energy into small wave numbers (i.e. small frequencies).

REFERENCES

1. Hundhausen, A.J., Space Sci. Rev. 8, 690 (1968).
2. Parker, E.N., "Interplanetary Dynamical Processes", Interscience Publishers, John Wiley & Sons (1963).
3. Kennel, C.F. and Sagdeev, R.Z., J. Geophys. Res. 72, 3303 (1967).
4. Kennel, C.F. and Scarf, F.L., J. Geophys. Res. 73, 6149 (1968).
5. Elsässer, K. and Schamel, H., MPI-PAE/Astro/44 (1971), to be published.
6. Jockers, K., Astron. & Astrophys. 6, 219 (1970).
7. Meyer, F., and Pfirsch, D., Kleinheubacher Tagung (1969).
8. Shapiro, V.D. and Shevchenko, V.I., Soviet Phys. JETP 18, 1109 (1964).
9. Elsässer, K., J. Plasma Phys. 5, 31 (1971).
10. Elsässer, K., J. Plasma Phys. 5, 39 (1971).
11. Elsässer, K., and Gräff, P., Annals of Physics, N.Y. (in press).

RESONANT DIFFUSION IN STRONGLY TURBULENT PLASMAS

T.J. Birmingham[†] and M. Bornatici[††]

[†] Goddard Space Flight Center, Greenbelt, Md.

[††] European Space Research Institute, Frascati

INTRODUCTION

Plasmas in general and cosmic plasmas in particular are frequently turbulent. The turbulence is often the direct result of the release of free energy in the plasma via an instability mechanism. The turbulence is evident in the random fluctuations of particle properties and the electric and magnetic fields which the plasma supports.

We are concerned here with the effect of turbulent fluctuations on the plasma particles, our ultimate goal being to derive equations which describe the evolution (as driven by the fluctuations) of macroscopic properties -- temperature, flow speed, etc. -- of the turbulent plasma.

As an initial step toward this goal we derive here for an unmagnetized plasma a diffusion equation for $\langle f \rangle$, the one particle distribution function averaged over an ensemble of plasmas. We use ensemble theory because the turbulent fluctuations are random and we are interested in average plasma properties. The ensemble may be thought of as consisting of realizations which differ only in the phases of the Fourier-decomposed microfields at some arbitrary instant.

The turbulence most strongly interacts with resonant particles, i.e., particles whose velocity matches the phase velocity of some component of the Fourier spectrum of the waves. Care must be exercised in applying a perturbation theory to the interaction of such particles with turbulence, for it is known that a straightforward expansion in $\delta\underset{\sim}{E}$ lead to secular deviation from linear trajectories. Even for plasmas in which the wave energy density $(\delta E_{rms})^2/8\pi$ is much smaller than the thermal energy density nT, the straightforward perturbation can describe the turbulence-particle interaction incorrectly. (In many situations of interest in the laboratory and in space the ratio $(\delta E_{rms})^2/8\pi nT$ is indeed small, being of $O(10^{-2})$ or less in observations of electrostatic turbulence associated with perpendicular collisionless shock experiments[1] and the earth's bow shock[2].

We here adopt the perturbation procedure proposed by Dupree[3] and further developed by Weinstock[4]. This technique avoids secularities by perturbing about zeroth order orbits which contain the effects of fluctuations in a statistical manner. The results of this statistical approach differ from those of a direct $\delta\underset{\sim}{E}$ expansion when $\delta\underset{\sim}{E}$ is large enough that a particle can diffuse at least one typical wavelength of the fluctuations in an interval τ_c, the correlation time of the fluctuations as observed by that particle moving on its straight line orbit. Such a diffusion may occur even when the energy density of the fluctuations is less than the thermal energy density, if the fluctuations are narrow-band.

RESONANT DIFFUSION BY STRONG PLASMA TURBULENCE

We consider an ensemble of three-dimensional plasmas with an approximately homogeneous and stationary distribution of random electromagnetic fluctuations. For each realization, the one-particle distribution function f satisfies the Vlasov equation. Denoting the average of a quantity over the ensemble by $\langle\rangle$ and its deviation from the average by δ, we obtain from the Vlasov equation an equation for the evolution of $\langle f\rangle$

$$\frac{\partial\langle f\rangle}{\partial t} + \underset{\sim}{v}\cdot\nabla\langle f\rangle + \langle\underset{\sim}{F}\rangle \cdot \frac{\partial\langle f\rangle}{\partial\underset{\sim}{v}} = - \langle\delta\underset{\sim}{F} \cdot \frac{\partial\delta f}{\partial\underset{\sim}{v}}\rangle \quad , \qquad (1)$$

and an equation for δf, the fluctuating part of the distribution function,

$$\frac{\partial\delta f}{\partial t}+\underset{\sim}{v}\cdot\nabla\delta f+(\langle\underset{\sim}{F}\rangle+\delta\underset{\sim}{F})\cdot\frac{\partial\delta f}{\partial\underset{\sim}{v}} = -\delta\underset{\sim}{F}\cdot\frac{\partial\langle f\rangle}{\partial\underset{\sim}{v}}+\langle\delta\underset{\sim}{F}\cdot\frac{\partial\delta f}{\partial\underset{\sim}{v}}\rangle \quad , \qquad (2)$$

with $\underset{\sim}{F}=\langle\underset{\sim}{F}\rangle+\delta\underset{\sim}{F}$ the force per unit mass on the element of plasma located at the phase space point $\underset{\sim}{x}$, $\underset{\sim}{v}$ at time t.

The next step is to eliminate δf from these two coupled equations and to obtain a single equation for $\langle f\rangle$ in terms of correlation functions of $\delta\underset{\sim}{F}$. We solve Eq. (2) in the weak coupling approximation to strong plasma turbulence, i.e., by iterating on the source term $\langle\delta F \cdot \frac{\partial \delta f}{\partial \underset{\sim}{v}}\rangle$ and retaining elements of the convective t term $\delta\underset{\sim}{F} \cdot \frac{\partial \delta f}{\partial \underset{\sim}{v}}$ in lowest order. The weak coupling approximation is discussed in detail by Birmingham and Bornatici[5], and those interested are referred to their paper and the references cited there for further details of this calculation. Substituting the lowest order solution of Eq. (2) into Eq. (1) and making standard assumptions result in a velocity-space diffusion equation for $\langle f\rangle$

$$\frac{\partial\langle f\rangle}{\partial t} + \underset{\sim}{v}\cdot\nabla\langle f\rangle + \langle\underset{\sim}{F}\rangle \cdot \frac{\partial\langle f\rangle}{\partial\underset{\sim}{v}} = \frac{\partial}{\partial\underset{\sim}{v}} \cdot \underset{\sim}{\underset{\sim}{D}}\cdot\frac{\partial\langle f\rangle}{\partial\underset{\sim}{v}} \quad . \tag{3}$$

The diffusion tensor is

$$\underset{\sim}{\underset{\sim}{D}}(\underset{\sim}{v},t-t_o)=\int_o^{t-t_o} d\tau\langle\delta\underset{\sim}{F}(\underset{\sim}{x},\underset{\sim}{v},t)\delta\underset{\sim}{F}[\underset{\sim}{x}^*(t-\tau),\ \underset{\sim}{v}^*(t-\tau),\ t-\tau]\rangle, \tag{4}$$

where $\underset{\sim}{x}^*$ and $\underset{\sim}{v}^*$ are the solutions of the characteristic equations

$$\frac{d\underset{\sim}{x}^*}{d\tau} = \underset{\sim}{v}^* \quad , \quad \frac{d\underset{\sim}{v}^*}{d\tau} = \underset{\sim}{F}(\underset{\sim}{x}^*,\underset{\sim}{v}^*,\tau) \quad , \tag{5}$$

with the boundary conditions $\underset{\sim}{x}^*(t)=\underset{\sim}{x},\underset{\sim}{v}^*(t)=\underset{\sim}{v}$. If $t-t_o>>\tau^*$, with τ^* the time characteristic of the decay of elements of the autocorrelation tensor $\langle\delta\underset{\sim}{F}(\underset{\sim}{x},\underset{\sim}{v},t)\,\delta F[\underset{\sim}{x}^*(t-\tau),\underset{\sim}{v}^*(t-\tau),t-\tau]\rangle$, the upper limit of integration in Eq. (4) may be extended to infinity, eliminating any dependence upon conditions at t_o,

$$\underset{\sim}{\underset{\sim}{D}}(\underset{\sim}{v},\infty)=\int_o^{\infty} d\tau\langle\delta\underset{\sim}{F}(\underset{\sim}{x},\underset{\sim}{v},t)\delta\underset{\sim}{F}[x^*(t-\tau),\underset{\sim}{v}^*(t-\tau),t-\tau]\rangle \quad . \tag{6}$$

The kinetic equation (3) is formally the same as the lowest order equation for $\langle f\rangle$ derived in the weak turbulence approximation. The diffusion tensor Eq. (6) differs, however, from the corresponding weak turbulence diffusion tensor in the fact that the correlation tensor in (6) is evaluated along the perturbed orbit of the particle.

An explicit evaluation of $\underset{\sim}{D}(\underset{\sim}{v},\infty)$ is performed for the case of electrostatic turbulence $\delta\underset{\sim}{F} = \dfrac{q\delta\underset{\sim}{E}(\underset{\sim}{x},t)}{m}$ and no zeroth order magnetic field. By expanding the $\delta\underset{\sim}{F}$'s in Fourier series and making a cumulant expansion we obtain

$$\underset{\sim}{D}(\underset{\sim}{v},\infty) = \sum_{\underset{\sim}{k}} \hat{k}\hat{k} < |\delta F_{\underset{\sim}{k}}|^2 > \int_0^\infty d\tau \, \exp - i\omega_{\underset{\sim}{k}}\tau \tag{7}$$

$$\exp\left\{-i\underset{\sim}{k}\cdot<\Delta\underset{\sim}{x}^{*}(t-\tau)> - \frac{\underset{\sim}{k}\underset{\sim}{k}}{2} : \left[<\Delta\underset{\sim}{x}^{*}(t-\tau)\Delta\underset{\sim}{x}^{*}(t-\tau)> - <\Delta\underset{\sim}{x}^{*}(t-\tau)> <\Delta\underset{\sim}{x}^{*}(t-\tau)>\right]\right\} ,$$

where $<|\delta F_{\underset{\sim}{k}}|^2>$ is the ensemble average square amplitude of the Fourier mode with wave vector $\underset{\sim}{k}$ and $\hat{k}$ is the unit vector along $\underset{\sim}{k}$. The term $i\underset{\sim}{k}\cdot<\Delta\underset{\sim}{x}^{*}(t-\tau)>$ contains the effect of the turbulence on the ensemble average phase space position of the particle at time $t-\tau$. Inclusion of the $\delta\underset{\sim}{E}$ effects of this term is a unique contribution of our theory. The term $\underset{\sim}{k}\underset{\sim}{k}:[<\Delta\underset{\sim}{x}^{*}\Delta\underset{\sim}{x}^{*}> - <\Delta\underset{\sim}{x}^{*}><\Delta\underset{\sim}{x}^{*}>]$ represents the statistical dispersion in position about the average. Equations (5) relate $\Delta\underset{\sim}{x}^{*}(t-\tau) \equiv x^{*}(t-\tau) - \underset{\sim}{x}$ to the fluctuations and we show that[5]

$$<\Delta\underset{\sim}{x}^{*}(t-\tau)> \simeq -\underset{\sim}{v}\tau - \frac{\tau^2}{2}\frac{\partial}{\partial\underset{\sim}{v}} \mathrm{Tr}\left\{\underset{\sim}{D}(\underset{\sim}{v},\tau)\right\} , \tag{8}$$

$$<\Delta\underset{\sim}{x}^{*}(t-\tau)\Delta\underset{\sim}{x}^{*}(t-\tau)> - <\Delta\underset{\sim}{x}^{*}(t-\tau)><\Delta\underset{\sim}{x}^{*}(t-\tau)> \simeq \frac{2}{3}\tau^3 \underset{\sim}{D}(\underset{\sim}{v},\tau) ,$$

where $\mathrm{Tr}\left\{\underset{\sim}{D}\right\}$ is the trace of the matrix of $\underset{\sim}{D}$. Plugging Eqs. (8) into (7) yields

$$\underset{\sim}{D}(\underset{\sim}{v},\infty)=\sum_{k}\hat{k}\hat{k}<|\delta F_{\underset{\sim}{k}}|^2>\int_0^\infty d\tau\ \exp-i\left[(\omega_{\underset{\sim}{k}}-\underset{\sim}{k}\cdot\underset{\sim}{v})\tau-\frac{\tau^2}{2}\underset{\sim}{k}\cdot\frac{\partial}{\partial\underset{\sim}{v}}\right.$$

$$\left.\mathrm{Tr}\left\{\underset{\sim}{D}(\underset{\sim}{v},\tau)\right\}\right]\quad\exp-\frac{\tau^3}{3}\underset{\sim}{k}\underset{\sim}{k}:\underset{\sim}{D}(\underset{\sim}{v},\tau)\quad. \qquad (9)$$

The non-asymptotic $\underset{\sim}{D}(\underset{\sim}{v},\tau)$ is given by the r.h.s. of Eq. (9) with τ replacing ∞ as the upper limit of integration. Since $\underset{\sim}{D}(\underset{\sim}{v},\tau)$ occurs in Eq. (9) in terms which drop off abruptly with τ, we can use the small τ expansion of $\underset{\sim}{D}(\underset{\sim}{v},\tau)$ in Eq. (9) and obtain[5]

$$\underset{\sim}{D}(\underset{\sim}{v},\infty)=\sum_{\underset{\sim}{k}}\hat{k}\hat{k}<|\delta F_{\underset{\sim}{k}}|^2>\int_0^\infty d\tau\ \exp-i\left[(\omega_{\underset{\sim}{k}}-\underset{\sim}{k}\cdot\underset{\sim}{v})\tau-\frac{\tau^5}{6}\underset{\sim}{k}\cdot\sum_{\underset{\sim}{k}'}\underset{\sim}{k}'\right.$$

$$\left.<|\delta F_{\underset{\sim}{k}'}|^2>(\omega_{\underset{\sim}{k}'}-\underset{\sim}{k}'\cdot\underset{\sim}{v})\right]\exp-\frac{\tau^4}{3}\underset{\sim}{k}\underset{\sim}{k}:\sum_{\underset{\sim}{k}'}\hat{k}'\hat{k}'<|\delta F_{\underset{\sim}{k}'}|^2> \qquad (10)$$

To carry out the τ-integration of Eq. (10) will in general require numerical work. It is, however, possible to proceed analytically when there is a clear differentiation of the time scales on which the integrand of Eq. (10) would converge if the three exponential factors occurred individually rather than in combination. The factor $\exp-i(\omega_{\underset{\sim}{k}}-\underset{\sim}{k}\cdot\underset{\sim}{v})\tau$ produces the usual weak turbulence convergence on the time scale $\tau_c=(\Delta\omega^*)^{-1}$, where $\Delta\omega^*$ is the frequency spread in the Fourier components of the turbulent spectrum as observed by the resonant particles moving with their unperturbed velocities. The factor $\exp-\frac{\tau^4}{3}\underset{\sim}{k}\underset{\sim}{k}:\sum_{\underset{\sim}{k}'}\hat{k}'\hat{k}'<|\delta F_{\underset{\sim}{k}'}|^2>=\exp-\frac{\tau^4}{3}\underset{\sim}{k}\underset{\sim}{k}:<\delta\underset{\sim}{F}(\underset{\sim}{x},t)$ $\delta\underset{\sim}{F}(\underset{\sim}{x},t)>$ produces the convergence due to the statistical spread in particle orbits on the time scale $\tau_2\cong\left(\frac{1}{k_o}\frac{m}{q\delta E_{rms}}\right)^{1/2}$, k_o being the characteristic wave number of the fluctuations. Finally, the term $\exp i\frac{\tau^5}{6}\underset{\sim}{k}\cdot\sum_{\underset{\sim}{k}'}\underset{\sim}{k}'<|\delta F_{\underset{\sim}{k}'}|^2>(\omega_{\underset{\sim}{k}'}-\underset{\sim}{k}'\cdot\underset{\sim}{v})$ produces the convergence associated with the acceleration of the ensemble average orbit on the time scale

$\tau_1 = \left(\frac{k_o}{\Delta k} \frac{1}{k_o^2} \frac{m^2}{q^2 \delta E_{rms}^2} \frac{1}{\Delta\omega^*}\right)^{1/5} \cong \left(\frac{k_o}{\Delta k} \tau_2^4 \tau_c\right)^{1/5}$. In strong turbulence τ_2 is less than or equal to τ_c. If we further consider the limit $\tau_2 \ll \tau_c$, Eq. (10) can be expanded and at lowest order in the small parameter $\left(\frac{\tau_2}{\tau_c}\right)^2$ we get

$$\underset{\sim}{D}(\underset{\sim}{v},\infty) = \frac{3^{1/4}}{4} \Gamma\left(\frac{1}{4}\right) \sum_{\underset{\sim}{k}} \frac{\hat{k}\hat{k} \langle |\delta F_{\underset{\sim}{k}}|^2 \rangle}{[\underset{\sim}{k}\underset{\sim}{k} : \langle \delta \underset{\sim}{F}(\underset{\sim}{x},t) \delta \underset{\sim}{F}(\underset{\sim}{x},t) \rangle]^{1/4}} , \qquad (11)$$

where $\Gamma\left(\frac{1}{4}\right) \cong 3.6$.

The diffusion process is described by Eq. (3) with the diffusion tensor (11) if the time scale τ_2 is much shorter than the time scale τ_c. For the electrons we have

$$\frac{\tau_c}{\tau_2^{(e)}} \cong \left[\frac{(\delta E_{rms})^2}{8\pi n T_e}\right]^{1/4} \left[2(k_o \lambda_D)^2\right]^{1/4} \left(\frac{m_i}{m_e}\right)^{1/2} \frac{\omega_{pi}}{\Delta\omega^*} , \qquad (12)$$

with $\lambda_D = \left(\frac{T_e}{4\pi q^2 n}\right)^{1/2}$ the electron Debye length. As an example we consider the ion sound turbulence present in collisionless perpendicular shocks[1] and in the earth's bow shock[2]. For this case $\frac{(\delta E_{rms})^2}{8\pi n T_e} \cong 10^{-2}$ and $(k_o \lambda_D)^2 = 0.5$. Equation (12), then, shows that $\tau_2^{(e)} \ll \tau_c$ for $\Delta\omega^* \ll 10\omega_{pi}$. This last condition is easily satisfied and, therefore, the electron diffusion is properly described by the diffusion tensor (11).

For the ions it is $\tau_2^{(i)} \cong \left(\frac{m_i}{m_e}\right)^{1/2} \tau_2^{(e)}$ and the condition $\tau_2^{(i)} \lesssim \tau_c$, rather than $\tau_2^{(i)} \ll \tau_c$, is likely to be satisfied. In fact $\tau_2^{(i)} \lesssim \tau_c$ for $\Delta\omega^* \lesssim \frac{\omega_{pi}}{\sqrt{10}}$.

Therefore, also for the ions the effects of turbulence on the orbits are significant and could play an important role in stabilizing the ion sound instability.

ACKNOWLEDGEMENT

This work was begun while M. Bornatici was NAS-NRC Resident Research Associate at Goddard Space Flight Center. Much of the work was done while M. Bornatici held an ESRO fellowship at the MPI Institut für Extraterrestrische Physik, Garching, Germany.

REFERENCES

1. Paul, J.W., Daughney, C.C. and Holmes, L.S. 1969, Nature 223, 822.

2. Fredricks, R.W., Crook, G.M., Kennel, C.F., Green, I.M., Scarf, F.L., Coleman, P.J. and Russell, C.T. 1970, J. Geophys. Res. 75, 3731.

3. Dupree, T.H. 1966, Phys. Fluids 9, 1773.

4. Weinstock, J. 1968, Phys. Fluids 11, 1977; 1969, Phys. Fluids 12, 1045.

5. Birmingham, T. and Bornatici, M., Goddard Space Flight Center, Preprint X-641-71-320 (submitted to Physics of Fluids).

THE STRUCTURE OF THE EARTH'S BOW SHOCK

B. Bertotti[†], D. Parkinson, K. Schindler
European Space Research Institute, Frascati

P. Goldberg
Richmond College, City University of New York

1. INTRODUCTION

The investigation which we shall briefly describe is an attempt to shed light on the structure of the earth's bow shock wave. Unlike gas dynamic shocks collisionless plasma shocks can dissipate directed flow energy in a number of different ways. In some of the models the shock structure is governed by plasma-magnetic field interaction (Kennel and Sagdeev, 1967a and b); in other cases the interaction is assumed to be predominantly electrostatic (Tidman,1967; Tidman and Krall, 1971). In the bow shock the experimental evidence (Fredricks et al., 1970; see also Bertotti and Schindler 1971 for a review) points to the presence of both magnetic and electrostatic fluctuations and no unambiguous picture exists, even on a qualitative basis, for the dissipation processes.

As Lindman and Drummond, 1971, we make no a-priori assumption about the nature of the interaction; this attitude arose after the failure of extensive attempts to construct predominantly magnetic models for strong shocks. A first indication was given earlier (Bertotti et al., 1970; Bertotti and Schindler, 1971) when it was suggested that, in the fluid picture, a viscous effect is probably dominant. We are now able to interpret

---

[†] Also Institute of Physics, University of Messina. Present address: Institute of Physics, University of Pavia.

this phenomenon in terms of electrostatic subshocks; it is interesting that Lindman and Drummond, 1971, have independently reached a similar conclusion with a different approach.

2. SUMMARY OF THEORETICAL ANALYSIS

We have studied the relevant properties of a number of different fluid models. In the simplest of all the electron and ion pressures are scalars and dispersion is provided by the inclusion of the Hall term; in addition, a scalar resistivity simulates the effect of small scale electrostatic fluctuations which heat the electrons:

$$\frac{\partial \rho}{\partial t} + \underline{\nabla} \cdot (\rho \underline{v}) = 0; \qquad \rho \frac{\partial \underline{v}}{\partial t} + \rho \underline{v} \cdot \underline{\nabla} \underline{v} + \underline{\nabla}(p_e + p_i) = \underline{j} \times \underline{B}$$

$$\underline{E} + \underline{v} \times \underline{B} - \frac{1}{en} \underline{j} \times \underline{B} + \frac{1}{en} \underline{\nabla} \underline{p}_e = \eta \underline{j}; \qquad \left(\frac{\partial}{\partial t} + \underline{v} \cdot \underline{\nabla}\right) \frac{p_i}{\rho^{\gamma_i}} = 0$$

$$\left(\frac{\partial}{\partial t} + \underline{v} \cdot \underline{\nabla}\right) \frac{p_e}{\rho^{\gamma_e}} = \eta \frac{j^2}{\rho^{\gamma_e}} (\gamma_e - 1) + \frac{1}{en} \underline{j} \cdot \underline{\nabla} \frac{p_e}{\rho^{\gamma_e}}; \quad \underline{\nabla} \times \underline{B} = \mu_o \underline{j}$$

Standard notation is used.

In other models, always based on the fluid picture (see Bertotti and Schindler, 1971), the ion pressure anisotropy was included; all models essentially lead to the same conclusions.

For upstream conditions corresponding to the central part of the earth's bow shock, with an oblique angle between the magnetic field and the shock normal, we find the following results.

a) Integrating the one-dimensional steady state equations out of the upstream singular point we find a whistler growing in the downstream direction (see Fig. 1). For sufficiently small amplitudes the density and velocity is almost unperturbed. Rather suddenly the wave ceases to be incompressive and at a certain point ($x \approx 26$ in Fig. 1) the velocity gradient develops a singularity.

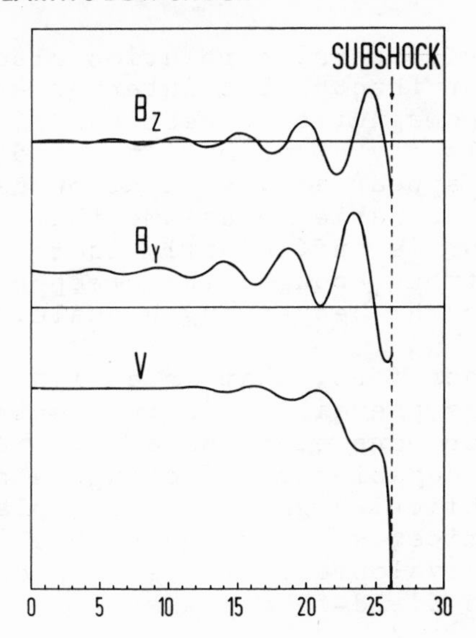

Figure 1: Model profiles for typical bow shock parameter (length unit $v_o/(M_A{}^2\Omega_i)$).

b) The time dependent problem was solved† for the case of a piston advancing into the plasma and showed a similar behaviour; the breakdown occurred after violent density oscillations.

c) In the above model (steady state) the singularity develops because of a vanishing denominator $\Delta = 1-(\gamma_i p_i+\gamma_e p_e)/\rho v_x^2$. We observe that $\Delta=0$ gives exactly the phase velocity of a sound wave standing in the shock in the absence of a magnetic field. At that point the electrons are substantially heated by the anomalous resistance.

d) Our model leads to the ordinary whistler dispersion relation for phase velocities exceeding the sound speed. With the scaling

$$\frac{\partial}{\partial t} \Rightarrow \frac{1}{\varepsilon}\frac{\partial}{\partial t},\ \underline{\nabla} \Rightarrow \frac{1}{\varepsilon}\underline{\nabla},\ \underline{B} \Rightarrow \underline{B}_o(\varepsilon\underline{r},\varepsilon t)+\varepsilon\underline{B}_1(\underline{r},t)$$

(corresponding to frequencies larger than the ion cyclotron frequency Ω_i, but to finite phase velocity), one can easily see that when $\varepsilon\to 0$ the magnetic field decouples and one is left with the momentum and mass equations for an ordinary fluid. This indicates, therefore, that if a whistler steepens sufficiently rapidly, it must eventually give rise to a shock embedded in a comparatively smooth magnetic field.

---

† With the help of the consulting firm A.R.S. of Milan.

e) The Vlasov dispersion relation also contains the ion sound branch for frequencies intermediate between Ω_i and Ω_e, with electromagnetic corrections of order Ω_i/ω. If the electron velocity distribution is not isotropic, dispersive effects appear at the electron Larmor radius scale. Is seems reasonable to assume that if the electron pressure tensor is sufficiently isotropic or if the initial pulse is strong enough, the steepening will go all the way down to the Debye length scale.

We conclude therefore, that in a high Mach number shock, in conditions prevailing in the central part of the bow shock,one or more electrostatic, small scale structures develop capable of affecting substantially the ion directed motion. Magnetic forces play very little role in their dynamics, so that the problem is reduced to investigating the development of a strong density gradient in a magnetic-field-free plasma.

The relevant theory is well developed for the cold ion cases ($T_e >> T_i$) and leads, if the Mach number is not too large, to stationary and laminar solutions, where the electrostatic potential exhibits non-linear oscillations on the Debye length scale (Moiseev and Sagdeev, 1963, Montgomery and Joyce, 1969, Mason, 1970, Tasso, 1969, Forslund and Shonk, 1970; see also the experimental evidence by Taylor et al., 1970). This type of picture, in which the temperatures based upon velocity averages over several Debye lengths show an increase, like in ordinary shocks, is not unreasonable in view of the electron heating in the whistler region.

Little is known when the ratio T_e/T_i decreases. Theoretical and experimental evidence (Forslund and Shonk, 1970; Mason, 1970; Taylor et al., 1970) seemsto indicate that this quenches the oscillations and increases the number of reflected ions. D. Biskamp (private communication) has suggested that one might have in the bow shock a potential barrier, large enough to reflect a sufficiently large fraction of the ions, which would subsequently be trapped and thermalized (at least on a gross scale) in magnetic wells.

There have been also attempts to construct theoretically turbulent electrostatic shocks (Tidman, 1967; Tidman and Krall, 1971), where the presence of a sufficiently strong electrostatic instability is the essential feature. It seems however interesting to note that both the current (Fredricks et al., 1970) and the two ion stream (Bertotti and Biskamp, 1969) instabilities are marginal if not irrelevant.

3. OBSERVATIONS

Evidence concerning the bow shock comes from the OGO-5 (Fredricks et al., 1970) and Vela-4 measurements (Montgomery et al., 1970).

a) The linear whistler dispersion relation in the high Mach number limit yields a wave length $\lambda=2\pi\cos\alpha(c/\omega_{pi})/M_A$ (Bertotti and Schindler, 1971) where α is the angle between the shock plane and the magnetic field;for typical bow shock conditions λ is about 70 km. If one takes into account the uncertainty in the velocity of whistlers (not necessarily stationary in the shock frame), it appears that the whistler interpretation is not inconsistent with the observed time scale of 1-5 sec of the precursor waves (Fredricks et al., 1968, Fredricks and Coleman, 1969, Fredricks et al., 1970, Montgomery et al., 1970). Therefore the interpretation of the observed scale in terms of c/ω_{pe} (Fredricks et al. 1970) does not seem to be the only possibility.

b) Neither magnetic fluctuations nor the observed electrostatic spectrum seem able to account by themselves for the strong dissipation and heating occurring in the bow shock. Proton thermalization due to magnetic fluctuations can occur only after several Larmor radii v_o/Ω_i, where v_o is the upstream velocity; experimentally, however, a scattering length of a few times $c/\omega_{pi}\approx v_o/M_A\Omega_i$) has been quoted (Ossakow and Sharp, 1970). Electrons thermalize at a much faster rate (in one case in .03 sec, Montgomery et al., 1970). Given the observed electrostatic spectrum we did not find a satisfactory way in which these trains can thermalize the 1 KeV protons of the solar wind. There are also theoretical reasons (Biskamp, 1970) to exclude dominance of dissipation due to ion sound turbulence, which would heat electrons and ions at the same rate, while experimentally $(dT_i/dt)/T_i \sim 10(dT_e/dt)/T_e$ (Montgomery et al., 1970).

c) Owing to the low frequency threshold of the TRW plasma wave detector used on board of OGO-5 (Fredricks et al., 1970) of about 500 Hz, predominantly laminar subshock oscillations are not observable if they move with respect to the spacecraft with a velocity less than 15 km/sec. On the other hand the absence of strong nonlinear fluctuations speaks against the possibility of a turbulent electrostatic subshock.

d) The important question remains, what is the role of the electrostatic structure in the dissipation process. In the conventional (laminar and oscillatory) model of an electrostatic shock the normal kinetic energy is transformed into thermal energy, the characteristic length of the fluctuations being the Debye length. The observed electrostatic turbulence can then perhaps complete the ion thermalization. One may wonder how this can be reconciled with the experimental observation of an enhanced proton flux along a direction making a finite angle with the main flow (Fredricks et al., 1970) occurring after a distance of roughly c/ω_{pi} (to wit, less than an ion Larmor radius). We point out that, independently of the weak turbulent diffusion, the magnetic field itself will turn the ions away from the original direction over a fraction of a gyro-period. A similar process may occur for the electrons, albeit on a smaller scale: magnetic bending will take place in a fraction of an electron gyro-radius. One can hope, therefore, to explain in this way the surprisingly small thermalization length claimed by Montgomery et al., (1970).

If the electrostatic structure is just a large potential barrier in a magnetic field stronger than upstream, reflection will disorder the Larmor gyration of particles, simulating thermalization (private communication from D. Biskamp). In this case a density peak occurs.

It is clear that these two alternatives can be mixed in any degree; moreover different dissipation mechanisms can prevail at different places.

e) The observed ion reflection (Asbridge et al., 1968; Montgomery et al., 1970) also fits in our model, since it is in fact a property of sufficiently strong electrostatic shocks (D.W. Forslund private communication).

4. CONCLUSION

In summary, our model suggests the presence of an electrostatic structure within the bow shock; the magnetic field does not play a relevant role in their profile, but determines their position. They produce directly (through electric oscillations) or indirectly (through reflection in a strong magnetic field) the main particle dissipation. The survival of this idea depends upon further observations, in particular low frequency electric field measurements and accurate scanning of the particle velocity distribution with a high time resolution.

We are very grateful to Dr. M.D. Montgomery for his most useful comments.

REFERENCES

Asbridge, J.R., Bame, S.J. and Strong, I.B., 1968, J. Geophys. Res. 73, 5777.

Bertotti, B. and Biskamp, D., 1969, "Collision-Free Shocks in the Laboratory and Space, ESRO SP-51, p. 41.

Bertotti, N., Goldberg, P., Parkinson, D. and Schindler, K., 1970, "On the gross magnetic structure of the bow shock of the magnetosphere", paper presented at the International Symposium on Solar-Terrestrial Physics, Leningrad, May 1970.

Bertotti, B. and Schindler, K., 1971, "The bow shock of the magnetosphere", invited paper at the VII International Symposium on Rarefied Gas Dynamics, Pisa, June 29-July 3, 1970, to be published.

Biskamp, D., 1970, J. Geophys. Res. 75, 4659.

Forslund, D.W. and Shonk, C.R., 1970, Phys. Rev. Lett. 25, 1699.

Fredricks, R.W. and Coleman, Jr., P.J., 1969, Conference on Plasma Instabilities in Astrophysics, Pacific Grove, October 14-17, 1968.

Fredricks, R.W., Crook, G.M., Kennel, C.F., Green, I.M. and Scarf, F.L., 1970, J. Geophys. Res. 75, 3751.

Fredricks, R.W., Kennel, C.F., Scarf, F.L., Vrook, G.M. and Green, I.M., 1968, Phys. Rev. Lett. 21, 1761.

Kennel, C.F. and Sagdeev, R.Z., 1967a, J. Geophys. Res. 72, 3303.

Kennel, C.F. and Sagdeev, R.Z., 1967b, J. Geophys. Res. 72, 3327.

Lindman, E.L. and Drummond, W.E. 1971, "Studies of oblique shock structure", preprint.

Mason, R.J., 1970, Phys. Fluids 13, 1042.

Moiseev, S.S. and Sagdeev, R.Z., 1963, Plasma Physics 5,43.

Montgomery, D. and Joyce, G., 1969, J. Plasma Physics 3,1.

Montgomery, M.D., Asbridge, J.R. and Bame, S.J., 1970, J. Geophys. Res. 75, 1217.

Ossakow, S.L. and Sharp, G.W., 1970, Paper presented at the International Symposium on Solar-Terrestrial Physics, May 1970, Leningrad.

Tasso, H., 1969, "Collision-Free Shocks in the Laboratory and Space", ESRO SP-51, p. 183.

Taylor, R.J., Baker, D.R. and Ikezi, H., 1970, Phys. Rev. Lett. 24, 206.

Tidman, D.A., 1967, J. Geophys. Res. 72, 1799.

Tidman, D.A. and Krall, N.A., 1971, "Shock waves in collisionless plasmas" (Wiley-Interscience).

EXPERIMENTAL STUDY OF ELECTRON AND ION HEATING IN HIGH-β PERPENDICULAR COLLISIONLESS SHOCK WAVES

M. Keilhacker, M. Kornherr,
H. Niedermeyer, K.-H. Steuer

Max-Planck-Institut für Plasmaphysik
Euratom Association, Garching

I. Introduction

Collisionless shock waves play an important part in various astrophysical phenomena, such as the heating of plasma in the solar corona, or in connection with flares in the solar atmosphere, or as interplanetary shocks within the solar wind. The best established example of a shock in an extraterrestrial plasma is the bow shock, which results from the interaction of the solar wind plasma with the earth's magnetic field.

This paper deals with laboratory experiments on perpendicular collisionless shock waves that resemble the bow shock in many respects. The shock waves are produced by fast magnetic compression of a plasma cylinder 14 cm in diameter using a theta pinch [1]. They propagate perpendicularly to a magnetic field B_1 into a high-β plasma formed by a theta-pinch preionization [2]. By changing the parameters of this initial plasma - and thus the magnetosonic velocity v_1 - or the amplitude of the piston field, almost stationary shock waves with magnetosonic Mach numbers $M = u_s/v_1$ ranging from 2 to 5 can be produced.

Table I compares data of two laboratory shock waves having Mach numbers of 2.5 [3] and 4.9 [4] with the corresponding bow shock data [5]. It is seen that both the bow shock and the laboratory shock waves propagate into a high-β plasma ($\beta_1 = 8\pi nk(T_{e1} + T_{i1})/B_1^2 \sim 1$). The electron and ion temperatures in front of the shocks are also of the same magnitude, but the electron-ion tem-

	LABORATORY SHOCKS		BOW SHOCK
Mach number			
M	2.5	4.9	5
Initial plasma			
T_{e1} (eV)	4	5	14
T_{i1} (eV)	18	20	7
β_{e1}	0.1	0.5	1.0
β_{i1}	0.6	2.0	0.5
Temperature jump			
T_{e2} / T_{e1}	27	5	4.5
T_{i2} / T_{i1}	4	17	27

Table I

perature ratios T_{e1}/T_{i1} are different, the ions being the hotter species in the laboratory plasma, while the opposite holds for the solar wind plasma.

Another interesting quantity is the increase in the electron and ion temperatures through the shock transition. Here one finds a striking correspondence between the bow shock data and the high Mach number laboratory shock wave: In both shocks the heating goes mainly into the ions and even the absolute values for the temperature jumps agree very well. In the low Mach number shock, on the contrary, most of the heating goes into the electrons. This result is well established by now [4,6] and can be attributed to a critical Mach number M_{crit}*): Below the critical Mach number the main dissipative mechanism is collective resistivity resulting in strong electron heating. Above M_{crit} resistivity fails to produce the necessary entropy change and some kind of turbulent viscosity that heats the ions becomes important.

The following part of this paper summarizes experimental results on electron and ion heating in high-β collisionless shock waves [2-4, 7,8].

---

*) M_{crit} is 2.8 for $\beta = 0$ and decreases with increasing β.

II. Electron heating in medium Mach number shock waves ($M < M_{crit}$)

The electron heating observed in medium Mach number shock waves ($M < M_{crit}$) implies an effective resistivity in the shock front which is about two orders of magnitude larger than the Spitzer value [2]. It is speculated that this anomalous resistivity results from microturbulence excited by the diamagnetic current within the shock front.

This notion is confirmed by ruby laser scattering measurements (for experimental details see ref. [3]) that reveal a strongly suprathermal level of fluctuations within the shock front [3,8]. As an example, figure 1 shows the total level of density fluctuations $n_e\ S(k,\omega)$ within the wave vector band $k.D = 0.8 \pm 0.16$ (D = Debye length) as a function of time for the M = 2.5

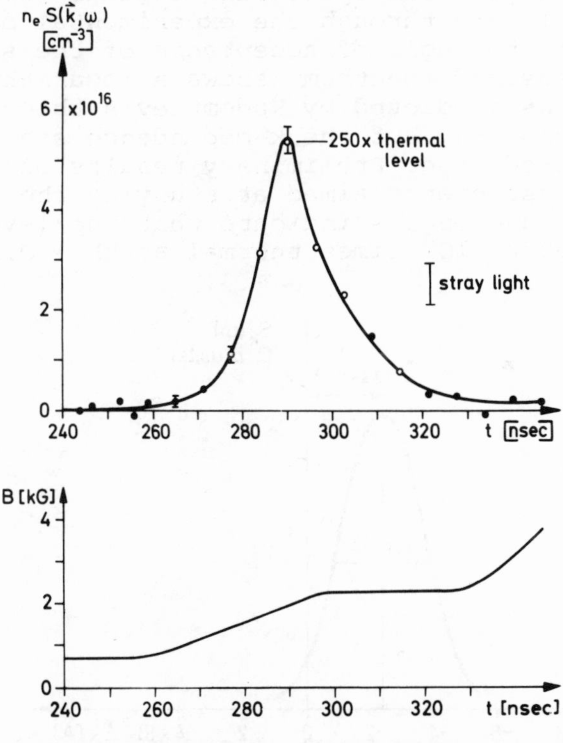

Fig.1 Intensity of density fluctuations $n_e\ S(k,\omega)$ and magnetic field B in a collisionless shock wave with Mach number 2.5 in deuterium. The fluctuations reach 250 times the thermal level.

deuterium shock wave of Table I. As a time comparison the magnetic field profile B is plotted, too. In the shock front the fluctuations reach about 250 times the thermal level.

Figures 2 - 4 show the frequency spectrum $S_k(\omega)$, the wave number spectrum $S(k)$, and the angular distribution $S(\varphi)$ (in the plane $\perp B_1$) of these fluctuations, measured at the time of maximum turbulence [3,7].

The scattered light spectrum (Figure 2) is shifted in frequency with respect to the laser line. This shift reverses sign if the current I_θ is reversed [3], thus indicating that the electron current drives the turbulence. The magnitude of the shift corresponds to scattering by plasma waves with frequency $\omega \sim 0.5\ \omega_{pi}$ (ω_{pi} = ion plasma frequency).

Figure 3 shows the short-wavelength part of the k-spectrum as obtained in the ruby laser scattering experiment by varying the scattering angle between 2° and 4°. Horizontal bars through the experimental points indicate the finite angle of acceptance of the scattered light. The measured spectrum shows a logarithmic cutoff for $k \geq 1/D$, as predicted by Kadomtsev's theory of ion wave turbulence [9], but the k-dependence seems to be weaker than predicted. Preliminary results of CO_2 laser scattering measurements aimed at studying the turbulence at longer wavelengths indicate that the level of fluctuations is about 10^4 times thermal at $kD \sim 0.12$.

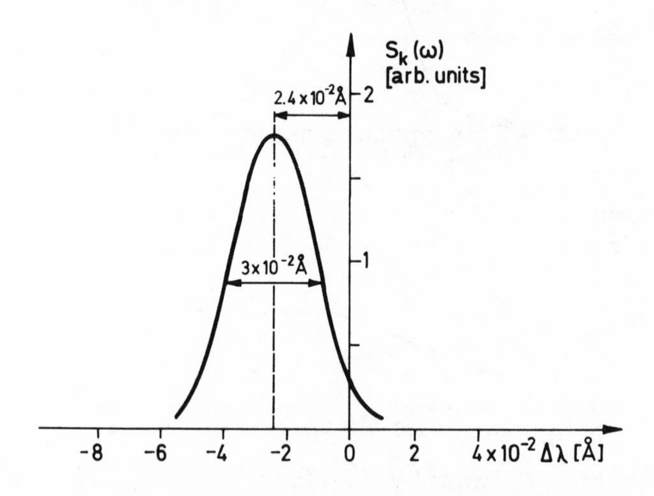

Fig.2 Frequency spectrum $S_k(\omega)$ of enhanced fluctuations

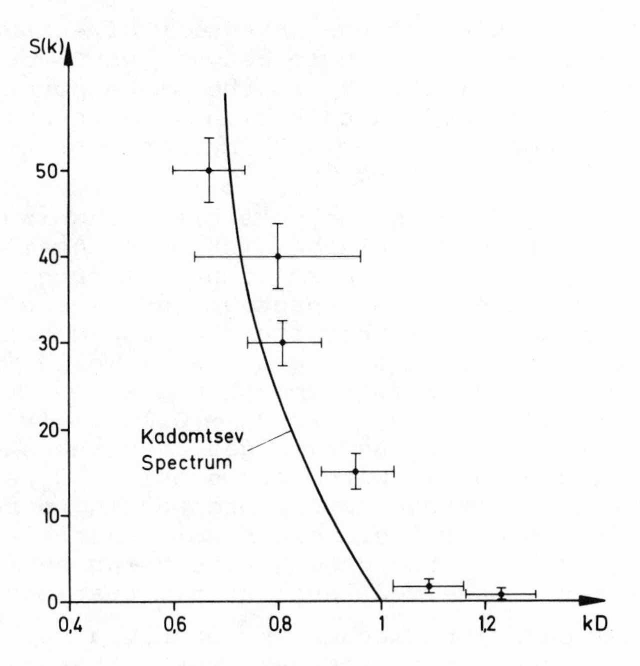

Fig.3 Wave number spectrum S(k) of enhanced fluctuations together with the form predicted by Kadomtsev

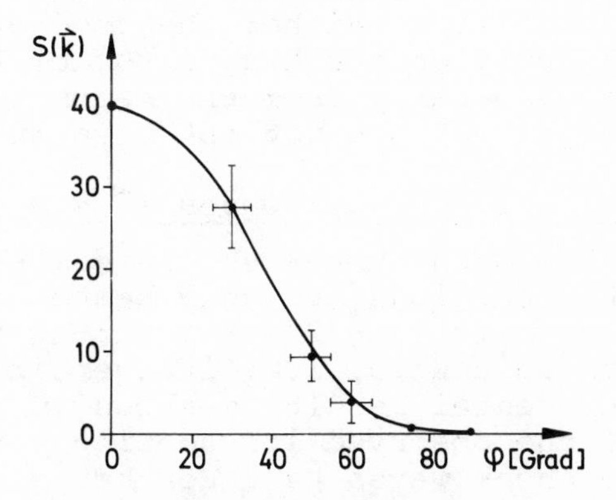

Fig.4 Angular distribution of enhanced fluctuations S(k) in the plane ⊥ B_1 (φ is the angle between wave vector k and drift velocity v_d)

Figure 4 is a plot of the intensity of enhanced fluctuations versus φ, the angle between wave vector k and electron drift velocity v_d in the plane perpendicular to B_1. It shows that in this plane the turbulence is spread out through a cone, with half-intensity at about 40° to the drift velocity.

We now briefly discuss the possible mechanism that leads to the observed turbulence (for more details see ref.[3] and [7]). The observation that the frequency shift of the scattered light spectrum reverses sign with reversal of current shows that the electron current drives the turbulence. The question therefore is which waves become unstable and grow fast enough under the conditions existing in the shock wave, viz. $v_d \sim 0.1\ v_e$ ($v_e = (2\ T_e/m_e)^{1/2}$) and $T_e \sim T_i$ as averages over the shock front. While ion acoustic waves are stable [10], electron cyclotron waves (Bernstein waves) propagating perpendicularly to the magnetic field can become unstable under these conditions [11], the growth rate being sufficient to account for the observed level of fluctuations [3,7].

The final question whether the observed turbulence can account for the effective resistivity that is calculated from the measured electron heating cannot yet be answered quantitatively. The model of stochastic electron heating proposed by Paul [12] gives too small a resistivity if one substitutes the measured cone and level of turbulence (using the Kadomtsev form of $S(k) \sim k^{-3}\ \ln\ (k\ D)^{-1}$ to extrapolate from the measured value at $k\ D = 0.8$ down to $k = 0$). So either the stochastic heating model does not apply or the k-spectrum of the turbulence deviates considerably from the Kadomtsev form in a wavelength range that has not yet been investigated.

III. Ion heating in strong shock waves ($M > M_{crit}$)

While in the low Mach number shocks discussed in the previous chapter the ions are only heated adiabatically ($T_{i2}/T_{i1} = (n_2/n_1)^{2/fi} \sim 2 - 4$), strong shock waves result in non-adiabatic, i.e. irreversible, ion heating. This experimental result is shown in Figure 5, which is a plot of the observed ion heating versus the Mach number of the shock waves [4]. For better comparison the actual ion temperature behind the shock T_{i2} is related to a temperature $T_{i2\ ad}$ calculated on the assumption of merely adiabatic ion heating. This temperature ratio is plotted versus the difference between the observed and critical Mach numbers $M - M_{crit}$. One clearly recognizes that for $M > M_{crit}$ the observed ion heating is well above the calculated adiabatic heating and

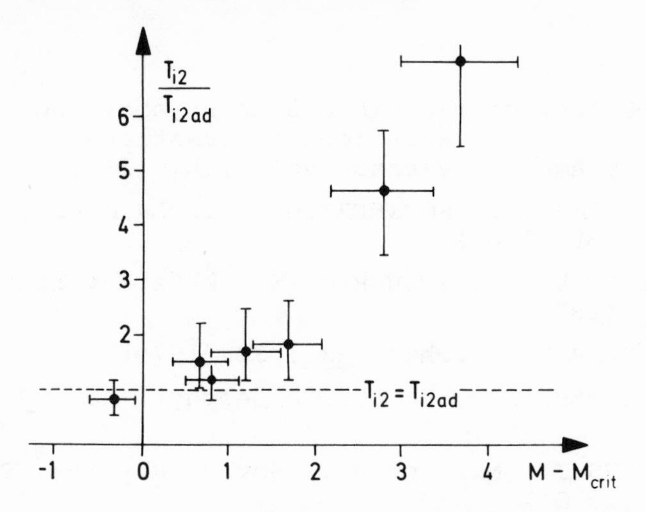

Fig.5 Dependence of ion heating on Mach number M ($T_{i2}/T_{i2\ ad}$ is the ratio of observed to merely adiabatic ion heating). Strong non-adiabatic ion heating for $M > M_{crit}$.

strongly increases with Mach number. For high Mach numbers the electron heating becomes negligibly small compared with the ion heating, and therefore the ion heating is proportional to M^2 for stationary shocks.

The ion temperatures T_{i2} used for Figure 5 were obtained by applying the Rankine-Hugoniot relations to the stationary shock waves. Recent direct measurements of the ion temperatures by light scattering [7] confirm the previous, more indirect findings.

The physical processes of ion heating in supercritical shock waves are only understood qualitatively as yet. As mentioned before, above the critical Mach number the steepening of the shock front can no longer be balanced by resistive dissipation of the diamagnetic current and the shock front "overturns". A numerical investigation of such shock waves which describes the ions by a Vlasov equation reveals that the ion distribution develops a two-stream structure [13]: fast ions that have been accelerated by the magnetic piston stream through the remaining part of the plasma that is still almost at rest. This two-stream structure of the ion flow can be subject to instabilities that randomize the ion energies.

References

[1] CHODURA,R. et al., in Plasma Physics and Controlled Nuclear Fusion Research, International Atomic Energy Agency, Vienna, Vol.I, 81 (1969)

[2] KEILHACKER,M., KORNHERR,M., STEUER,K.-H., Z.Phys. 223, 385 (1969)

[3] KEILHACKER,M., STEUER,K.-H., Phys.Rev.Lett. 26, 694, (1971)

[4] KORNHERR,M., Z.Phys. 233, 37 (1970)

[5] MONTGOMERY,M.D. et al., J.Geophys. Res. 75, 717 (1970)

[6] ALINOVSKII, N.I. et al., Soviel Physics JETP 30, 385 (1970)

[7] KEILHACKER,M. et al., Proceedings of 4th Conference on Plasma Physics and Controlled Nuclear Fusion Research, Madison 1971,Paper CN-28/J-10, IAEA 1971

[8] STEUER,K.-H., KEILHACKER,M., Bull.Amer.Phys.Soc. 15, 1410 (1970)

[9] KADOMTSEV,B.B., Plasma Turbulence, Academic Press, London (1965)

[10] STRINGER,T.E., Nucl.Energy, Part C 6, 267 (1964)

[11] FORSLUND,D.W., MORSE,R.L., NIELSON, C.W., Phys.Rev. Lett. 25, 1266 (1970); LASHMORE-DAVIES,C.N., J. Phys.A 3, L 40 (1970) and Phys.Fluids 14, 1481 (1971); GARY,S.P., BISPAMP,D., J.Phys.A 4, L 27 (1971)

[12] PAUL,J.W.M et al., Nature 223, 822 (1969); GARY,S.P., PAUL,J.W.M., Phys.Rev.Lett 26, 1097 (1971)

[13] CHODURA,R., v.FINCKENSTEIN,K., Institut für Plasmaphysik, Garching, Report IPP 1/113 (1970)

NONLINEAR THEORY OF CROSS-FIELD AND TWO-STREAM INSTABILITIES IN THE EQUATORIAL ELECTROJET

André Rogister

European Space Research Institute

Frascati (Rome)

Experimental studies of the equatorial ionospheric E region have revealed the presence of an assortment of non-thermal plasma motions in the equatorial electrojet referred to as Type I and Type II irregularities[1,2]. The linear theories[3] of Farley and Buneman describe an instability mechanism which can plausibly explain the generation of Type I irregularities: the current flowing in the equatorial ionosphere in the East-West direction can drive longitudinal ion-acoustic waves whenever the electron drift velocity in the wave direction exceeds the velocity of sound in the ionospheric plasma. The linear theory[4] of Rogister and D'Angelo appears to explain well the origin of Type II irregularities: a universal-like ("crossfield") instability[5] sets in in an inhomogeneous plasma immersed in crossed electric and magnetic fields.

It has been shown[6] that at the dip equator, the secondary (vertical) electric field E_s is related to the primary (in the East-West direction) field E_p, in the absence of turbulence, by the relation

$$E_s = (\sigma_2/\sigma_1)E_p \quad ,$$

which is arrived at by requiring that the vertical electron and ion fluxes in the electrojet be equal. σ_1 and σ_2 are respectively the Pedersen and the Hall conductivities[7]. In the presence of turbulence, the electrojet self-consistency condition will be[8]

$$\bar{n}\bar{v}_{ex}+\langle\delta n\delta v_{ex}\rangle=nv_{ix}+\langle\delta n\delta v_{ix}\rangle$$

($\hat{i}_x$ in the vertical direction) where the brackets are the correlations of the density and velocity turbulent fields.

It can be shown that $\langle\delta n\delta v_{ix}\rangle$ is negligible compared to $\langle\delta n\delta v_{ex}\rangle = -(\hat{s}\cdot\hat{i}_y)^2 \frac{\nu_i}{\Omega_i} (1-\frac{\nu_e\nu_i}{\Omega_e\Omega_i})^{-1}\bar{v}_{ey}\langle\delta n\delta n\rangle/\bar{n}$

where $\Omega_e(\Omega_i)$ and $\nu_e(\nu_i)$ are the electron (ion) Larmor and collision frequencies (with neutrals) respectively; $\hat{s}\cdot\hat{i}_y$ is the cosine of the angle between the direction of wave propagation and the East-West axis.

With $\langle\delta n\delta v_{ex}\rangle/\bar{n}\bar{v}_{ex}$ negative, the vertical ion flux $\bar{n}\bar{v}_{ix}$ must decrease. Hence the secondary electric field $\bar{E}_s=m_i\nu_i\bar{v}_{ix}/q_i$ and the electrojet electron current $q_e\bar{n}\bar{v}_{ey}=c\bar{E}_s/B$ also decrease.

The resistive two-stream instability, with growth rate proportional to $\frac{(\hat{s}\cdot\hat{i}_y)^2 v_{ey}^2}{(1-\nu_e\nu_i/\Omega_e\Omega_i)^2} - c_s^2$, will thus be quenched when $v_{ey}=c_s(1-\nu_e\nu_i/\Omega_e\Omega_i)$. The results we find in this way agree reasonably well with what is available of experimental evidence on the following points:

a) the turbulence level is in agreement with the figure quoted in Ref. 1,

b) the electron drift velocities in the turbulent state are appreciably lower than those computed on the basis of the equilibrium theory[6] of the equatorial electrojet,

c) the phase velocities are independent of the electron drift velocities, in contrast to the prediction of linear theory, and are of order of the sound speed c_s,

d) the drift velocities are consistently larger than the phase velocities.

The cross-field instability, on the other hand, cannot be quenched by the above quasi-linear mechanism alone since its growth rate is proportional to

$$-\frac{\nu_i}{\Omega_i}\bar{v}_{ey}(\hat{i}_y\cdot\hat{s})^2\frac{1}{\bar{n}}\frac{\partial}{\partial x}\bar{n}+\frac{\nu_e}{\Omega_e\Omega_i}k^2c_s^2\quad(v_{ey}\frac{\partial}{\partial x}\ell n\bar{n}<0;\Omega_e\Omega_i<0),$$

when electron inertia is negligible $(v_{ey}<c_s)$.

A nonlinear theory has thus been developed[9] in which the energy is transferred by mode coupling from large *unstable* to small *stable* wavelengths where it is absorbed by linear damping (classical diffusion); this process naturally leads to the formation of an overall marginally stable state. Although the asymptotic turbulence level has not yet been evaluated, it is already clear that the origin of the observed[10] oscillations with wavelengths corresponding to linear stability can readily be explained by the present theory.

The coupling process is also relevant to the theory of the resistive two-stream instability in the sense that it will determine the *form* of the spectrum $\delta n_k \delta n_{-k}/\bar{n}\bar{n}$.

REFERENCES

1. Bowles, K.L., Balsley, B.B. and Cohen, R., 1963, J. Geophys. Res. 68, 2485.

2. Balsley, B.B., 1969, J. Geophys. Res. 74, 2333.

3. Farley, D.T., Jr., 1963, J. Geophys. Res. 68, 6083, and Buneman, O., 1963, Phys. Rev. Lett. 10, 285.

4. Rogister, A. and D'Angelo, N., 1970, J. Geophys. Res. 75, 3879.

5. Simon, A., 1963, Phys. Fluids 6, 382, and Hoh, F.C., 1963, Phys. Fluids 6, 1184.

6. Sigiura, M. and Cain, J.C., 1966, J. Geophys. Res. 71, 1869.

7. Chapman, S., 1956, Suppl. del Nuovo Cimento 4, 1385.

8. Rogister, A., to appear in J. Geophys. Res. Nov. 1971.

9. Rogister, A., to be published.

10. Prakash, S., Gupta, S.P. and Subbaraya, B.H., 1971, Lett. to Nature 230, 170.

FERMI ACCELERATION IN INTERPLANETARY SPACE

G. Wibberenz and K. P. Beuermann

Institut für Reine und Angewandte Kernphysik

Universität Kiel

1. THE PHYSICAL MODEL

Since Fermi's original idea that charged particles may on the average be accelerated by randomly moving "magnetic clouds", this mechanism has been applied under a variety of circumstances. Recently, the question of possible Fermi acceleration in interplanetary space has been raised anew. Murray et al.(1971) have observed that a characteristic break in the energy spectrum during the decay phase of a solar proton event moves towards lower energies with a time rate markedly slower than expected on the basis of adiabatic cooling of these particles in the expanding wind. Jokipii (1971) estimates that the observed time scale could be accounted for if the adiabatic deceleration is partly cancelled by Fermi effects. It should be noted, however, that the evidence for the occurrence of Fermi acceleration is not conclusive. The observed behaviour can also be explained by a solution of the full transport equation including spatial diffusion. Part of the observed variation of the MeV proton spectrum could be due to a spatial feature in the particle distribution transported across the observer, without taking into account Fermi effects (see Forman, 1971; Gleeson, personal communication, 1971).

Nevertheless, it is of interest to study the modifications of low-energy cosmic-ray transport theory, once the existence of waves in the interplanetary medium is established. Waves or spatial structures moving with respect to the solar wind will cause energy changes of individual particles. The theoretical treatment of Fermi acceleration has to be modified as compared to the original model of moving magnetic clouds. In interplanetary space, the particles gyrate around the average interplanetary magnetic field and undergo continuous pitch-angle scattering due to irregularities superimposed on the average field. A stochastic treatment relates the pitch-angle scattering and the resulting spatial diffusion coefficient with the power spectrum of magnetic field fluctuations (see Jokipii, 1966, 1967; Hasselmann and Wibberenz, 1968, in the following referred to as paper I).

In paper I, a rigorous treatment of Fermi acceleration has been given. It was assumed that the electromagnetic field in interplanetary space is obtained by superimposing a set of fields, each member of which is derived from the same time-independent magnetic field by different Lorentz transformations. The magnetic field contains only fluctuations axisymmetric around and transverse to the average magnetic field direction (denoted as model (a) in paper I). This is exactly the situation which arises if the irregularities observed in the interplanetary magnetic field stem from Alfvén waves moving in both directions along the magnetic field. In this case, the Fokker-Planck coefficients describing changes in energy are obtained by a linear transformation from the pitch-angle diffusion coefficient. For a pitch-angle distribution close to isotropy, averaging over all pitch-angles yields two additional terms in the particle transport equation which is now written as

$$\frac{\partial \varrho}{\partial t} + \frac{\partial}{\partial x^i}\left(V^i \varrho - K^{ij}\frac{\partial \varrho}{\partial x^j}\right) + \frac{\partial}{\partial \gamma}\left[\left(\frac{d\gamma}{dt}\right)_{ad}\varrho\right] + \frac{\partial}{\partial \gamma}\left(A_F \varrho - D_F \frac{\partial \varrho}{\partial \gamma}\right) = 0 \qquad (1)$$

where V^i is the solar wind bulk velocity vector and K^{ij} is the spatial diffusion tensor. The equation without the Fermi terms containing A_F and D_F was first introduced by Parker (1965) and has since been extensively used to describe the propagation of cosmic rays in interplanetary space by spatial diffusion, convection, and adiabatic deceleration.

The coefficients for the mean acceleration and the energy diffusion are related by

$$A_F = \frac{D_F}{\gamma}\left(1 + \frac{1}{\beta^2}\right) \qquad (2)$$

with

$$D_F = V_A^{\,2}\left(\frac{e}{mc}\right)^2\left(\frac{\Omega}{ck_0}\right)^q \frac{\pi\, M\beta^{-q+1}\gamma^{-q}}{2(2-q)(-q)c^3} \qquad (3)$$

(see equation (8.12) in paper I). Here e, m, βc, and Ω are the charge, rest mass, velocity, and cyclotron frequency of the particles under consideration, $\gamma = (1-\beta^2)^{-1/2}$ the Lorentz factor.

The spectral density of the interplanetary magnetic field fluctuations is taken as a power law for all wave numbers, $f(k) = M(k/k_o)^q$. V_A is the Alfvén velocity.

In relating the Fermi effects to the measured magnetic field power spectrum, we shall assume that all power is contained in Alfvén wave motion and that for a given wave number interval equal power is contained in waves travelling inward and outward along the average magnetic field, from now on referred to as "bi-directional waves". The numerical results obtained in this way give an upper limit to the Fermi effects actually occuring in interplanetary space.

2. CHARACTERISTIC TIME CONSTANTS

It is the purpose of this paper to describe the energy changes which ~1 MeV protons undergo locally by adiabatic deceleration and Fermi acceleration. Effective time constants for the energy changes are defined by expressing the last two terms of (1) as

$$\frac{\partial}{\partial\gamma}\left(\left(\frac{d\gamma}{dt}\right)_{ad}\rho\right) + \frac{\partial}{\partial\gamma}\left(A_F\rho - D_F\frac{\partial\rho}{\partial\gamma}\right) \equiv \frac{\partial}{\partial\gamma}\left(\frac{\gamma-1}{T}\rho\right)$$

where

$$\frac{1}{T} = \frac{1}{T_{ad}} + \frac{1}{T_F} = \frac{1}{\gamma-1}\left(\frac{d\gamma}{dt}\right)_{ad} + \frac{1}{\gamma-1}\left(A_F - D_F\frac{1}{\rho}\frac{\partial\rho}{\partial\gamma}\right). \quad (5)$$

T_{ad}, T_F, and T are the time constants for adiabatic deceleration, Fermi acceleration, and for the net energy change. In defining a characteristic time constant T_F, the effects of diffusion in energy which actually influence the whole energy spectrum are converted to an apparent energy gain of individual particles. Since in interplanetary space Fermi acceleration is important only for sufficiently small particle energies, we shall use the non-relativistic limits of the relevant formulas from now on. For a constant and radial solar wind with speed V, the time constant for the adiabatic energy loss is given by

$$\frac{1}{\gamma-1}\left(\frac{d\gamma}{dt}\right)_{ad} = \frac{1}{T_{ad}} = -\frac{4V}{3r} \quad (6)$$

Time constants T_F are evaluated for particle kinetic energy spectra of the power law type and of the exponential type. We obtain

$$\frac{1}{T_F} = \frac{V_A^2}{c^2}\frac{1}{\tau}(1-2n)\ \beta^{-q-3} \tag{7}$$

for a particle density spectrum $\rho(\gamma) \propto (\gamma-1)^n$, and

$$\frac{1}{T_F} = \frac{V_A^2}{c^2}\frac{1}{\tau}\left(1+\beta^2\frac{mc^2}{E_0}\right)\beta^{-q-3} \tag{8}$$

for a particle density spectrum $\rho(\gamma) \propto \exp(-(\gamma-1)mc^2/E_0)$.
The time constant τ defined by

$$\frac{1}{\tau} = \left(\frac{e}{mc}\right)^2\frac{\pi M}{c(2-q)(-q)}\left(\frac{\Omega}{ck_0}\right)^q$$

depends upon the magnetic field parameters, but is independent of the particle energy and spectrum. Two points are worth noting:

(1) Contrary to the adiabatic energy loss, the rate of Fermi energy gain is energy dependent. The functional form of this energy dependence varies with the spectral shape of the particle energy spectrum and with the spectral distribution $f(k)$ of the magnetic fluctuations.

(2) The two additive terms in $1/T_F$ derive from A_F and D_F in equation (5), the terms describing acceleration and energy diffusion, respectively. It is seen that, except for a flat particle spectrum, the diffusive term is by far the dominant one. Consequently, the apparent energy gain of particles of a given energy will strongly depend on the particle spectral shape $\rho(\gamma)$.

Finally, we note that Fermi energy gains will be of importance, in particular, in highly disturbed regions where the relaxation time is small. It is of interest, therefore, to relate the expressions for spatial and energy diffusion. For the model of purely axisymmetric transverse magnetic fluctuations with zero polarisation, the value for the spatial diffusion coefficient is given in paper I (in relativistically correct notation) as

$$K_\parallel = \frac{2}{\pi M}\left(\frac{mc}{e}\right)^2\left(\frac{\Omega}{ck_0}\right)^{-q}\frac{c^3\ \gamma^{q+2}\beta^{q+3}}{(q+2)(q+4)} \tag{9}$$

For the case that all spectral power is contained in bi-directional waves, we may replace τ in (7) by $K_\parallel$. In the non-relativistic limit, this yields

$$\frac{1}{T_F} = \frac{2\ (1-2n)}{(-q)(2-q)(2+q)(4+q)}\frac{V_A^2}{K_\parallel} \tag{10}$$

The numerical factor in (10) is of the order of 1 for q=-1.5 (see below) and n varying between -2 and -4. Relating $K_{\parallel}$ to the pitch-angle relaxation time τ_{rel} by $K_{\parallel} = v^2 \tau_{rel}/3$, we obtain

$$T_F \approx \frac{v^2}{V_A^2} \frac{\tau_{rel}}{3}$$

which closely resembles the original Fermi result (e.g., Jokipii, 1971).

Relation (10) may be profitably used to study the variation of the effect with radial distance r from the sun. No reliable estimates exist for the dependence of the spectral power M on r. However, model calculations for solar flare particle propagation and successful fits to the observed intensity-time profiles seem to indicate that the radial dependence of the diffusion coefficient is $K_{\parallel}(r) \propto r^m$ with m = 0..1. Since the Alfvén velocity V_A varies approximately as r^{-1}, this leads to a radial dependence of the Fermi time constant for a given particle energy $T_F(r) \propto r^{2+m}$. If our assumptions concerning the magnetic field fluctuations were valid throughout interplanetary space, we would expect Fermi acceleration processes to occur predominantly in highly disturbed regions as, e.g., colliding stream regions, and to become of increased importance as the radial distance to the sun decreases. (With respect to equation (10) it should be added that this relation is of limited use for computational purposes. As discussed in paper I, the above expression for $K_{\parallel}$ develops an "escape hole" singularity for q = -2 which is not characteristic of real diffusion coefficients. In combination with observationally determined values for $K_{\parallel}$, (10) should be used with care while being correct together with $K_{\parallel}$ as given by (9).)

3. RESULTS

Numerical values for the time constants T_{ad} and T_F have been calculated from (6) to (8) for protons with energies between 1o keV and 10 MeV. For the average magnetic field a value of B = 6.5γ, for the solar wind number density a value of 10 cm^{-3} is taken.

Fig.1 illustrates the magnitudes of adiabatic deceleration and Fermi acceleration near the orbit of the earth for moderately disturbed times. The curves are based upon the Mariner 4 spectral power distribution reported by Siscoe et al. (1968). This spectrum is fitted by the above power law with M=0.83 $Gauss^2$cm at k_0=1.57 10^{-10} cm^{-1}, corresponding to f_0=10^{-3} Hz, and q=-1.5. The spectrum is assumed to retain this shape up to infinite k. The relative energy loss by adiabatic deceleration for a solar wind speed of 400 km/sec is indicated by the dotted line. (Note that for purposes of comparison the value of $-1/T_{ad}$ is plotted.) The solid and dashed curves represent the relative energy gain by Fermi acceleration according to (7) and (8) for the two values of the power spectral index n and the e-folding energy E_0 indicated in the figure. These values of n and E_0 have been chosen as typical for a steady state solar particle flux (e.g., Gleeson et al., 1971). It is seen that in moderately disturbed regions near the orbit of the earth adiabatic deceleration dominates over Fermi acceleration for proton energies above 100 keV. Near 100 keV, Fermi acceleration may balance the adiabatic energy loss if the proton energy spectrum is sufficiently steep. Exponential energy spectra with $E_0 \gtrsim 0.5$ MeV are too flat and do not lead to a net acceleration for the power level chosen here.

There exist, however, highly disturbed regions where the spectral power may be increased by an order of magnitude and the spectral slope is near $q=-1.5$ or even near $q=-1$ (Coleman, 1966). Fig. 2 shows $1/T_F$ for different values of q, the <u>same</u> spectral power M at 10^{-3} Hz as in Fig. 1 and for moderately steep proton energy distributions. It is apparent that the $q=-1$ spectrum with its increased high frequency wave power may result in a net acceleration for protons with energies up to ~ 0.3 MeV. While no quantitative information on the relative power contained in waves is available so far, it is clear that for the highly disturbed regions with spectral power $\sim$10M a relative contribution of only 10 % in suitable waves would suffice to yield an effect similar to that depicted in Fig.2. In quiet regions, on the other hand, with spectral power of the order of 0.1 M and spectral indices of $q=-1.7$ or below, Fermi acceleration will be totally negligible.

It is seen from Fig.2 that the functional energy dependence of the effect is strongly dependent on the spectral shapes of (i) the proton kinetic energy distribution and (ii) the wave power distribution via q.

Apart from the magnitude of the effect close to the orbit of the earth, it is of interest to study possible radial variations. Since no reliable estimates exist of the relative power contained in waves suitable for Fermi acceleration, we shall assume, for illustrative purposes, that this relative contribution does not vary with distance from the sun. The radial variation of $1/T_F$ may then be profitably related to the radial dependence of $K_{\parallel}$, applying equation (10). Starting from the curves labeled $q=-1.5$ in Fig.2 which may reflect conditions near the orbit of earth in disturbed regions, we obtain rather drastical effects extrapolating back to 0.3 AU

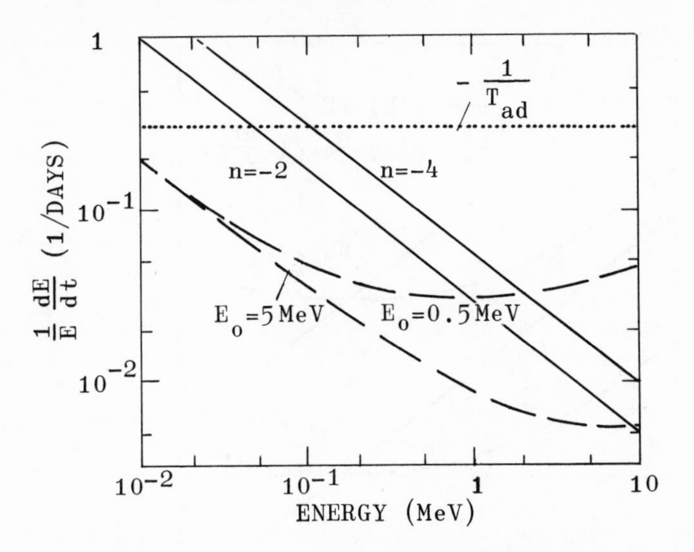

Fig.1 Rate of energy gain of protons by Fermi acceleration for power law spectral density $f(k) \propto k^{-1.5}$. Results are given for proton kinetic energy spectra of the power law and the exponential type. Dotted line indicates the level of energy loss by adiabatic deceleration at 1 AU for a constant solar wind velocity of 400 km/sec.

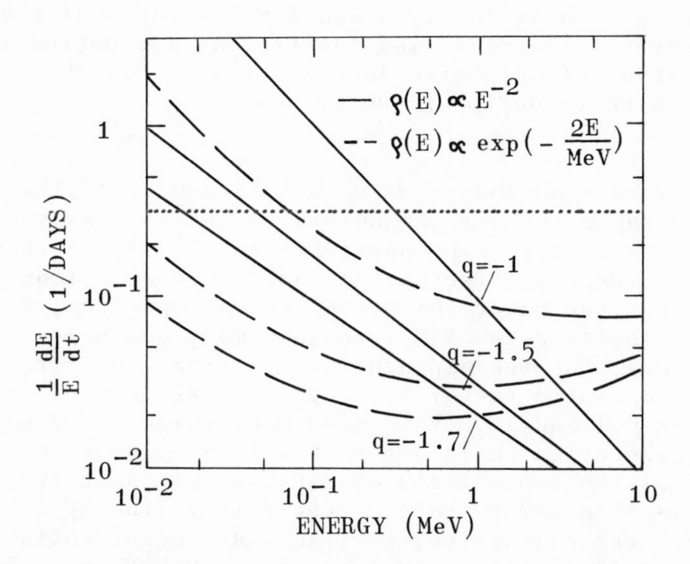

Fig.2 Same as Fig.1 for various values of the power index q of the spectrum of magnetic fluctuations and for proton kinetic energy spectra proportional to E^{-2} and to $\exp(-E/E_o)$ with E_o=0.5 MeV.

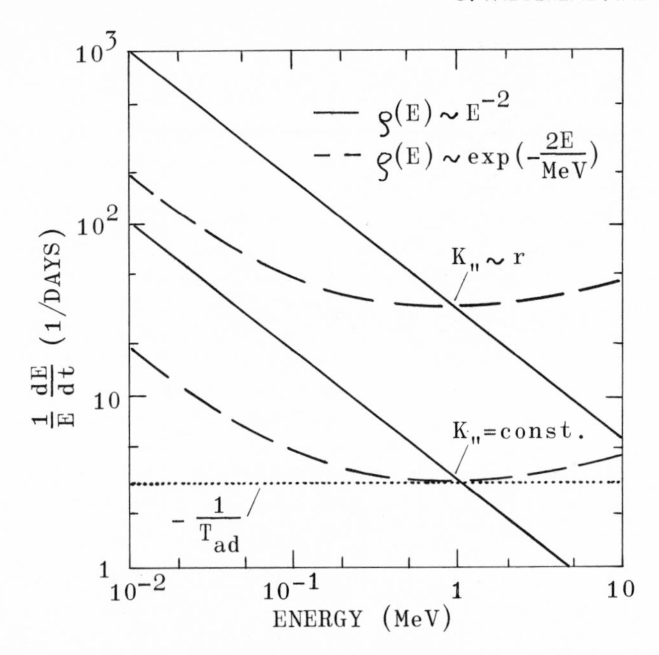

Fig.3 Rate of energy gain of protons by Fermi acceleration at a radial distance from the sun of 0.1 AU, for the same spectral density as in Fig.1. Results are given for the same proton kinetic energy spectra as in Fig.2 and for two forms of the radial dependence of the diffusion coefficient $K_{\parallel}$. Dotted line indicates the level of adiabatic deceleration at 0.1 AU for a constant solar wind velocity of 400 km/sec.

and 0.1 AU, if we assume that $K_{\parallel} \propto r^m$ with r = 0 or 1. The ratio of the Fermi energy gain over the adiabatic energy loss varies as r^{-1-m} since the latter is proportional to r^{-1}. Fig.3 shows the magnitudes of the two effects at a radial distance from the sun of 0.1 AU. For the power law proton energy spectrum, Fermi acceleration dominates below 1 MeV in case m=0, and below 20 MeV in case m=1. For the exponential spectrum, the Fermi effect and the adiabatic energy loss approximately cancel for energies between 0.1 and 10 MeV in case m=0, whereas for m=1 rapid Fermi acceleration would occur at all energies. The relative magnitude of the two effects at a radial distance from the sun of 0.3 AU is obtained by raising the dotted line by a factor of three. For comparison, at 1 AU a net acceleration is formally obtained for proton energies below ~40 keV only. This result appears reasonable, in particular, for highly disturbed regions. On the other hand, the result for 0.1 AU and $K_{\parallel} \propto r$ is certainly far from the real physical situation, implying that one or more of the underlying assumptions are in error.

4. DISCUSSION

The Fermi effect may be of considerable importance for low-energy particle transport in interplanetary space. An understanding of its influence, however, requires a detailed knowledge of the wave properties of the interplanetary magnetic fluctuations. This knowledge just begins to accumulate. Belcher and Davis (1971) have shown that Alfvén waves have a characteristic pattern of association with the large scale structure of the solar wind. From their work, one may conclude that the most likely regions for pronounced Fermi acceleration are the compression regions at the leading edges of high velocity streams, which seem to contain significant amounts of Alfvénic structures. However, it is not clear as yet which part of the structures will effectively contribute to Fermi acceleration. The relation of the numerical values presented here to the real conditions in interplanetary space will be discussed in an extended version of this paper.

Particles moving along field lines with large superimposed irregularities will have small mean free paths. Accordingly, diffusive effects have been neglected in some models of particle propagation. Gleeson et al. (1971), for example, treat the purely convective transport for >0.3 MeV protons. In these models, the particles suffer considerable adiabatic energy losses. We propose that Fermi acceleration may partly cancel the adiabatic losses. Accounting for the Fermi effect may, therefore, significantly alter the solutions of low-energy particle transport theory under convective conditions.

For a magnetic field power spectrum $f(k) \propto k^q$ with $q > -3$, one obtains formally the result that the net acceleration becomes arbitrarily large for sufficiently low energies. Following a group of low-energy particles which are convected outward with the solar wind, we should expect an energy spectrum which is gradually steepening at the low-energy end. Anderson and Lin (1971) have recently found a stable low-energy particle component in interplanetary space, possibly protons, extending with an energy spectrum $\propto E^{-3}$ down to 30 keV. It will be interesting to see how this component extends to still lower energies and how its properties relate to the wave power spectrum at the high-frequency end. It was noted that the Fermi effects may be more important closer to the sun, so that observations near the orbit of earth, reflecting the net effect on the particles during the propagation outward, are not easily interpreted in terms of wave observations performed close to the earth.

References:

Anderson, K.A., and Lin, R.P.: 1971, Conf. Papers 12th Internat. Conf. on Cosmic Rays, Hobart, Tasmania (Paper SOL-36).
Belcher, J.W., and Davis, L., Jr.: 1971, J. Geophys. Res. 76, 3534.
Coleman, P.J.: 1966, J. Geophys. Res. 71, 5509.
Forman, M.A.: 1971, Conf. Papers 12th Internat. Conf. on Cosmic Rays, Hobart, Tasmania (Paper SOL-26).
Gleeson, L.J., Krimigis, S.M., and Axford, W.I.: 1971, J. Geophys. Res. 76, 2228.
Hasselmann, K. and Wibberenz, G.: 1968, Z. f. Geophys. 34, 353 (quoted as "paper I").
Jokipii, J.R.: 1966, Astrophys. J. 146, 480.
Jokipii, J.R.: 1967, Astrophys. J. 149, 405.
Jokipii, J.R.: 1971, Phys. Rev. Letters 26, 666.
Murray, S.S., Stone, E.C., and Vogt, R.E.: 1971, Phys. Rev. Letters 26, 663.
Parker, E.N.: 1965, Planet. Space Sci. 13, 9.
Siscoe, G.L., Davis, L., Jr., Coleman, P.J., Smith, E.J., and Jones, D.E.: 1968, J. Geophys. Res. 73, 61.

THE GALACTIC COSMIC RAY DIURNAL VARIATION AS A STREAMING PLASMA INTERACTION BETWEEN GALACTIC AND SOLAR CORPUSCULAR RADIATION

V.J. Kisselbach

Max-Planck-Institut für Aeronomie

Lindau/Harz

Despite considerable experimental and theoretical attention over the years, the cosmic ray diurnal variation still poses many questions.

It occurred to the present author some time ago that some of the discrepancies between the characteristics of the cosmic ray diurnal variation which are predicted by corotation theory and the experimentally observed cosmic ray diurnal variation might possibly be explained by additional sources of cosmic rays located outside the orbit of the earth, i.e., by a spatial anisotropy of the primary cosmic radiation. If such an anisotropy does indeed exist, then one observed discrepancy, i.e., the fact that amplitudes of the diurnal variation in excess of the 0.8% maximum predicted from corotation are observed, would have a natural explanation. The resultant diurnal variation observed at earth would have to be considered to be superposition of the diurnal variation due to corotation and that due to the primary anisotropy.

On inquiring into the experimental observations to be expected on this model, it was indicated that an excellent time of the year to look for this effect should be at the autumnal equinox. At this time, for example, the 23°°h R.A. direction is almost parallel to the earth sun line.

Diurnal time series recordings of muon coincidence data appropriate to a basic data acquisition interval of

0.1 h in rectascension were initiated at Lindau in September 1969. The coincidence selection device consists of a double, twofold circular-disc, plastic scintillation counter vertically oriented in the laboratory whose magnetic asymptotic response has been given (1). The apparatus itself has been described in detail elsewhere (2, 3).

A detailed analysis of diurnal variations recorded at the autumnal equinoxes of 1969 and 1970 yields the following results:

1. The occurrence of "micro-structures" of the order of ca. 1 h fwhm in the diurnal variation, which display annual recurrence.

2. Two main, separately congruent sub-intervals in the diurnal variations which are solar- and sidereal-associated, respectively.

It is concluded that the diurnal variation time series of the autumnal equinox-proximate cosmic ray muon intensity data displays two annually recurrent subintervals, one evidencing pronounced solar association, the other displaying sidereal invariance.

We conclude that the diurnal variation is a resultant of the interaction of the now widely studied solar wind plasma stream and streams of galactic beams whose particle energies are above 10^{11} eV.

REFERENCES

1. Hatton, C.J. and D.A. Carswell, Deep River AECL Document No. CrGP-1165

2. Kisselbach, V.J., Mitteilungen aus dem Max-Planck-Institut für Aeronomie, No. 40 (S), 1970.

3. Kisselbach, V.J., TE-24 Proceedings of the Eleventh International Conference on Cosmic Rays, Budapest (in press).

THE INTERPLANETARY CONDITIONS ASSOCIATED WITH COSMIC RAY FORBUSH DECREASES

L.R. BARNDEN

Laboratorio Plasma Spaziale del CNR, Roma

This paper presents new evidence concerning the interplanetary phenomena which produce the Forbush decreases observed in the cosmic ray intensity on Earth (1-60 GV). It is generally accepted that Forbush decreases are associated with interplanetary disturbances which originate in active regions on the Sun, and effectively sweep cosmic rays as they propagate through interplanetary space. The drop in cosmic ray number density behind the disturbance is recorded as a Forbush decrease as it passes the Earth.

An interplanetary disturbance will sweep cosmic rays if it has an effective cosmic ray transmission coefficient < 1.0, and it is propagating relative to the cosmic ray "gas", which is essentially fixed with respect to the Sun. Parker (1963) has suggested that these conditions may exist (a) at interplanetary shock waves, (b) in magnetic tongues which are drawn out by flare ejecta (after Gold, 1960), or (c) in extended regions of magnetic field irregularities. However, because of the complexity of the observed Forbush decreases it was not possible to reliably deduce the spatial structure of the cosmic ray decrease, nor its arrival time at Earth. Thus it was not possible to establish which, if any, of these 3 features produces Forbush decreases. This Forbush decrease complexity is illustrated in Figure 1, which shows the event of Jan 26-27, 1968, as recorded at 3 different stations. Deep River and Leeds view equatorially but are separated by 100° in longitude while Thule views almost due North.

For this event it is seen that (a) the apparent arrival time of the decrease differs by 3-4 hours at Deep River and Leeds, (b) at any one station the decrease extends smoothly over 2-4 hours, and (c) there is a modulation before and after the decrease at the equatorial viewing stations.

Recently a technique has been developed (Barnden 1971 a,b) which permits the calculation of the cosmic ray variations at Earth corresponding to any arbitrary cosmic ray number density distribution in space. This, the "origin-of-scatter" technique, computes the response of a given detector to the number density in regions of space remote from the Earth. Figure 2 shows the response, at 4 different times of day, of the Deep River neutron monitor to the cosmic ray number density at heliocentric radius r (and close to the spiral field line which passes "through" the Earth). During hour 12 UT

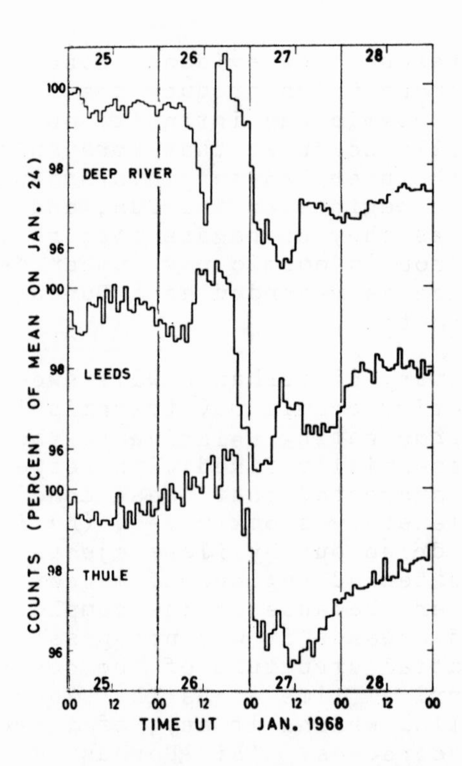

Fig. 1. The Forbush decrease of Jan 26-27, 1968.

Deep River looks inwards and parallel to the interplanetary magnetic field, and so on.

This technique was applied (Barnden 1971 b) to the extreme case of a discontinuous step in cosmic ray number density which moves outwards past the Earth. It was found that the general features of the observed Forbush decreases (illustrated in Fig. 1) could be reproduced. It was concluded therefore that the observed Forbush decreases are consistent with a cosmic ray decrease in space which extends over 1 hour (0.01 AU) or less. Further results have suggested a quick method to determine the arrival time of this cosmic ray step (to within 1-2 hours for suitable events). This method will be described elsewhere. We can therefore start from this point and use the cosmic ray step arrival times to elucidate the interplanetary features which produce the observed cosmic ray effects.

We first examine the relationship between the cosmic ray step arrival time and the shock arrival time. The Forbush decreases used in this study were chosen such that (a) the step arrival time could be determined reliably and without ambiguity, and (b) an interplanetary shock was observed by a satellite in Earth (or Moon) orbit. Most of the shock arrival times were taken from the papers listed in Hundhausen (1970). Twenty-two events from 1965-1968 were used. It was found that:

1) in only 1 out of the 22 events was a large (>2%) Forbush decrease directly associated with the interplanetary shock;

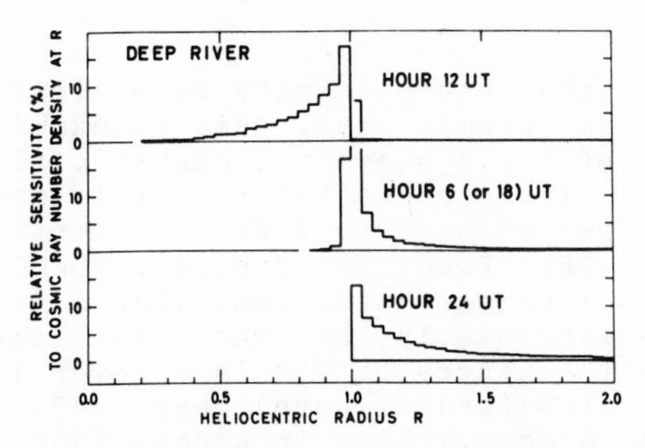

Fig. 2. The relative response of the Deep River detector to the cosmic ray number density at R, for 4 times of day.

2) 5 events exhibited a double cosmic ray step. The first and smallest step could be associated with the shock while the major step occurred several hours later;

3) for the remaining events the shock preceded the cosmic ray step by 2-17 hours.

The distribution of the delay between the shock and the cosmic ray step is shown in Fig.3. Where a double step was present, the delay was calculated using the largest (second) step. The average delay is $7\frac{1}{2}$ hours. It is seen that even allowing for a systematic error of 3 or 4 hours, most Forbush decreases cannot be directly associated with interplanetary shocks, but they occur after the passage of the shock.

To attempt an identification of the features which produce the cosmic ray effects, we now examine the interplanetary data for one particular event, namely that of Jan. 26-27, 1968, shown in Fig.1.

A shock was observed by Explorers 33 and 35 at 1430 UT, Jan.26, and a sudden commencement at 1441, Jan.26. Lepping (1971) has deduced that the normal to the shock surface was directed almost radially. For this particular event the station at Thule was deduced to show most truly the cosmic ray conditions in the immediate vicinity of Earth. The arrival time of the cosmic ray step was taken to be 2400 $\pm$ 2 hours UT on Jan. 26, i.e. 9 hours after the shock. This is supported by the observation of a Forbush decrease in >12 MeV cosmic rays on Pioneer 8 near 0200 UT, Jan.27 (see Lockwood and Webber, 1969). This 2 hour delay is consistent with the 2 hour delay between the sudden commencement at Earth and the shock at Pioneer 8.

Since only the interplanetary magnetic field can directly influence cosmic rays, this is examined first. The upper half of Fig.4 shows the magnetic field observed by Explorer 33. It is seen that the field magnitude and direction remained quiet from 2 hours after the shock until 2200 UT Jan.26. From 2200 Jan.26 to 0200 Jan.27, large directional changes occurred. The field magnitude did not change appreciably, however. In these 4 hours there were smooth changes of 90° in θ (over 1 hour) near 2230, and 180° in ϕ (over 1 hour) near 2330. There was a 90° directional discontinuity in θ near 2300, and 70° and 180° directional discontinuities in ϕ at 0030 and 0130. Quenby and Sear (1971) have established that both rotational and tangential discontinuities have a cosmic ray transmission coefficient <1.0. In addition, it is easy to imagine either of the observed smooth large amplitude directional changes having an important effect on cosmic rays, since, whatever its incident pitch angle, a cosmic

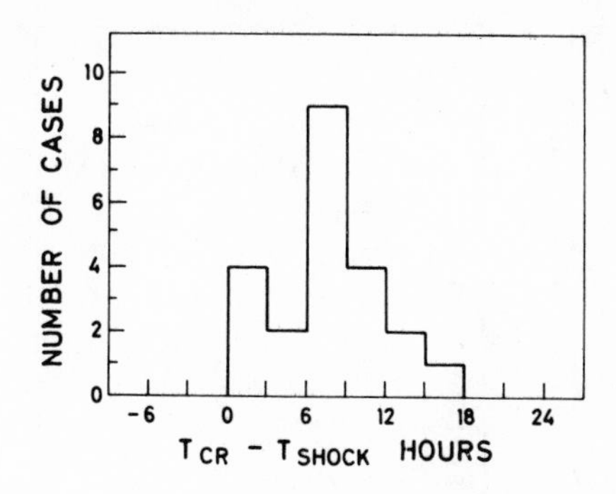

Fig. 3. The distribution of the delay at Earth between the shock arrival time T_{SHOCK}, and the cosmic ray step arrival time T_{CR}.

ray would, at some point in this region, find its pitch angle near 90° - the condition for mirroring. It would appear therefore that any of these features could be expected to sweep cosmic rays, leaving a region of depleted number density in its wake. For simplicity we will also refer to these smooth (over ∿1 hour) directional features as discontinuities.

Such magnetic field discontinuities are relatively common, of course, but very few produce Forbush decreases. It is proposed that the special feature of the discontinuities which do cause Forbush decreases is that they extend over a vast surface. Clearly any cosmic ray depletion caused by a discontinuity of small extent will quickly be filled by cosmic rays diffusing from nearby regions of normal cosmic ray number density. Since the magnetic field is "frozen-in" to the solar wind, such extended directional discontinuities can be expected at the boundary between shells or streams of different solar wind plasma regimes. These boundaries will be marked by sharp changes in one or more of the plasma parameters. The lower half of Fig.4 shows the plasma density and bulk velocity from Explorer 33. A sharp decrease in plasma density was observed near 2300 UT Jan. 26, and in bulk velocity near 2330 UT Jan. 26. These features therefore support the contention that the cosmic rays were swept by a magnetic field discontinuity which extended over the surface of a new plasma regime.

It has been reported recently (e.g. Ogilvie and Wilkerson, 1969; Hirshberg et al, 1970) that the solar wind He/H density ratio increases sharply several hours after some shocks. Hirshberg et al interpret this to

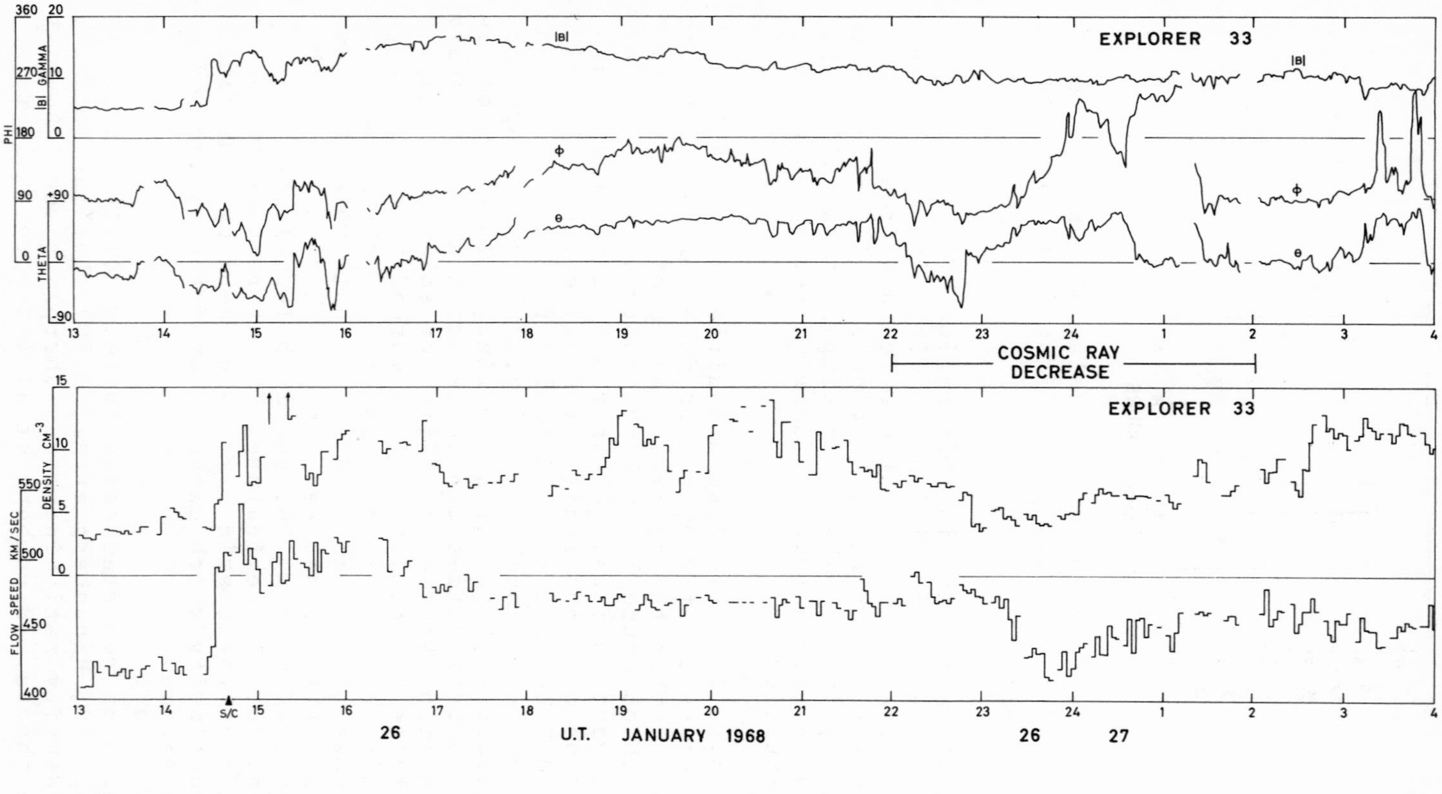

Figure 4. Interplanetary data from 1300 UT, January 26, to 0400 UT, January 27, 1968, from Explorer 33.
TOP: The magnetic field components in solar equatorial coordinates.
BOTTOM: The solar wind plasma number density and bulk velocity.

indicate the arrival of the actual flare ejecta or driver gas. In 2 of the 5 examples described by these authors the He/H enhancement coincided (within 2 hours) with a cosmic ray decrease. Thus it appears that in some cases cosmic rays are swept by interplanetary features associated with the surface of the driver gas. Indeed, the driver gas could be expected to define an expanding magnetic tongue with a tangential discontinuity extending in a homogeneous fashion over its whole surface - ideal conditions for the sweeping of cosmic rays.

At first sight it would appear that the directional discontinuities associated with the shock should extend over a large surface and therefore satisfy the condition to effectively sweep cosmic rays. However the shock is propagating relative to the ambient solar wind, and, if its magnetic features are essentially the ambient magnetic features compressed, this directional discontinuity surface will not have the homogeneous nature of a surface which separates two different gas regimes. It is suggested that, at any one time, the shock surface will have regions where the directional discontinuities are weak, which previously corresponded to uniform conditions in the ambient solar wind. Such areas will effectively be "holes" for cosmic rays and will greatly reduce the sweeping capability of the shock-directional discontinuity surface as a whole.

We have concluded therefore that most Forbush decreases are not directly produced by shocks. Evidence from 1 event supports the proposal that they are produced by a magnetic field directional irregularity which extends over the total surface of a new plasma regime. It is not clear whether this directional irregularity is a discontinuity or a large amplitude change which extends smoothly over about 1 hour (0.01 AU). There is further evidence that in some cases this directional irregularity may be coincident with the boundary of the flare driver gas.

ACKNOWLEDGEMENTS

I would like to thank Drs. J.H. Binsack and H.C. Howe Jr. for the Explorer 33 plasma data, and the Ames Research Center for the Explorer 33 interplanetary magnetic field data. I would also like to thank Dr. J.K. Chao for many informative discussions on the nature of interplanetary shocks.

REFERENCES

Barnden, L.R., 1971 a, Solar Phys., 18, 165.
Barnden, L.R., 1971 b, Unpublished Thesis, University of Adelaide.
Gold, T., 1960, Astrophys. J., Suppl. 4, 406.
Hirshberg, J., Alksne, A., Colburn, D.S., Bame, S.J. and Hundhausen, A.J., 1970, J. Geophys. Res., 75, 1.
Hundhausen, A.J., 1970, Rev. Geophys. Space Phys., 8,729.
Lepping, P.R., 1971, Goddard Space Flight Center, pre-print X-692-71-138.
Lockwood, J.A. and Webber, W.R., 1969, J. Geophys. Res. 74, 5599.
Ogilvie, K.W. and Wilkerson, T.D., 1969, Solar Phys. 8, 435.
Parker, E.N., 1963, Interplanetary Dynamical Processes, Interscience Publishers, Inc., New York.
Quenby, J.J. and Sear, J.F., 1971, Planet. and Space Sci., 19, 95.

THE DIURNAL EFFECT OF COSMIC RAYS AND ITS DEPENDENCE ON THE INTERPLANETARY MAGNETIC FIELD

E. Bussoletti, N. Iucci (*)

Laboratorio Plasma Spaziale del CNR, Roma

(*) Istituto di Fisica, Università di Roma

Immediately after the discovery, made by Wilcox and Ness (1), of the interplanetary magnetic field sector structure, Ryder and Hatton (2) used Deep River neutron monitor data for the period December 63 - February 64 to study the diurnal variation in view of a possible connection with this magnetic structure of space.

In this period, the sectors were each of a width of about 8 days and were stable for several solar rotations, while the diurnal waves had, in each sector, the same typical behaviour. In the central period (4 days) the normal corotation shape was observed while in days near the sector boundary, besides the corotation wave, additional waves were found with a "free space" amplitude of about 0.3 - 0.4% and a phase of the maximum approximately centered along the field line of $\underline{B}$. Moreover, the phase of the maximum of these additional waves was, in the initial period, towards the sun, while in the final period it was in the opposite direction. This suggests a possible "streaming" of cosmic rays (C.R.) along the magnetic field lines.

This behaviour has been interpreted by the authors following Parker (3). In fact, if the magnetic sectors can be considered as "closed boxes" without any C.R. flux through their boundaries, the "net streaming velocity" $\underline{u}$ of the C.R. can have a radial component. Its direction depends on the ratio V/K between the solar wind velocity and the diffusion coefficient of the C.R.

The method described in Bussoletti _et al_. (4) has been applied to data of 14 neutron monitors for the period 15 October 1965 - 30 June 1966. This method is particularly suitable for the study of the day by day changes in the diurnal variations. The following results were obtained:

1) in the central period the diurnal waves showed the typical corotation shape for phase and amplitude;
2) in the final and initial periods, additional waves whose direction was "systematically" along the sun and antisun line, as it happened between December 1963 and February 1964, were never found.

On the other hand in most of the cases large fluctuations in amplitude of the anisotropy were observed in days before and after the change of polarity of $\underline{B}$, i.e. near the sector boundaries. Therefore, it can be said that in 1965 and 1966, years in which the solar activity was rising, there was no "steady streaming" of C.R. along the interplanetary magnetic field at the edges of the magnetic sectors.

We then analysed the trend of the diurnal variation wave at the sector crossing taking periods of three days centered around the day in which there is the change of polarity of $\underline{B}$. We analysed 17 cases; 14 showed an abrupt change in the percent amplitude of the wave, while considerable phase changes did not occur. These variations showed no systematic tendency to rise or diminish, nor were they correlated with the polarity of the field.

At the same time, from magnetic data of the Imp C satellite we must note that in the period investigated when $\underline{B}$ is changing its polarity (see Fig. 1):

1) the direction of $\underline{B}$ takes a fixed value, positive or negative, only after a time interval of several hours during which the vector $\underline{B}$ changes direction many times;
2) the hourly variance of $\underline{B}$, σ_B, during the same time interval, is higher, by about a factor of two or three, than its mean value in any magnetic sector.

This trend of $\underline{B}$ suggests that we are dealing with magnetic irregularities of the interplanetary field in both its micro- and meso-structure. Such inhomogeneities can act as "scattering centers" for the C.R. whose gyroradius is smaller than or comparable with their scale length L. As the irregularities are observed for a

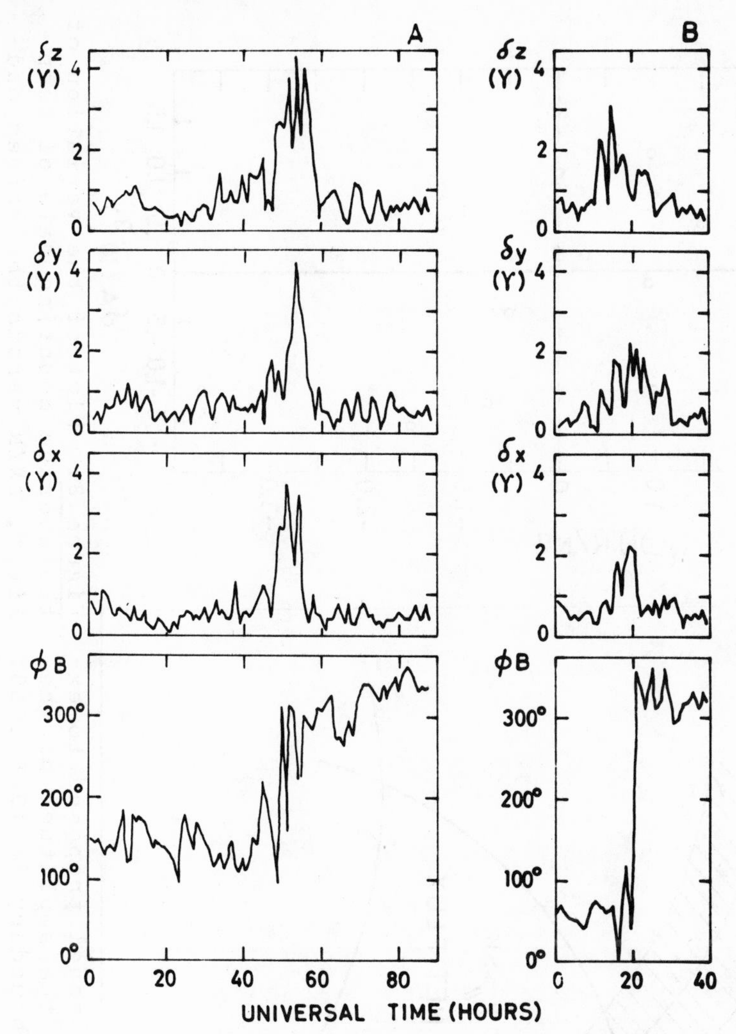

Figure 1 A, B: The typical trend of Φ, azimuthal direction of B, and the variance δ_x, δ_y, δ_z, of the three field components at two different sector boundaries.

period T≃ 6-12h, L≃10^6-10^7km, and so they can affect C.R. with rigidities up to 10-20 GeV which largely contribute to the counting rate of high latitude neutron monitors.

Diffusion phenomena of the C.R. across B can occur, and therefore if C.R. gradients are present between two sectors, transverse fluxes of particles are possible (see Fig. 2).

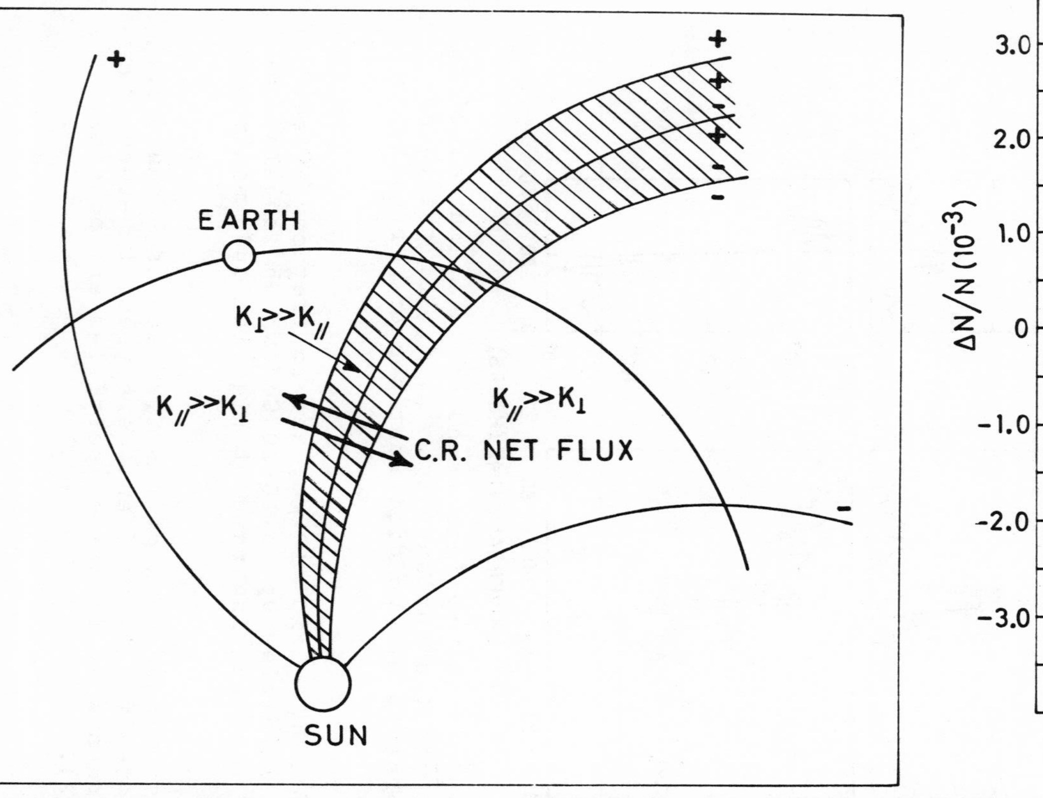

Figure 2: Phenomenological model proposed to explain the abrupt changes in the amplitude of the diurnal wave at the sector boundary in 1965-1966. The dashed region represents scattering centers at the sector boundary.

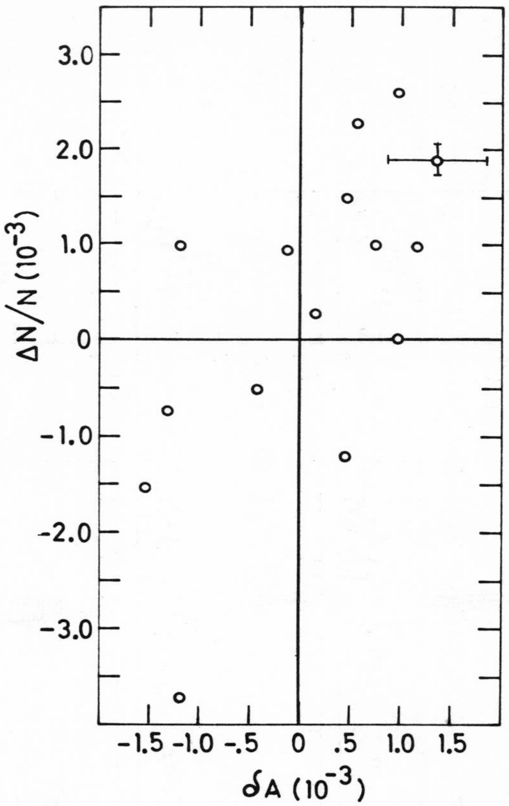

Figure 3: Plot of the variation of the average counting rate of Deep River, $\Delta N/N$ versus the corresponding change in the mean diurnal variation δA, taken between days before and after sector boundaries.

This mechanism well explains the discontinuities seen in the percent amplitude of the diurnal wave at the crossing through a sector edge. In fact, when the net diffusion flux is in the same direction as the corotation flux the two fluxes sum and the amplitude rises. When the diffusion flux is in the opposite direction the two fluxes subtract and there is a lowering of the amplitude of the diurnal variation.

In order to test this mechanism we have taken the daily intensity changes of the C.R., measured by the Deep River super monitor, as a good index of the tangential gradient of the C.R. Then we have done a regression analysis between the variation of the anisotropy amplitude δA and the percent average counting rate variation $\Delta N/N$, both taken between the day before and the day after the inversion of the polarity of $\underline{B}$ (see Fig. 3); we found a correlation coefficient of $r_t = 0.7 \pm 0.1$ and a slope $\alpha = \delta AN/\Delta N = 0.30 \pm 0.07$.

In this case, for the C.R., a diffusion equation can be written

$$\underline{F} = N\underline{u} = N\underline{V} - K\nabla N \tag{1}$$

Because of the tangential diffusion, Eq.(1) becomes

$$N\underline{u} = - K_{\perp}\nabla N \tag{1'}$$

so the net streaming velocity is

$$\underline{u} = - K_{\perp} \frac{\nabla N}{N} \tag{2}$$

Utilizing the Compton-Getting formula an estimate of the value of $K_{\perp}$, at our energies, is possible:

$$\frac{1}{2}\,\delta A = (2+\mu)\frac{|u|}{c} \cong \frac{K_{\perp}}{c}(2+\mu)\frac{|\nabla N|}{N} \tag{3}$$

where c is the velocity of light and μ is the exponent of the differential energy spectrum of the primary C.R.: $D(E) = AE^{-\mu} = AE^{-2.7}$. The factor 1/2 comes from the definition of δA, as a difference between two amplitudes.

Therefore we obtain

$$K_{\perp} = \frac{c}{2}\,\delta A / \{(2+\mu)\frac{|\nabla N|}{N}\} \tag{3'}$$

Calling N_1 and N_2 the average daily counting rates

of the Deep River neutron monitor, taken respectively the day before and after the sector edge,

$$\frac{|\nabla N|}{N} = \frac{N_1 - N_2}{N} \cdot \frac{1}{L}$$

where L is now the distance in km covered by the transition zone in one day.

With L=400km/sec × 3.6 × 10^3 × 24.2 sec we have

$$K_\perp = \frac{\alpha L c}{2(2+\mu)} = (5.5 \pm 1.5) \times 10^{21} \text{ cm}^2/\text{sec}. \quad (4)$$

This value of $K_\perp$ has been obtained at the crossing of the sector boundary and becomes roughly equal to the value of $K_\parallel$ in the same region in the simple model of $K_\perp = K_\parallel/1+(\omega\tau)^2$ (ω is the Larmor frequency and $1/\tau$ is the collision rate). Because $K_\perp$ is obtained in a highly perturbed period, its value must be considered as an upper limit with respect to its value during quiet periods.

It is now clear that with the changes in magnetic structure in space from 1963 - 1964 to 1965 - 1966, the morphology of the diurnal variation has also changed. In 1963 -1964 the magnetic sectors were stable and could be considered as "closed" sectors also for the C.R., so that streaming of particles along the lines of force occurred at the boundaries. On the other hand in the period that we have analysed, the magnetic structure of space was no longer stable in time, but continuously evolving. The polarity inversion regions were turbulent and could act as "scattering centers" for C.R. Therefore, diffusion processes across these regions occurred while the streaming along the lines of force at the boundary did not.

ACKNOWLEDGMENT

The authors are indebted to Dr. N.F. Ness for making available the Imp C data which made the analysis possible.

REFERENCES

(1) N.F. Ness, J.M. Wilcox, Science 148, 1592, 1965

(2) P. Ryder, C.J. Hatton, Can. J. Phys. 46, 999, 1968

(3) E.N. Parker, Planet. Sp. Sc. 12, 735, 1964

(4) E. Bussoletti, N. Iucci, G. Villoresi, to be published in Il Nuovo Cimento.

SUBJECT INDEX